Textile Science and Clothing Technology

Series Editor

Subramanian Senthilkannan Muthu, SgT Group & API, Hong Kong, Kowloon,
Hong Kong

This series aims to broadly cover all the aspects related to textiles science and technology and clothing science and technology. Below are the areas fall under the aims and scope of this series, but not limited to: Production and properties of various natural and synthetic fibres; Production and properties of different yarns, fabrics and apparels; Manufacturing aspects of textiles and clothing; Modelling and Simulation aspects related to textiles and clothing; Production and properties of Nonwovens; Evaluation/testing of various properties of textiles and clothing products; Supply chain management of textiles and clothing; Aspects related to Clothing Science such as comfort; Functional aspects and evaluation of textiles; Textile biomaterials and bioengineering; Nano, micro, smart, sport and intelligent textiles; Various aspects of industrial and technical applications of textiles and clothing; Apparel manufacturing and engineering; New developments and applications pertaining to textiles and clothing materials and their manufacturing methods; Textile design aspects; Sustainable fashion and textiles; Green Textiles and Eco-Fashion; Sustainability aspects of textiles and clothing; Environmental assessments of textiles and clothing supply chain; Green Composites; Sustainable Luxury and Sustainable Consumption; Waste Management in Textiles; Sustainability Standards and Green labels; Social and Economic Sustainability of Textiles and Clothing.

More information about this series at http://www.springer.com/series/13111

Subramanian Senthilkannan Muthu
Editor

Leather and Footwear Sustainability

Manufacturing, Supply Chain, and Product Level Issues

 Springer

Editor
Subramanian Senthilkannan Muthu
SgT Group & API
Hong Kong, Kowloon, Hong Kong

ISSN 2197-9863   ISSN 2197-9871   (electronic)
Textile Science and Clothing Technology
ISBN 978-981-15-6298-3   ISBN 978-981-15-6296-9   (eBook)
https://doi.org/10.1007/978-981-15-6296-9

This Springer imprint is published by the registered company Springer Nature Singapore Pte Ltd.
The registered company address is: 152 Beach Road, #21-01/04 Gateway East, Singapore 189721,
Singapore

This book is dedicated to:

The lotus feet of my beloved Lord Pazhaniandavar;

My beloved late father;

My beloved mom;

My beloved wife Karpagam;

My beloved Daughters Anu and Karthika;

My beloved brother Raghavan;

Last but not least

To everyone working in Leather and Footwear

Sector to make it Sustainable.

Contents

About the Editor

Dr. Subramanian Senthilkannan Muthu is currently Head of Sustainability at the SgT Group and API in Hong Kong. He holds a Ph.D. from The Hong Kong Polytechnic University. He is a leading expert in the areas of environmental sustainability in textiles and the clothing supply chain, product life cycle assessment (LCA) and product carbon footprint (PCF) assessment in various industrial sectors and has extensive industrial experience in these fields. He has published more than 75 research papers, written numerous chapters and authored/edited over 80 books.

Environmental and Chemical Issues in Tanneries and Their Mitigation Measures

P. Senthil Kumar and G. Janet Joshiba

Abstract In spite of the fact that the tanning industry is known to be one of the main financial parts in numerous nations, there has been an expanding ecological concern with respect to the arrival of different hard-headed contaminations in tannery wastewater. The releasing and dumping of squanders close to the water bodies without treatment make it nearly look like a territory which is lying under the cover of contamination. It is a developing issue for all the living organisms and the wellness of the environment. This chapter elaborates the various environmental and chemical issues' cause due to the tanneries and their impact towards the environment. In addition, it also clearly explains the mitigation measures followed in the tannery industry to reduce its pollution level.

Keywords Tannery · Contaminations · Wellness · Mitigation · Environment

1 Introduction

In current scenario, the generation of wastes from various domains such as agriculture, transport, mining, industry and energy sectors is of prime environmental concern, and also it is considered to be one of the greatest threats to the wellness of the environment. Once more, among all the modern squanders tannery effluents are positioned as the most elevated toxins. On account of the generally modest expense of work and materials, over a large portion of the world's tanning business happens in predominantly in the developing nations. It has been reported that during the period of 1970–1995, the production of light leather materials progressively increased from 35 to 56% and the production of heavy leather materials progressively increased from 26 to 56% [2].

The leather industries are one of the important industries which play a vital role in the lifestyle of human beings. Basically, the raw animal skins are processed by converting the animal skin and hides into leather which can be used further for

P. Senthil Kumar (✉) · G. Janet Joshiba
Department of Chemical Engineering, SSN College of Engineering, Chennai 603110, India
e-mail: psk8582@gmail.com; senthilkumarp@ssn.edu.in; senthilchem8582@gmail.com

© The Editor(s) (if applicable) and The Author(s), under exclusive license
to Springer Nature Singapore Pte Ltd. 2020
S. S. Muthu (ed.), *Leather and Footwear Sustainability*, Textile Science
and Clothing Technology, https://doi.org/10.1007/978-981-15-6296-9_1

1

manufacturing various essential things for humans. The waste skin materials from the slaughter house are used as the key material in the tannery industries; further, the tanneries lessen the burden of disposal of slaughter house by utilizing the skin waste which is not suitable for the edible usage. Several reports have claimed that the tannery is one of the highest pollution-creating industries because of the strong and toxic chemicals used in the processing of the animal skins. The implementation of sustainable and cleaner wastewater treatment technologies in the tanneries helps in lessening the negative impacts of the tannery wastewaters towards environment. In addition, with consistently expanding interest for new water and the water assets getting rare, the decrease of soil fruitfulness because of the defilement by the saline wastewater from the enterprises are significant issue in Indian states, for example, Tamil Nadu [1]. As per the investigation of Blacksmith Institute, it is reported that around 75% of the chromium-polluted sites are present in South Asia and their nearby countries. In addition, the major source of chromium pollution occurs from the industrial sources such as tannery, metallurgy and mining industries. Among these industries, the tanneries present in South Asian countries contribute greatly to a higher amount of generation of chromium into the environment [2].

The tannery wastewater is considered to be one of the highly toxic pollutants because of its toxic composition and its deleterious effects. The colour, higher oxygen demand, toxic metals, dyes and other poisonous chemicals make the tannery effluent more complex to degrade, and also it is highly difficult to treat using the wastewater treatment technologies. In different stages of preparation of leather such as soaking, tanning and post-tanning, various chemicals such as wetting agents, ammonium sulphide, sodium sulphite, ammonium chloride, soda ash, CaO and enzymes are involved in the tannery industry. The disposal of the above-mentioned tannery waste into surface and groundwaters without proper treatment results in negative impacts towards the environment causing health disorders and environmental effects [3]. This chapter clearly explains the various environmental and chemical effects of the tanneries, and also it includes the various mitigation measures to combat the negative impacts of the leather industry. It also focuses on the various treatment technologies used to eliminate the toxic compounds from the tannery wastewater.

2   Overview of Tannery Wastewater

The tannery wastewater is composed of various compounds such as pH, BOD, nitrogen, sulphides, suspended solids, settleable solids, gross solids, fats, oils and grease, heavy metals, dyes and toxic chemicals [4]. The significance of ecological gauges in India's fare was first experienced when Germany, one of the significant merchants of cowhide and calfskin merchandise from India, restricted the import of cowhide items containing in excess of 5 mg/kg of pentachlorophenol (PCP) in 1990. This was trailed by a German prohibition on the import of cowhide and materials treated with various Azo colours in 1994. In India, the significant bunches of tanning businesses are Ambur, Pallavaram, Vaniyambadi, Pernambut, Dindigul,

Table 1 Characteristics of tannery wastewater

No.	Characteristics	Average concentration in mg/l (before treatment)	Average concentration in mg/litre (after treatment)
1.	BOD	1850	700
2.	COD	4500	3000
3.	Suspended solids	3750	1500
4.	Chloride	5500	1200
5.	Chromium	165	38

Source [7]

Ranipet and Trichy in Tamil Nadu; Kolkata in West Bengal; Kanpur in Uttar Pradesh; and Jalandhar in Punjab. Around 45% of the all the tanneries in India are situated at Tamil Nadu. In Tamil Nadu, the significant tannery bunches are situated at Vellore district [1]. The characteristics of the tannery wastewater are clearly explained in Table 1.

The generation of effluent containing chromium is one of the major threats to the environmental issues. Chromium is a profoundly lethal compound, and the dumping of chromium containing material is in many nations limited to a couple of uncommon dumping grounds. Outflows into the air are essentially identified with vitality use, yet additionally the utilization of natural solvents and colour causes discharges into the air. The production of fresh animal hides has been calculated at about 8–9 million ton per year. Nearly 1.4 million tonnes of soil waste are generated during the production of hides. For each ton of crude conceal prepared, the measures of strong waste and results might be delivered. Assortment and safe removal of strong waste, particularly chrome containing strong waste and slime, are ordinarily checked by natural specialists and related with costs. In any occasion, decrease of waste is basic so as to satisfy needs for diminished contamination load from tanneries [5].

Around 15% of solid waste flushing out of the tanneries is a mixture of both solid and liquid materials in which 29% in the liquid state. Several chemicals such as formic acid, chromium oxide, sodium chloride, ammonium sulphate and sulphuric acid are utilized in the processing of leather. In addition, the HCl, caustic soda, formic acid, chromium, sodium arsenite, soda ash, sulphuric acid and arsenic sulphate are utilized in the various tanning process such as soaking, liming, deliming and tanning processes. After every consequent operation, water is used to wash and clean the chemicals and machineries. A huge part of solid waste is generated from the trimmings, shavings, buffing, curing and packaging of the animal skins or hides.

In 2015, Kavitha and Ganapathy have investigated about the environmental impact of tannery process in the Vellore district, Tamil Nadu. The studies showed that the water utilization for the creation of skin from one tone of crude covers up is around 15,000–40,000 l and 110–260 l for every sheep skin. Quality synthetic substances and ideal dosing in each procedure can lead to downing the contamination level, and along these lines diminish the ecological effect. Cutting-edge innovations are followed in the tanneries based on reusing and recuperating procedures that lead to the improvement of calfskin quality.

The wastewater from the tanneries been treated in the emanating treatment plants utilizing reverse assimilation (RO) innovation and changed over into reusable water for the tannery forms and the reject from the RO treatment being vanished either by nearby planetary group or by mechanical dissipation and changed over into salt. In this manner, the ecological effect of tannery industry in Tamil Nadu, India, has radically diminished by accomplishing Zero Liquid Release framework [1]. Tinni et al. have investigated the impingement of tannery effluent on the environment of Dhaka city corporation. They concluded that the solid and liquid of around 56% waste flushing out of the tanneries are in black colour and causes noxious smell in the environment. In addition, the disposal of tannery effluents without proper treatment causes health disorders such as skin disease, diarrhoea, dizziness, asthma, fever and respiratory issues. It also affects the aquatic life and other living organisms by creating various negative effects on the environment [6]. Rahaman et al. have conducted a comparative study of the various concentrations of heavy metals in different layers of vicinity soil and agricultural soil. The studies showed that heavy metals' concentration was found to be higher in the aquatic region than the normal agricultural land. This metal focus in soil is mindful for the advancement of lethality in horticultural items. In this way, the human well-being and condition are influenced by these territories [8].

3 Process Involved in Tanneries

Tanning is the process of conversion of protein present in the raw animal skin into a strong material which can be used for wide variety of applications. The tanned leather changes into a flexible form, and it does not get rancid when it is wetted back with water. The leather tanning process involves various stages of processing of raw skin. Initially, the animal skin is prepared using some preparatory stages such as:

- Preservation—The animal skin is subjected to a preservation treatment to protect it from decomposition
- Soaking—Water is used in this method to wash or hydrate the skin
- Liming—Elimination of undesirable protein and opening up is achieved
- Unhairing—Elimination of animal hair from the skin
- Fleshing—Removal of subcutaneous material
- Splitting—The animal skin is cut into two or mo
- Rehorizontal layers
- Reliming—Further elimination of protein and opening-up process
- Deliming—Removal of liming and unhairing chemicals from the skin
- Bating—Further proteins are removed using the proteolytic proteins and undergoes softening process
- Degreasing—Removal of natural fats, oils and grease from the skin/hide

- Frizing—Physical removal of fat layer inside the skin
- Bleaching—Chemical modification of darker pigmented skins into light coloured pelt
- Pickling—Lowering of pH value to acidic region
- Depickling—Increasing of pH to acidic region.

After the preparatory stages, the prepared skin is subjected into various stages such as tanning, dyeing and finishing procedures. The tanning process is of two types such as vegetable tanning and chrome tanning [4]. In the pretanning process, the prepared skin is subjected into pretanning operations which include soaking, fleshing, trimming, bating and pickling processes. In the chrome tanning process, the chromium is fixed by slowly increasing the pH by addition of base. The cross-linkage of chromium ions with carboxyl groups results in the chrome tanning process. In the vegetable tanning process, the skin is subjected to series of vats and agitated accompanied with progressive increase of tanning liquor. The vegetable tannins are basically made up of polyphenolic compounds which are divided into hydrolysable tannins and condensed tannins.

In the finishing process, the chromium/vegetable tanned hides are frequently retanned to incorporate some desirable properties and textures including their colour, smoothness and filling in the hides. In addition, the excess amount of water is removed using the drying process, and consequently, it is cooled with cooling water. After retaining and drying, the crust is exposed to various completing tasks. The reason for these activities is to make the shroud milder and to veil little slip-ups. The cover-up is treated with a natural dissolvable or water-based colour and varnish. The completed final result has somewhere in the range from 66 to 85 wt% of dry issue [5].

4 Environmental and Chemical Concerns of Different Tannery Process

The tannery procedure is majorly into three main stages such as pretanning, tanning and post-tanning process. In the pretanning process, several procedures such as soaking, liming, deliming, pickling and degreasing are done to the raw skin/hides. In this process, several chemical compounds such as pesticides and detergents are gushed out to the environment with the wastewater generated from the pretanning process. Further, the tanning process is divided into three major types such as vegetable tanning, chrome tanning and combined tanning. Various harmful salts such as chromium, zirconium and aluminium are used in the tanning process which are capable of creating various deleterious health effects and environmental concerns. Also, the formaldehyde, fish oil, silica and Calgon are gushed out of the wastewater generated by the tanning process. The oils are one of the main ingredients of the leather wastewater, and they are applied in the treated skins or hides subsequently. A combination of vegetable oil, sulphated oil, mineral oil, Epsom and glucose is used in the tanning process which is highly difficult to degrade in the tannery wastewater treatment. In the post-tanning process, the tanned leather is trimmed and conditioned into a fine leather. Furthermore, the final step of the leather processing is finishing method. Several resins, pigments, waxes and binders are utilized in the finishing process which are one of the hazardous waste in the environmental criteria [1].

Leather industries are one of the most elevated dangerous producers of toxic effluent per unit of output. The process of conversion of skin into leather is an intensely concentrated procedure using around 130 synthetic substances. The principle synthetics utilized in the different preparing stages incorporate lime powder, sodium sulphide, chromium sulphate, ammonium sulphate, sulphuric corrosive, formaldehyde, sulphonated and sulphated oils and colours. In the processing stages such as pretanning, tanning, wet finishing, drying and finishing process, the effluent gushing out of these processes is in the form of air, liquid and solid. Hydrogen sulphide and alkali are the significant gases discharged into the environment. Be that as it may, laboratory results indicated outflows lower than the national ecological quality measures.

A large portion of the strong squanders are reused. The drums, containers and synthetic packs are obtained for reuse. Fleshing, crude cutting and buffing dust are purchased by cowhide board or poultry feed makers. Chromium is one of the most important chemical residues released in this tannery effluent which are highly carcinogenic and capable of causing various deleterious effects to human beings. Animal shavings are utilized as modest fuel in ovens causing the arrival of chromium into nature. The staying strong squanders are generally wrongfully dumped around the processing plant territory on unutilized lands. These strong squanders incorporate metal substance, for example, chromium, aluminium and zirconium, which detrimentally affect plant development. Over the span of preparing of cover-up into cowhide, approximately 50–150 l of water was utilized per one kilogram of changed over cowhide. In this manner, effluents released from tanneries are voluminous, profoundly hued, and contain an overwhelming residue load including lethal metallic mixes, synthetic concoctions, naturally oxidizable materials and enormous amounts of rotting suspended issue. Tannery effluents, with no pretreatment, are released unpredictably into water bodies or open land, coming about in tainting of surface just as subsurface water. The smelling salts outflow during the deliming cause bothering of the respiratory tracts. Other negative impacts of the alkali outflows incorporate the misfortune of land efficiency, impediment of the germination of plants and seeds, migraines, stomachaches, discombobulation, night visual impairment, sickness, dermatitis and other skin issues. Calfskin dust results in sensitivities and tumours that harm local people around the tanneries [9].

The research finding of Garai at 2014 is about the environmental aspects and health risks of leather tanning industry: A study in the Hazaribag area showed that the exploration discoveries uncover that the tannery business creates an unhygienic and poisonous condition in the Hazaribag region. Harmful substance and other lethal squander results of the tannery business make air, smell, water and soil contamination, and so on which all seriously sway living conditions. The discoveries additionally show that 33% of respondents asserted a boundless and harsh scent as the primary issue; 37.5% referenced that grimy streets are the fundamental issue in the examination territory, on account of the business releasing squanders in an ill-advised way. Concerning untreated waste discharge, 35.6% said that the these spread different sorts of sickness (e.g., jaundice, the runs, skin infections), while 39.3% asserted that the severe smell in the region is added to an unhygienic condition for everyone [10].

Skin disease is one of the main problems caused due to the contamination of tannery wastewater in the water and soil. Based on the toxicity of the tannery wastewater, the extent of skin disease varies from minor rashes and itches to major skin cancer. More than 8000 labourers in the tanneries of Hazaribag experience the ill effects of gastrointestinal, dermatological and different maladies, and 90% of this populace kicks the bucket before the age of 50. The nearness of arsenic in the groundwater expanding shortage of crisp drinking water causes skin injury, kidney, liver inconvenience, malignant growth and so forth. The tannery workers did not wear any gloves, cover and uncommon shoes instead they work in exposed feet.

The tannery squanders were arranged in spontaneous manner. These squanders fall in the nearby stream, and the ghettos' individuals utilize this contaminated water and get various illnesses in them. These squanders secured the encompassing territory which makes noxious. The putrid condition damaged human well-being. The tannery wastewater negatively influenced the other sectors such as livestock growth, fisheries production and agricultural sectors causing various health disorders and damaging their lives. Shakir et al. in 2012 investigated the ecotoxicological risks associated with tannery effluent wastewater. The results showed that the hexavalent chromium and tannery wastewater have huge ecoharming potential, and significant levels of chromium are representing an impressive hazard to the human populace, aquaculture and agrarian industry that can pulverize biological system encompassing the tanneries [11].

5 Mitigation Measures Followed by the Tanneries

Tannery effluents are viewed as an unconventional type of dirtied wastewater since they fluctuate across tanneries in both volumes just as in contamination load. 62 As such, every tannery presents it possess profluent issue. Along these lines, in any event, for a unique kind of cowhide, it is hard to figure a standard plan for emanating treatment. The strategies being used for the profluent treatment might be of a physical, concoction or organic nature, utilized either alone or in mix. A short record of a portion of these techniques utilized in the nation has just been accounted for above. Like all other modern wastewater treatment, the medication cost can be significantly decreased by embracing great in-house rehearses, measures identified with squander decrease at source and utilizing greater condition amicable advancements [9].

From the findings of Azom and other researchers in 2012, some of the sustainable strategies are enunciated which can be used as effective mitigation measures in controlling the harmful effects of the tannery wastewater. They are as follows:

- Transferring the tanneries from centre of the city to outskirts to avoid its negative impacts on the environment
- Implementation of proper effluent treatment plant to eliminate the undesirable harmful toxins present in the tannery wastewater
- Proper monitoring of the safety of the workers by providing protective aids such as masks, shoes and gloves
- Examining the impacts of tannery effluent on the quality and richness of the soil with respect to physical and biological studies
- Utilization of necessary chemicals in prescribed limit according to the Department of Environment
- Proper management of the operation hours of the industry to reduce the negative effects of the tannery towards the habitat
- Proper disposal of solid wastes consisting of toxic chemicals and undesirable solid wastes
- Squanders utilized as poultry nourishment ought to be inspected previously as it is given to the poultry ranch, strong waste containing chromium that must not be utilized as poultry nourishment
- Considering the financial part, minimal effort coagulant, for example, alum, lime and ferric chloride can be picked for the treatment of tannery effluents
- Effective Environmental Management Plan (EMP) ought to be presented for most extreme contamination reduction.

At long last, one might say that satisfactory preventive measures ought to be taken in tannery modern exercises with the end goal of guaranteeing sheltered, sound and sound condition for the more noteworthy advantage of our ecosystem [2]. Stream isolation is the underlying advance in executing in-plant controls. Because of the distinction in wastewater attributes from beamhouse (high pH and sulphides), tanning and retanning (low pH and chromium) tasks, increasingly proficient control could be accomplished through the utilization of a treatment procedure explicitly intended for the related contamination. Besides, the isolation could prompt the reuse or reusing of spent mixers and the recuperation of materials. Wastewater treatment is one of the major mitigation measures which can be used to reduce the negative impacts of the tannery effluent on the environment. Basically, three important treatment methods such as physical, chemical and biological treatment methods are applied in the elimination of hazardous compounds from the tannery industrial wastewater. The coagulation-flocculation is a first treatment step required to expel particulate material and different contaminations and also as chromium (VI), which hinders natural treatment. Essentially, to apply isolation of waste streams approach in tannery contamination anticipation is the most significant advance in spite of the fact that it is not broadly and essentially applied on the planet. Association of tanneries in the mechanical regions is another normal methodology which causes a lot to subside the contamination in corresponding to the reinforcing release limits. Anyway incorporated contamination counteraction procedure of the EU and the greening economy which incorporates the moving synthetic substances with the common ones, water minimization advancements and water reusing this segment will keep on spending endeavours for comprehending ecological issues [12].

6 Green Technology

Benzene- and naphthalene sulphonates are generally applied in tanning of cover-up and sulpho-subordinates of naphthalene are seen in tannery wastewater at convergences of 0.1–30 mg L^{-1}. Naphthalene sulphonates and their subbed analogues have been accounted for to be inadequately degradable which was associated with either their atomic structure or their immediate harmfulness to potential microbial degrader. The sulphonated mixes in wastewater are known to be debased by a few bacterial consortia, yet the vast majority of the xenobiotic organosulphonates are dependent upon desulphonation. Ultrasound has been practised as an elective innovation to diminish synthetic concoctions use and calfskin quality. Synthetic adjustment of chromium tanning salt is one of the choices for improving the take-up of chromium. Manufactured tanning material dependent on chromium improved essentially (90%) chromium take-up. Chromium was complexed utilizing multifunctional polymeric grid. The substitution of ammonium salts in the deliming forms via carbon dioxide and the reuse of wastewater and synthetic compounds after layer filtration of the deliming/bating alcohol can be used as an effective alternative [12].

7 Conclusion

Tanneries are seemed to be one of the highly pollution industries among the other commercial industries. In the conversion of raw skin to leather, various processing stages are involved such as preparation, pretanning, chemical tanning, vegetable tanning, wet finishing, drying and finishing procedures. Every processing stage of the tanneries consumes a huge amount of water and also releases toxic effluents with harmful chemicals and solvents. The disposal of these toxic tannery effluents without proper treatment into the environment can cause deleterious skin and respiratory issues, and also it causes tumours and organ damages. In environmental perspective, the release of tannery effluent into soil decreases the richness of the soil and also causes accumulation of toxic metal ions in the soil which spoils cause impotency of the soil nutrients. In addition, it also affects the livestock, fisheries and human beings in a larger scale. Several advancements have been made in the tannery wastewater treatment methodologies to obtain a sustainable and effective treatment of this effluent. Apart from the conventional methodologies, many new advancements such as membrane process, photocatalysis are implemented in the tannery effluent treatment. The minimization of water usage, reuse and recycle of wastewater in the various sectors of the leather processing stages can reduce the intensity of tannery effluents. Further implementation of green chemicals and solvents can reduce the toxicity of the tannery effluents and ease the treatment methodology. Thus, this chapter has clearly explained the various environmental and chemical issues of the tanneries, and it also clearly explains some of the mitigation measures which can be undertaken to reduce the severity of the tannery effluents.

References

1. Kavitha PR, Ganapathy GP (2015) Tannery process and its environmental impacts a case study: Vellore District, Tamil Nadu, India. J Chem Pharmaceut Sci
2. Azom MR, Mahmud K, Yahya SM, Sontu A, Himon SB (2012) Environmental impact assessment of tanneries: a case study of Hazaribag in Bangladesh. Int J Environ Sci Dev 3(2)
3. Ahmed IM (2015) Hazaribagh tanning area and pollution management. Department of Environmental Science, State University of Bangladesh, Dhaka
4. Jamal MT, Rahman S, Tasnim G, Farooq M, Zaki M (2017) Tannery industry
5. http://www.fao.org/3/X6114E/x6114e05.htm
6. Tinni SH, Islam MA, Fatima K, Ali MA (2014) Impact of tanneries waste disposal on environment in some selected areas of Dhaka City Corporation. J Environ Sci Nat Resour 7:149–156
7. Technical EIA guidance manual for skin/leather/hide processing industry. The Ministry of Environment and Forest, India
8. Rahaman A, Afroze JS, Bashar K, Ali MdF, Hosen MdR (2016) A Comparative study of heavy metal concentration in different layers of tannery vicinity soil and near agricultural soil. Am J Anal Chem 7:880–889
9. Khan SR, Khwaja MA, Khan AM, Ghani H, Kazmi S (1999) Environmental impacts and mitigation costs associated with cloth and leather exports from Pakistan. A report on trade and sustainable development submitted by Sustainable Development Policy Institute and IUCN-P to IISD Canada for the IISD/IUCN/IDRC project on building capacity for trade and sustainable development in developing countries, Islamabad
10. J. Garai (2014) Environmental aspects and health risks of leather tanning industry: a study in the Hazaribag area, Chinese Journal of Population Resources and Environment
11. Shakir L, Ejaz S, Ashraf M, Qureshi NA, Anjum AA, Iltaf I, Javeed A (2012) Ecotoxicological risks associated with tannery effluent wastewater. Environ Toxicol Pharmacol 34:180–191
12. Lofrano G, Meriç S, Zengin GE, Orhon D (2013) Chemical and biological treatment technologies for leather tannery chemicals and wastewaters: a review. Sci Total Environ 461–462:265–281

Leather in the Age of Sustainability: A Norm or Merely a Cherry on Top?

Mukta Ramchandani and Ivan Coste-Maniere

Abstract In the area of fashion and luxury industry, there is a growing trend in consumer awareness when it comes to leather sustainability. The current chapter aims to understand and discuss for major leather users what is considered to be the major criteria for sustainable leather consumption. For example, vegan consumers refrain from buying animal-based leather but might indulge in using PU or PVC leather or other alternatives which could be ranking low in the sustainability criteria. The current chapter uses the primary data with an interview and secondary data from the industry assessments and industry cases which highlights what could be misleading for the buyers and producers. In addition, what are the perspectives of the manufacturers to develop newer sustainable methods of leather production, recycling needs and wastage are also discussed.

Keywords Leather sustainability · Sustainable luxury · Sustainable fashion · Footwear sustainability · Sustainable consumption · Decision making · Sustainable production

1 Introduction

How leather became synonymous with the luxury and fashion industry can be understood from its consumption patterns. The limited sourcing of the animal hides in the past, available only in the hands of the few, to the royals and the rich elites made it

M. Ramchandani (✉)
UIBS, Zurich, Switzerland
e-mail: muktaramchandani@gmail.com

I. Coste-Maniere
Luxury and Fashion Management SKEMA Business School, Sophia Antipolis, France
e-mail: ivan.costemaniere@skema.edu

Luxury and Fashion Management SKEMA Business School, Suzhou, China

Global Luxury Management SKEMA Business School, Raleigh, USA

Luxury Retail, LATAM, Florida International University, Miami, USA

S. S. Muthu (ed.), *Leather and Footwear Sustainability*, Textile Science and Clothing Technology, https://doi.org/10.1007/978-981-15-6296-9_2

11

distinctively a symbol for differentiation and exclusivity. But currently, we live in the age of over-consumption which has driven the planet and its beings to the brims of extinction and irreversible damages. From the lens of sustainability, the description in the leather sector cannot just be limited to the animal vs. non-animal-based hides. But needs to be seen with a larger holistic viewpoint. It must as well include the aspects of humane ways of animal hides sourcing, production in limited quantities, bi-product of the food industry, the tanning of the raw hides to the final stage and the conditions of the workers coming in contact with the production. The current chapter presents the various points of assessments for the sustainability in the leather sector. Both manufacturer and consumer views of leather are included in the current chapter as we believe these must be considered for a deeper understanding of the subject as it can help determine the future. Our research was developed using various secondary and primary sources including an interview with a leather manufacturer in India.

The manufacturing processes and the environmental impacts are a decision-making factor for companies using leather and also the consumers buying the leather products. Generally, the distinction made these days is on the basis of the tanning procedure, i.e., chromium-based tanning or vegetable tanning. The production processes of leather include curing, storage, sorting and trimming, dehairing and liming, fleshing, deliming and bating [1]. Post-tanning processes include the softening of the leather with oils and tanning agents to improve the feel and handling properties of the leather which further helps in coloring, lubricating and adding other characteristic properties like water repellence and oleophobicity [2]. The research by Laurenti et al. [2] suggests that when the life cycle assessment (LCA) is compared between the chromium-based leather and vegetable-tanned leather, it is important to consider that the vegetable tannins come from the trees (a renewable resource), whereas the chrome is mined and used once in leather. This aspect of how a fair comparison can be made between renewable natural materials and materials that come from fossil fuels or from one-time use of a mined resource is a major area which needs consideration. Furthermore, the impact categories need to be evaluated for such comparisons with toxicity. However, another consideration from our perspective should also be the effects of the production procedures on the workers working at these tanning factories and the hygiene practices, which could be different because the two tanning processes have different levels of the carcinogens and skin irritant chemicals. In the next sections of this chapter, we include the background on animal-based leather and non-animal-based leather, greener paradigms for consideration by manufacturers and brands, interview with a leather manufacturer, and key decision-making factors for leather consumers.

2 Methodology

In order to understand the overall sustainable practices in the production and consumption, we gathered our literature from exploring various scientific research articles, reports and other publications, which included assessments of the previous literature as well. In the primary data, we have conducted an interview with a leather manufacturer based in Kanpur, India, to research in depth their view on key sustainable issues, pitfalls faced by a leather manufacturer and suggestions for the future.

3 Background

Producing raw materials for the fashion and luxury industry comes along with damages to the environment, eco-systems and people. A major challenge in the race to be more profit-oriented fashion company and being sustainable at the same time stems from the fact that the textile industry is highly dependent on the use of chemicals, which are primary reasons for pollution and causing environmental degradation. It is not a new knowledge that a step toward a better sustainable future inclines toward the usage of sustainable long-lasting materials which cause minimal emissions in the environment. But when companies mislead the customers in the name of sustainability, it reflects poorly on the industry as such. Moreover, with the latest developed smart textile materials like bio-fabricated leather or synthetic leather look-alike, the quantitative data on the long-term environmental impacts is limited. Several claims of sustainability have lead users to believe that a company is sustainable. But what part of sustainable goals are these companies fulfilling is the question to be asked?

Considering, for example, the usage of water as a resource in the textile industry, a heavy metal laden water bi-product used in the dying or washing process can be of adverse damage not only to the workers but also to the soil and ground-water basin. Affluent treatment plants would consider filtration of the water before discharging from the factories, but it requires better invigilation systems and checks. The next sections highlight some of the differences in the animal-based leather and non-animal-based leather.

3.1 Animal-Based Versus Non-animal-based "Leather"

When it comes to leather, the basic norm in the consumption historically is related to the animal-based leather from the raw skin hide tanned to the usable conditions. In the past, the animal-based leather has been sourced from cattle-based animals like cows, goats and pigs as a subset of the bi-product from the meat industry. From the luxury sector, the animals not part of the popular food consumption innovated to exclusivity with the deriving leather from the animals like alligator, crocodile, snakes and ostrich. With the rampant unsustainable methods of sourcing animal hides, like torturous methods of killing the animals, the attention from luxury and exclusivity shifts more and more toward sustainability.

- **Oxymoron terms**

  A report from Henkel [3] shows that the German leather industry (VDL) has been taking into consideration the misleading of the term "vegan leather" used by certain companies producing shoes, bags and furniture. As the term leather itself means coming from the animal, it cannot be used as a terminology for synthetic materials made of plastic petroleum based like PU and PVC leather. Vegan leather does not mean it is genuine leather, but what does it signify to the consumers? The VDL has accelerated their process of bringing into law the clear terminology for non-genuine leather. One good example is of Pinnatex which produces leather look-alike materials from the pineapple skins, setting a step forward in honest terminology. Obviating the use of the honest terminology and "synthetic leather" where ever applicable, adds to the greenwashing marketing elements of the consumers in the name of sustainability.

- **Cactus Leather**

  One of the latest innovations in the non-animal-based leather sector has been the development of the cactus leather named **Desserto**. As per their website [4], Desserto is a highly sustainable plant-based vegan leather made from cactus, distinguished by its great softness at touch while offering a great performance for a wide variety of applications and complying with the most rigorous quality and environmental standards. Their aim is to offer cruelty-free, sustainable alternative, without any toxic chemicals, phthalates and PVC. It is also partially biodegradable and has the technical specifications required by the fashion, leather goods, furniture and even automotive industries. The company claims of using little amount of water for growing the cactus plantations using mainly the rainwater irrigation systems and every 6–8 months having a new harvest in the same plantation. The process induces less energy by using only the mature leaves of the cactus plant which are dried under the sun for few days to achieve the desirable humidity levels and does not use any herbicides or pesticides.

- **Fish leather**

  The usage of the fish leather was traditionally by the indigenous heritage of the Arctic people, specifically the Inuit, Yup'ik and Athabascan of Alaska and Canada, in Siberia with the Nivkh and Nanai, the Ainu from the Hokkaido Island in Japan, Sakhalin Island Russia, the Hezhe from northeast China, and the people from Iceland [5]. The usage has now been taking into consideration for its sustainable features as more and more alternatives for sustainable leather are being researched. In particular, the highlights of the fish leather are that it does not require the resources or leave the carbon footprint associated with raising cattle and does not use endangered species that could threaten biodiversity. Mainly the salmon or cod from Iceland [6] is used or the pirarucu fish from Brazil, which is the staple food for the locals, and the skin is the bi-product of the food consumption. Its latest adaptation in the luxury sector can be seen in the Rick Ovens S/S20 collection [5].

4 Production Procedure and Supplier Norms in the Leather Industry

The tanning processes includes the chemical and mechanistic processes to transform the raw hides from the animals into a refined material used widely in different products like shoes, bags and upholstery furniture. But with respect to the environmental wastage, there exists bans and international restrictions for the export and import of the leather products. For instance, the European Union has banned the use of toxic chemicals in textiles and in leather with chromium VI. They promote the eco-friendly dyeing processes meaning that the chemicals that have the properties of being carcinogenic, mutagenic and toxic for reproduction or have metal complexes must be removed. Printing pastes cannot contain more than 5% volatile organic compounds, and chlorine bleaching is also prohibited [7].

The supplier stability is also a limiting factor for small- and medium-sized business in the leather sector. A lack of knowledge on the newer materials for the manufacturers could delay them in their supply. Suppliers are mostly driven by the demand in the market if the demand is for faster production of large quantities of the leather goods generated by the fast fashion brands. The suppliers do not have the capacity to cater to the sustainable leather like eco/vegetable tanned. Due to the fact that it takes longer to tan them compared to the chrome tanned leather. Additionally, when the sustainable leather does become available, the prices charged are much higher in comparison. Which also leads to the consumers lacking alternatives in the market for more sustainable consumption.

In the shoes and bags production as well, it is not just about the type of leather used. In the shoes and bags production as well, the sustainability criteria must not be just limited to the type of leather used or the leather tanning procedure. But consideration of even the glue used in assembling the parts of the product must be sustainable and free from harmful chemicals. One problem is that the glue used when inhaled in humid or hot weather can be harmful to the worker's health.

5 Greener Paradigms for Sustainability from Manufacturer/Brand Perspective

It is crucial to understand where the sustainable actions from the brands and companies end up. To begin with this purpose, it must be clear to the brands what are the visions and goals for the future. The responsibility lies in the production and consumption phases. A brand can aim to be producing more sustainably but may be lacking in maintaining the sustainable consumption pattern, or vice-versa. Both of which can shape the path for benefits or destruction of the natural resources.

In the first part of the paradigms, we describe the production, the second part describes the consumption, and the third part enriches the reader about the biodegradability aspect so that the leather product does go to landfills and minimize wastage.

Additionally, there can be many facets to include in these paradigms like supply chain and it can be extended to further parameters, but our focus in the current chapter is to help the readers with some starting points in their complex decision-making factors for manufacturing.

- **Production**

  Companies and brands producing materials and textiles for fashion must be responsible for their production processes to achieve their sustainability goals. Parameters like energy waste, harmful chemical emissions, wastage of water and other resources should be minimized. The use of harmful substances harms not only the workers involved in the production process but also the environment. For example, the use of heavy metals in textile dyes leads to the higher levels of toxicity. Other chemicals including alkylphenols, phthalates, formaldehyde, amines, etc., are hazardous to the environment [8].

  In the production of new age smart raw materials, the life cycle assessments (LCA) must be considered in order to predict the long-term use of the leather and the resources required for its production.

- **Consumption**

  It starts from the time the products are bought and starts to be utilized by the consumer. The duration of usage of the materials can last long or short depending on the frequency of usage, the quality and durability of the material, that is, for how long the produced material can be used and to what extent it will be intact in the long-term usage? The durability of natural materials like genuine leather is widely known to be long lasting in terms of different weather conditions, protection gear, atmospherics in the retail and storage environments. Therefore, the luxury and fashion brands must inform the consumers regarding the repairability of the materials. The consumption pattern for luxury consumers is different, where a high-end expensive material like silk or genuine leather is kept for decades and does not lose its value over time. Similarly, such long-lasting materials are the need of the hour to maintain the value of money and resources spent on producing textiles and the final product. As a company or a brand, it is important to also communicate effectively and be transparent about the sustainable practices so the end-user consumption can benefit and become more satisfied with the products purchased. As the demand for the new products will not be curbing down all of a sudden, the consumption to be sustainable should be a slow-paced process which needs to be dealt with inclusive of chemicals and processes that are organic and natural.

- **Biodegradability**

  The aspect of biodegradability is very important one for the sustainable paradigms. Especially when the oceans are predicted to be having more plastics in the future than the fishes and the other marine organisms, it is vital to reduce the waste generated. Also as seen in certain reports about the fashion industry, the pre-consumption wastage of unsold textile materials leading to the landfills every year is a major problem to be solved. Despite retail stores, giving high discounts and overseas shipping solutions, the problem still remains. One perspective that emerges as a solution is to tackle the need of newly produced materials and curbing the consumption levels in the society with the opportunities of recycled, upcycled and secondhand materials. But considerate to the fact that consumers especially who are fond of the luxury sector need more latest designs and exclusivity will not be leading the way for secondhand materials and their demand for new clothes.

6   Interview with a Leather Manufacturer

The city of Kanpur in India is situated on the banks of the river Ganges and famous for its leather tanneries and exports to the American and the European markets. The leather industries in Kanpur were set up since the British colonial times in India to cater to the leather demands. As a result, boosted the economy of the city and the state but ended up being one of the highly polluted cities in the world and polluting the Ganges. We conducted an interview with one of the leather manufacturers in Kanpur Mr. Zafar Iqbal, to understand how the production side of leather aims to tackle the sustainability issue; below are the questions and answers from the interview conducted.

Q1. *What types of leather do you produce in your factory?*
We produce several types of leather: Vegetable-tanned leather used in the sole. Belting leather used in the production of belts and the watch sector. The third type is the industrial leather used in machines which is vegetable tanned. Next is the chrome tanned leather using fat liquors, syntans like phenolic, sulfone, protein, maleic, melamine, dicyandiamide, polypeptide and acrylic, glycerin, fish oils for softness mellowness, stretchable and color fastness specifically used in the footwear industry.

Q2. *Is vegetable-tanned leather used widely in shoes?*
Vegetable-tanned leather is used mainly for thinner and soft linings like the inner of the women's shoes.

Q3. *What is in the sustainability/eco-friendly sector for the shoes sector?*
Following the government standards, we have to use the primary effluent treatment plant (ETP) at our factory meaning that the any non-diluted waste has to be stopped by us and not be discharged into the sewage. Our PH levels of the water initially are between 11 and 13 which is re-treated to be neutralized and reduced to the levels of 7–9. Then, from the ETP, the wastewater is discharged to common effluent treatment plant (CTP) which is from the government of India, where it again gets treated before finally being discharged.

Q4. *Yet why is it so polluted?*
It is not because of the leather sector that the Ganges is polluted, but because of the other industries which do not currently have as strict norms as the leather sector in Kanpur. The drainage system does not cater to the huge population of the city of 7 million people.

Q5. *What do you think of the latest innovative leather such as the pineapple skin-based leather, mushroom leather or the cork leather?*

In my opinion, it is merely for fashion shows and the trend. These types of leather cannot be for long-term practical use. Cork, for example, has been mainly used in the sole of the shoes as it has soft tissues. It is a different thing that they could use the cork on the shoe uppers in parts by gluing or other methods but for the entire shoe it cannot be as durable as the genuine leather. Practically for walking on streets or nature, the cork leather will not be able to withstand the roughness. It is merely a marketing and sales campaign for attraction to the cork leather in the shoes sector.

Q6. *What is your view of the recycling of genuine leather?*

The recycling in genuine leather or the recycled leather is no longer a genuine leather for me as the process converts it into something like a man-made leather which cannot be equivalent to the real form of the genuine leather. In textiles sector, it is possible as the raw materials can still be close to its original form but not in leather.

Q7. *What is the future of leather? Specifically for catering to the wastage? Like overproduction and overconsumption of leather goods?*

Natural resource like genuine leather should be used, and there should be a strict norm for amending the use of natural leather only in the shoes sector. As the animals are in the limited supply, the availability of the skins is scarce, which will make it more expensive and costlier making it into something more preserved and longer lasting. Consequently, it will minimize the consumption and people will use it with care to make it long lasting with the help of polish etc. This cannot be possible for a cheap pair of shoes. As the manufacturers keep producing the fake synthetic leather for shoes using much more harmful chemicals for processing. The preservability aspect of it besides the long-lasting impact will not be of importance to the consumers.

7 Environmental Assessments and Tests

The life cycle assessment (LCA) is a tool used to evaluate the environmental impacts of products over their full life cycle, including resource extraction, production, use, transport and end-of-life (EoL) stages [9]. The certifications and test for leather are done for the assessment of the hazardous chemicals or substances like testing allergenic disperse dyes, nickel in metallic parts in contact with the skin and dimethyl fumarate, detecting banned azo dyes, incl. P-aminoazobenzene, pentachlorophenol (PCP), chromium VI and cadmium and for footwear test methods for the assessment of ecological criteria (EN 14602).

According to Brugnoli and Král [10], there are numerous ways of calculating the LCA and product carbon footprint (PCF) in the leather sector but a harmonized way is necessary. Their research proposes that first the leather should be distinguished into one of the three considerations from the milk/meat industry:

- co-product,
- by-product,

- waste.

The methodology used in their report [10] included identification and analysis of existing standards and publications, selection of the reference standard (ISO DIS 14067) and analysis of the requirements of the reference standards (analysis of the different sectorial approaches and proposed harmonized methodology).

In their analysis and detailed work, some of the key results included different actors of the leather values chain and guidelines for obtaining reliable leather PCF data. Furthermore, Brugnoli and Král [10], shared the following key elements which serve as the recommendation for the convergence and harmonization of the finished leather LCA and PCF:

- **Functional Unit**: The functional unit in general should be measurable consistently, and it shall correspond to the basic unit that the tannery uses for trading the finished leather it produces; the proposal is to use, as functional unit, 1 m^2 of finished leather (1 kg in the case of sole leather), including an indication of the thickness of the material.
- **System Boundaries**: For the LCA and PCF of finished leather deriving from hides and skins which have been raised mainly for milk and/or meat production, it must start in the slaughterhouse, where the treatments are carried out to prepare the hides used in tanning and end at the exit gate of the tannery.
- **Quantification**: The harmonized methodology proposed in order to obtain Kg of CO_{2e}/m^2 of finished leather, lies in the quantification of CO_{2e} content of all the different products and material entering the tannery (upstream processes), adding CO_{2e} produced in the tannery itself (core processes), as well as CO_{2e} emanating from water and air purification and waste recycling/disposal (downstream processes).
- **Allocation**: In the leather making process, allocation shall be avoided whenever possible and, if unavoidable, it should be made according to the physical relationship within the single process under consideration.

8 Decision-Making Factors by Consumers

According to an interview reported in the research by Streit and Davies [11], the consumers for luxury goods do not consider "ethics" to be the priority factor for purchases. In addition, if a luxury product is ethical it is perceived more like a bonus factor from the brand. Their research also reported that luxury brands in general are perceived to be ethical by consumers, but it could be a beneficial factor for the consumers to know more if the brands would share more information about the material productions and supply chain. Several other researches have been done in understanding the types of consumers for the purchase of sustainable luxury fashion goods such as prosocial behavior, status consumption, costly signal theory, eco-conspicuous and eco-conscious consumption [12].

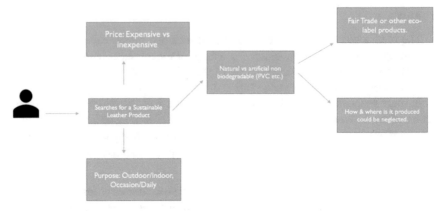

Fig. 1 Search and decision-making factors for purchase of sustainable leather products

From our research and experience in this sector, we propose that the following factors (Fig. 1) will help in understanding the consumer's search for sustainable leather product and decision making for the purchase:

1. **Price**: The leather product shoes/bags/apparel, etc., are subjective to price sensitivity level of each individual consumer. However, the more sustainably compliant the brand is, the prices charged tend to get higher. From the standpoint of financial consideration of purchasing leather, customers searching for high-quality genuine leather also associate the high prices for the quality and longevity. However, when it comes to bio-fabricated substitutes of leather, the prices do not justify the quality for majority of customers.
2. **Purpose**: If the product is bought for the purpose of outdoors or indoors. For example, the leather safety shoes worn by workers everyday ascertain the safety regulation vs. leather shoes meant for occasional use or indoor work.
3. **Type of leather**: Natural genuine animal leather or bio-fabricated synthetic leather look-alike which may mislead the consumers to an extent. Scarce leather sourced from crocodile, snakes, ostrich and others qualify for the higher prices due to the limited supply.
4. **Certifications and tests**: The buyers of the raw materials from factories-based overseas should not just solely rely on certifications which are given to the large firms and factories but must try to inculcate the small manufacturers and small suppliers overseas. Sometimes the certifications can be misleading as not all firms practice and adhere to the norms laid by the certificating companies, and one example is the case of Rana Plaza collapse where the manufacturing unit also claimed to have had certifications for better working conditions. By inculcating small manufacturer those who cannot afford the hefty fees of certifications, it becomes more a responsibility to further check and invigilation in terms of keeping regular and maintaining records of the safe working conditions of employees at the production houses, the hygiene standards, the wages paid and

impact on the people. Which naturally could be a time-consuming process but yields better sustainable results in the long term.

5. **Place of production**: Once the primary decisional considerations are made by the consumer based on the above factors, the place of production can play an important role. For example, there is a general notion in the European and American consumers of products being of inferior quality if made in Asia vs. made in Europe or USA [13]. Albeit many of the leather products claiming to be made in Europe are in fact partially or fully produced in Asia.

9 Discussion

Conclusively, from our research, we find that the leather production cycles must be limited for supply of quantities but higher in quality following the environmental regulations, be inclusive of having the fair price model than a mere high profit-oriented model. In addition, the manufacturers need to be more proactive on taking a lead for innovation of sustainable production methods of high-quality leather and not be dependent just on the market demand, i.e., being more a market driving company and taking the lead than be driven just by the market. They must no longer be ignorant of the high need for more sustainable reforms and LCA assessments of their different stages of the leather production. There is a higher possibility for the leather industry to emerge as a role model sustainable material due to its longevity, repairability and biodegradability compared to the other materials. In addition, the brands should not mislead the consumers and use the honest terminology. For the consumers, the more the awareness and education on the materials used by the brands, the more sustainable consumption can emerge, which is a necessary step to promote the slow fashion and a holistically sustainable maintenance and innovation of the leather industry. Nevertheless, it is difficult and complex to have a perfect system of sustainable leather production and consumption, but the future lies in the hands of those who can create the innovative technologies which is less evil and pave the way for better environmental preservation.

References

1. Black M, Canova M, Rydin S, Scalet BM, Roudier S, Sancho LD (2013) Best available techniques (BAT) reference document for the tanning of hides and skins. JRC 83005. EUR 26130 EN. ISBN 978-92-79-32947-0 (pdf). European Commission; Joint Research Centre; Institute for Prospective Technological Studies, Seville, Spain
2. Laurenti R, Redwood M, Puig R, Frostell B (2017) Measuring the environmental footprint of leather processing technologies. J Ind Ecol 21:1180–1187. https://doi.org/10.1111/jiec.12504
3. Henkel R (2019) Is the 'vegan leather' label misleading consumers? https://fashionunited.uk/news/fashion/are-manufacturers-misleading-consumers-using-the-label-vegan-leather/2019102845933
4. Desserto (2020) https://desserto.com.mx/desserto

5. Mallon J (2019) Is fish skin the new frontier for eco-friendly fashion? https://fashionunited.uk/news/fashion/is-fish-skin-the-new-frontier-for-eco-friendly-fashion/2019110546042
6. Atlantic Leather (2020) http://www.atlanticleather.is
7. European Commision Eco-Label. https://ec.europa.eu/environment/ecolabel/digital_toolkit.html
8. Rozas AC (2017) Textile toxicity: what lurks in your clothes. Retrieved from https://fashionunited.uk/news/fashion/textile-toxicity-what-lurks-in-your-clothes/2017061424828. Retrieved on 2 Aug 2019
9. ISO (International Organization for Standardization) (2006) ISO 14040:2006: environmental management—life cycle assessment—principles and framework. International Organization for Standardization, Geneva, Switzerland
10. Brugnoli F, Král I (2012) Life cycle assessment, carbon footprint in leather processing (review of methodologies and recommendations for harmonization). In: Eighteenth session of the leather and leather products industry panel, UNIDO. https://leatherpanel.org/sites/default/files/publications-attachments/lca_carbonfootprint_lpm2012.pdf
11. Streit CM, Davis IA (2013) Sustainability isn't sexy an exploratory study into luxury fashion. In: Gardetti MA, Torres AL (eds) Sustainability in fashion and textiles: values, design, production and consumption. Greenleaf Publishing Limited, Sheffield
12. Ramchandani M, Coste-Manière I (2018) Eco-conspicuous versus eco-conscious consumption: co-creating a new definition of luxury and fashion. https://doi.org/10.1007/978-981-10-8285-6_1
13. DW (2020) https://www.dw.com/en/luxury-behind-the-mirror/av-51115721

Blockchain Technology in Footwear Supply Chain

Hao Cui and Karen K. Leonas

Abstract This research investigated how adopting blockchain technology will strategically place footwear companies in a competitive market. The article attempted to answer three central research questions: (1) Should a footwear company adopt blockchain technology in its supply chain? (2) What can footwear companies gain from the adoption? (3) Where can the blockchain position the footwear companies in the competitive market if the implementation of the technology is in the supply chain? First, we explored the business model and the features of the footwear industry and examined the blockchain technology, which provided the information from the characteristics of the industry and the advantages of the technology as the inputs of the TOWS matrix. Next, the author highlighted what benefits the footwear companies could get from the adoption of the technology based on the TOWS matrix analysis. Also, the author analyzed where the blockchain would position the company if the adoption of the technology is in the supply chain by investigating the definition and characteristics of the disruptive technology and Roger's diffusion of innovations theory. Finally, the author concluded that technology still holds for promise though the implementation of the technology has several barriers and challenges, given the potential of the technology. In terms of the scope of this research, this chapter focused on catering to the needs of footwear giants (i.e., Nike, Adidas) due to their urgent needs of creating a transparent and traceable supply chain.

Keywords Blockchain · Footwear · Supply chain · TOWS matrix · Diffusion of innovations · Disruptive technology · Transparency and traceability · Competitive position

H. Cui
Wilson College of Textiles, North Carolina State University, Raleigh, NC, USA
e-mail: hcui6@ncsu.edu

K. K. Leonas (✉)
Textile and Apparel, Technology and Management, Wilson College of Textiles, North Carolina State University, Raleigh, NC, USA
e-mail: kleonas@ncsu.edu

© The Editor(s) (if applicable) and The Author(s), under exclusive license to Springer Nature Singapore Pte Ltd. 2020
S. S. Muthu (ed.), *Leather and Footwear Sustainability*, Textile Science and Clothing Technology, https://doi.org/10.1007/978-981-15-6296-9_3

23

1 Introduction

1.1 Background

Consider this scenario: When you walk into a fashion store, you pick up one Rounded-Toe Pump made in Vietnam. You like the style and color and decide to bring it back to your shoe rack. If someone told you that this product is a knockoff and was manufactured in a sweatshop that has been discharging tons of hazardous wastewater into the local river, would you still purchase this product? This billion dollar question bothers most of the footwear companies because of the customers' concerns about buying the counterfeit product and increasingly awareness of the importance of sustainable/responsible purchases. According to Muthu [36], the production of the footwear products poses a negative impact on the environment in several ways, given the resources used, energy consumed, and wastes generated in the life cycle of the product. Doorey [11] further indicated that social factors such as working conditions in the manufactures could affect consumers' decisions over their apparel and footwear. In response to these impacts and concerns from the consumers, most of the apparel and footwear brands have committed to follow up United Nations Sustainable Development Goals (UNSDGs), which could support brands to mitigate social and environmental impact throughout their value chain. Some sustainability pioneers even have gone beyond this commitment and have been starting setting up science-based targets (SBTi) and goals such as zero discharge of hazardous chemicals (ZDHC) to tackle more environmental issues such as climate change and the toxicity released from the environment. However, scholars and sustainability practitioners have reprimanded that these apparel and footwear brands are just considering sustainability as one of the elements in companies' public relations strategies [29]. That is to say, the brands and retailers are just making commitments and setting up goals to be in favor of their Greenwashing Public Relation (PR) strategies and therefore boost their sales accordingly. Apart from the social and environmental issues, the apparel and footwear industries are also urgently working on addressing the fake product issue given the products are easily being imitated. Carpenter and Edwards [6] found out that the counterfeit products did threaten the manufacturers and retailers of originator products. According to [31], the amount of counterfeit goods and fake products has skyrocketed thanks to the trends of globalization and online shopping. It is no doubt that fake products could undermine the brands' intellectual property rights and even the brands' reputation by compromising the quality of the knockoff products. Although brands are using a variety of product authentication verification methods (i.e., hidden printed messages, color-shifting inks, etc.) and adopting new product tracking and tracing technologies (i.e., radio-frequency identification, barcodes, etc.), the knockoff manufacturers, however, could always find a way to avert these methods and technologies. The recent "breakup" between Amazon Inc. and Nike Inc. is another solid evidence of the severities of the knockoff issue.

In response to the sustainability and counterfeit issues, what actions footwear giants such as Nike Inc. and Adidas AG should take to convince their customers that they are buying the authentical products and making conscious decisions? According to Doorey [11], information disclosure could mitigate the social sustainability impact from the supply chain, address the forced labor issues, and improve labor practices. James and Montgomery [21] also identify the consumers' demand for the sharing of information throughout the supply chain from apparel and footwear retailers [21]. In the meantime, the industry is foreseeing a trend to go beyond auditing, look beyond the first-tier manufacturers, integrate sustainability to core business practices, and bring transparency to the supply chain [23]. Several papers and researches also back up the trend and customers' needs for information disclosure, highlighting the importance of creating transparent and traceable apparel and footwear supply chain [24]. There is one shared keyword in these researches: information disclosure. Information disclosure renders the brands to build a transparent and traceable supply chain and therefore convinces customers that they are making a sustainable and conscious decision. Information disclosure is not only mitigating social and environmental impact throughout the supply chain but also addressing the counterfeit product issues by creating a traceable supply chain. Li [31] identified the potential of adopting tracking technology to disclose the production information throughout its entire supply chain and therefore secure the authenticity of the products. Due to the important role of information disclosure throughout the supply chain, this research will explore how footwear giants could unveil the "opaque cover" of their supply chain in the rest sections of this article. This chapter started off with the background of the importance of creating a transparent and traceable supply chain for the footwear industry and chose the technology that has the potential to address the issue. Next, we discuss the objectives and methods we used in this chapter. Then, we presented a brief introduction to the footwear industry and blockchain technology. In the following section, we analyzed what benefits could bring to the supply chain and what competitive position could bring to footwear companies by adopting the technology. The section following this investigates the application cases and barriers of the technology. The final section wrapped up with the conclusion and future suggestions on the adoption.

1.2 Available Technologies and Tools

The researchers will highlight three different technologies and tools: radio-frequency identification (RFID), Higg index, and blockchain to answer the question on creating a transparent supply chain. After the brief introduction on these three technologies and tools, the researchers will further compare them from different perspectives and find out what technology would fit in the footwear supply chain and address the emergency of information disclosure.

1.2.1 Radio-Frequency Identification (RFID)

Radio-frequency identification (RFID) could automatically identify and track tags where digital information is electronically encoded [51]. According to Kwok and Wu [28] and Li and Zhao [30], RFID makes the textile supply chain—from raw material production to garment production—more transparent and traceable and improves the productivity throughout the supply chain.

Kwok and Wu [28], Li and Zhao [30], and Bhagwat [2] listed several benefits and challenges of adopting the technology throughout the supply chain. In terms of the benefits, the technology could provide real-time inventory data and reduce the error and cost in terms of the wrong data-based decision. However, the technology has one significant flow in terms of data security, where the data generated from the technology is mutable and easily tampered by either external or internal parties.

1.2.2 Higg Index

Higg index is a tool to measure the sustainability performance of brands and manufacturers in the apparel and footwear industry based on the explanation of the Sustainable Apparel Coalition (SAC),[1] the creator of the Higg index. Lou and Cao [32] also provided a more comprehensive explanation regarding the Higg index and stated that it could help the participants to compare their sustainability-related performances among the peers in the industry and guide the customers to make the eco-conscious and social-conscious choice. When it comes to the supply chain and production, Higg material index and Higg facility index could help the decision-makers in the retailers to capture the social and environmental impact of a product from raw material extraction to final product manufacturing.

1.2.3 Higg Material Sustainability Index

Higg material sustainability index analyzes the environmental impacts that are categorized into different classification and renders the target audiences look at the eco-related issues derived from the materials from a different perspective. The Higg material index will walk the audiences through five main environmental impacts, most related impacts related to the apparel and footwear industry, namely global warming, eutrophication, water scarcity, abiotic depletion potential, and chemistry (see Table 1). The Higg material sustainability index could help the decision-makers to choose the right materials based on their sustainability agenda. For example, brands who are prioritizing tacking the chemistry and climate change issues could select

---

[1]Sustainable Apparel Coalition: is the apparel, footwear, and textile industry's leading alliance for sustainable production.

Source: https://apparelcoalition.org/the-sac/.

Table 1 Explanation and examples of the five environmental impact categories

Environmental impact	Definitions
Global warming	The effect leads to climate change
Eutrophication	Excessive richness of nutrients in a lake or other body of water
Water scarcity	Excessive use of water resources leads to the lack of the water
ADP	Depletion of non-renewable resources, i.e., fossil
Chemistry	The hazardous chemicals used in the production processes

Sources Adapted from https://apparelcoalition.org/the-sac/

the materials with less impact on chemistry and global warming potential for their material extraction processes.

1.2.4 Higg Facility Index

Higg facility index provides a tool for manufactures (i.e., tanneries, material mills, and shoe manufacturers) to evaluate and quantify the sustainability performance of their manufacturing units. The manufacturers are required to conduct self-assessment by filling out the questionnaire from the Higg facility module covering both environmental and social topics. The environmental module covers topics such as energy and water usage and environmental management system while the social part includes topics in terms of recruitment and hiring, wage and working hours, and health and safety, etc. The normalized results will be used as a benchmark for manufacturers to compare the social and environmental performance against those from their peers.

1.2.5 Blockchain

Blockchain is a digital ledger with blocks of information linked together. Each party on a blockchain has access to all the information, and no single person or entity controls a blockchain. Because each block of data (transactions) is linked to all previous ones ("forming the chain"), unlike the radio-frequency identification (RFID), the blockchain is immutable [50]. Although this technology is deployed mostly in the financial sector, there is a huge potential of adopting this technology in the supply chain to address transparency and traceability issues in the supply chain.

Accordingly, blockchain offers several advantages for sustainability performance and tracking in supply chain management with some accompanying barriers [26].

Benefits of blockchain:

- Recording and validating transactions efficiently and permanently (e.g., self-executing contracts);
- Auditing of all transactions (an audit trail that cannot be altered);
- Tracking time and location of actions;
- Reducing the risk of data loss or fraud;
- Exerting pressure on supply chain partners toward greater accountability; and
- Providing immediate access to transactions occurring in the supply chain.

Barriers of blockchain:

- Differences of laws, regulations, jurisdictions, and institutions across a global supply chain;
- Collaboration among all relevant parties; and
- Implementation of technology-based solutions for suppliers, both in developed and developing countries.

1.3   Comparison Among Different Technologies

In this section, the research compares the three technologies and methods from six perspectives: (1) transparency (2) traceability (3) speed (4) security (5) compatibility (6) cost.

1. Transparency: The information evaluates how the data could be disclosed transparently as per request. In other words, the original data is available when needed.
2. Traceability: The data throughout the supply chain could be tracked down and verified. Consumers could dig into the details about how the product is made (i.e., the raw materials, manufacturers' info).
3. Speed: Real-time data could be captured; it also assesses the time needed for interactions and communications . How long the data could be processed and obtained throughout the system?
4. Security: This information shows if the technology could secure the cybersecurity of the data. That is to say; the data is not being able to be hampered and hacked.
5. Compatibility: This perspective evaluates if the technology is easily integrated into the system. Are there any skillsets workforce should have if the new system is introduced?
6. Cost: The cost of the installation of the system and of maintenance and operation for the system.

Table 2 analyzes the performance of these six perspectives for the three technologies. Following literature provided the author with the hints of creating this table [2, 28, 26, 30, 32, 50] and the author's frontline working experience in the supply chain.

Matrix A presents the measurement of performance from these six perspectives for the three technologies based on above analysis. i represents the number of the

Table 2 Performance dimension of these technologies in the supply chain

Performance dimension	Blockchain	RFID	Higg index
Transparency	Information could be uncovered, and the original data is accessible when needed	Information could be uncovered, and the original data is accessible when needed	Information could be uncovered, and the original data is accessible when needed
Traceability	The data throughout the supply chain could be tracked down and verified. Consumers could dig into the details about how the product is made (i.e., the raw materials, manufacturers' info)	The data throughout the supply chain could be tracked down and verified. However, the data needs to be verified through addition steps given the data is easily being hampered at facility level	The Higg index is not able to track down all the information through the entire life cycle of the product due to the data collection methods. In addition, the data needs to be further verified and audited
Speed	Real-time data could be captured. With the integration of IoT devices, the time for interactions and communications will be brief. The data could be processed and obtained immediately throughout the system as per request	Real-time data could be captured through RFID devices and chips. However, it will take time for data to be communicated between the RFID hardware and the ERP system the facility is using. The physical limitation, such as Internet bandwidths, will also have an impact on the quality and speed of the interaction	Real-time data could not be captured through the Higg index. The data will be submitted manually by the facilities manager
Security	The data is immutable. The hackers are rarely able to hack the data given the existing computing power most hackers have	The data is easily being hampered. Without the protection mechanism, the data is also easily being hacked	The data is easily being hampered and hacked (Given all the information is stored in the server provided by SAC)
Compatibility	With the Internet of things (IoT) devices and blockchain platform, the technology is easily integrated within the existing supply chain system. However, the full deployment of the technology needs the workforce who has the knowledge of blockchain platform and IoT devices. The compatibility of the blockchain is quite similar to other two technologies	Most of the wireless communication hardware for RFID could be seamlessly integrated into the production line. However, the system needs technicians who could work on extracting the data from the software and understand how to maintain the RFID devices and chips	No additional devices will be integrated into the system. However, in some cases, it will require the facilities to have sensors or electronic devices to collect the data. The system needs experienced experts who can conduct the self-assessment for the facility

(continued)

Table 2 (continued)

Performance dimension	Blockchain	RFID	Higg index
Cost	Although the cost of generating blockchain digital certification by using the existing platform will be low (i.e., IBM and AWS blockchain platform), the cost for installment and maintenance and operation of the IoT devices and sensors which provide the data to the digital ledger could be high. Thus, the cost is comparatively higher than RFID and Higg index	The cost of the RFID consists mostly of hardware such as wireless communication devices and chips and software. Given the development of technology, the cost of the hardware has been reduced these years, and the cost will be lower than the blockchain	Higg index does not require any hardware devices, though the self-assessment does need the educated workforce to finish the Higg index. In addition, the self-assessed data needs to be verified through external audits. The cost of human resources and verification of the results will make it much higher than RFID, maybe in some cases higher than blockchain

row of the matrix which refers to technologies (blockchain, RFID, Higg index) while *j* represents the number of the column of the matrix which refers to the six aspects (transparency, traceability, speed, security, compatibility, and cost). a_{ij} stands for the measurements of the performance for assigned technology in particular aspects (i.e., a_{11} represents the measurement of the performance for blockchain from transparency perspective). All the measurements are on a scale from 1 to 5. 1 means very bad, while 5 indicates very good performance in the aspect. Y_i is the sum of all the measurements for a particular technology without assigning any weight to each perspective.

$$A = \begin{bmatrix} 5 & 5 & 5 & 5 & 4 & 1 \\ 5 & 4 & 3 & 2 & 4 & 3 \\ 5 & 3 & 1 & 1 & 4 & 2 \end{bmatrix}$$

Matrix A is the measurement of the performance

$$Y_1 = \sum_1^6 a1_j = 25 \tag{1}$$

$$Y_2 = \sum_1^6 a2_j = 21 \tag{2}$$

$$Y_3 = \sum_1^6 a3_j = 16 \tag{3}$$

Equations (1)–(3) represent the calculation methods of the measurement of the performance of three technologies based on the six evaluation metrics mentioned above. Based on the above calculation, the blockchain stands out among three technologies. Thus, next, this research focused on exploring the blockchain technology and figure out how blockchain could be used in the context of the footwear supply chain.

2 Objective and Method

2.1 Objective

In this chapter, the author attempts to answer a central question: Should footwear companies adopt blockchain technology in its supply chain? Two sub-questions revolved around the central question are raised and explained in the following sections, including (1) What can the footwear companies gain from the adoption? (2) Where can the blockchain position these companies in the competitive market if the implementation of the technology is throughout the companies' supply chain? The explanation of questions could facilitate the audiences to understand why blockchain technology could address the issues that footwear companies are facing and how early adopters could stand out among their peers and competitors.

2.2 Method

To elucidate these research questions, the authors analyzed the overview of the footwear industry, including the industry characteristics, production processes of footwear, and the supply chain structures and the challenges throughout the supply chain. In the next section, the research delved into the topics on what is the blockchain technology and how blockchain technology could address the issues throughout the supply chain. The study then looked into the key terms and popular blockchain services and frameworks. The research concluded this section by analyzing what benefits this technology could bring to the supply chain. After exploring the blockchain technology, authors cracked the gains from the adoption of the technology by analyzing outputs from the threats, opportunity, weakness, and strength (TOWS) matrix. Following the TOWS analysis, this study defined the strategic position in which the footwear companies will be by exploring the definitions of disruptive technology and theories of diffusion of innovations. The authors then investigated the applicable area of this technology and picked three examples of how this technology has supported supply chains in other industries. Finally, the authors explored the challenges most of the footwear companies will be facing and the potentials of this technology if these companies adopt the technology before concluding this study with the author's recommendation on whether footwear giants should adopt this technology throughout the supply chain and what this technology could bring to the company.

3 Footwear Industry Mapping

3.1 Footwear Industry Overview

Apparel and footwear industry is a globalizing and myriad industry that employs millions of people internationally [17]. A market report from Credence Research Inc. revealed that global footwear market value was at US 222.4 Bn in 2017 and expected to grow by about 31% in the following decade [8]. 222 Bn is almost equivalent to one-sixth of the Australian GDP in 2017. This report shows the increase in demands in both athletic and non-athletic shoes. The report also indicated there are several global footwear market leaders, including Adidas AG, Nike Inc., Bata Limited, Puma SE, VF Corporation, New Balance Athletics, Inc., The Columbia Sportswear Company, etc. [8]. Given the companies' size and global influences, this sector compared three key payers among the worldwide footwear market leaders: Nike Inc., Adidas AG, and VF Corporation (see Table 3).

These three companies are good examples of globalizing and myriad businesses mentioned in the literature. Specifically, these three companies took up 40% revenue of the footwear market, hired around 170,000 employees globally, and established a decentralized and globalized supply chain. Although all the companies

Table 3 Comparison among three market leaders

	Nike Inc.	Adidas AG	VF Corporation
Revenue 2018 (USD) (Bn)	36.39	23.78	13.8
Employees 2019	73,100	57,016	50,000
Headquarters	Beaverton, Oregon, USA	Herzogenaurach, Germany	Denver, Colorado, USA
Global supply chain	Yes	Yes	Yes
Focus area in sustainability	Transparent supply chain and circular economy design	Circular economy design	Traceable and transparent supply chain
Climate commitment	Science-based targets set	Science-based targets committed	Science-based targets set

Source NIKE Revenue 2006–2018 I NKE Retrieved from www.macrotrends.net
Adidas AG annual report 2019, VF Corporation annual report 2019, Nike Impact report 2019, VF corporation sustainability report 2019, Adidas AG sustainability report 2019

have committed to mitigating their social and environmental impact, each has very different sustainability agendas. For example, Nike committed to creating a traceable and transparent supply chain with the aid of big data. Adidas AG has focused on designing eco-friendly products with the introduction of the novel plant-based material. VF also allocated resources to make their supply chain more traceable and transparent; however, they are paying more attention to the social realm topics such as forced labor and fair wages.

When it comes to the spectrum of the retailing side, the footwear industry is also experiencing growth in demand due to the "fast fashion" effect. The idea of "fast fashion" was originally derived from the "Just-in-time" manufacturing philosophy and quick response strategies [35]. According to Morgan and Birtwistle [35], "fast fashion" companies could bring new styles from design to shop floor within as short as two weeks. Although "fast fashion" giants such as INDITEX Group (parent company of ZARA) and H&M Group never explicitly admit they are "fast fashion" company, they are the pioneers of adopting "Just-in-time" and "Quick Response Strategies." The pioneers of the fast fashion industry embrace this idea to satiate the appetite of fashion-hungry consumers wanting to buy items in new trends as they appear [35]. "Fast fashion" retailers, such as H&M, Zara, Primark, and TopShop, provide products that are designed to be worn less than ten times in the life cycle of the product at very competitive prices in the market. This characteristic of "fast fashion" has directed consumers to purchase and disposing of ever larger quantities of apparel and footwear [33, 35]. The low-price strategy and huge demand require a decentralized manufacturing and marketing strategy. The traditional footwear companies (Nike, VF, Adidas, etc.) have been catching up with their retailing competitors by adopting this "fast fashion" idea. Besides the footwear companies embracing the concept of "fast fashion," the industry also sees a trend of sustainable consumption. Morgan and Birtwistle [35] declared that the sustainability concerns of the apparel and footwear industry have come to public awareness since the 1990s. Unlike the older generation, the younger generation—especially Gen Y and Gen Z—has been paying more attention to both social and environmental sustainability other than only focusing on the styles, quality, and price of the products. In response to customers' preference over sustainability, most of the apparel and footwear companies have made commitments to mitigate their impact (i.e., setting up science-based targets) and introduce the new materials and designs which are considered to be beneficial to our ecosystem.

Finally, Nash [39] recently conducted a study showing how social media platforms are influencing the purchase decision of the generation Y and generation Z—target customers of apparel and footwear products. Given the increasing power of the social media platform, even scandals (i.e., Bangaladesh building collapse and Chinese worker strikes) in the remote supply chain may have the ripple effect of the entire brand. 2013 Dhaka garment factory collapse has led to several boycotts toward some US and EU retailers because this information has been spread and gone viral on the Internet and social media. Thus, footwear retailers should prevent these scandals from happening throughout their entire supply chain. Otherwise, the power of this new media could corrupt the reputation and image of the company and, therefore, debase the profit.

3.2 Footwear Production

Have you ever wondered how your shoes that you will be wearing every day are made? Given the mixture of types and styles of shoes, there are hundreds of different production processes due to the different designs and materials. But, in this study, we will focus on figuring out the production processes for sneakers and fashion shoes, given our target companies for this study (i.e., Nike Inc. Adidas AG). Normally, one pair of sneakers started his life cycle from preparation processes. During this process, the shoe last (shoe shape) will be made of plastic, wood, or metal, and the shoe pattern will be designed. Then, this pair of shoes will move to the cutting process where all the pattern parts for the shoe upper will be cut in a variety of ways (computer-controlled cutter, laser cutter, or hand cutting, etc.). Once all parts and components are prepared, the stitching lines—usually consists of 30–50 workers—will stitch and assemble the parts all together and attach the shoe upper onto the sole. Before this pair of sneakers ending up lying inside the package, the shoes will experience the last process where the workers will pull the upper from the last (shoe shape) made previously and make it the real shape of a shoe. Finally, the pair of sneakers will be quality checked before being packaged in the shoe box.

Subic et al. [47] summarized these production processes and classified all the production processes into two categorizations: tier 1 manufacturing and tier 2 manufacturing. Most of the components and accessories will be prepared in the tier 2 process (mold injection, shoelace production, etc.). Once all the components are being prepared, all these parts will be cut, stitched, assembled, finished, and being packaged for final shipping in the tier 1 manufacturers.

Like the apparel sector, footwear production also poses social and health threats to workers and environmental perils to our ecosystem. Rajnarayan [43] illustrated several concerns, such as occupational injuries, child labor, chemical hazard, and forced labor that are related to the footwear production processes. For example, the hazardous chemicals used in the gluing process will cause the severe health issues of the workers and even the life of them in some cases. Irresponsible suppliers were using child labor and forced labor to produce shoes for well-known western footwear companies. All these types of scandals will no doubt cost the reputation damage of these companies due to the misconduct or wrongdoings from their suppliers.

3.3 The Footwear Supply Chain

As mentioned earlier, the trend of "pursuit of fast fashion" requires a decentralized and globalized manufacturing strategy for the industry. Also, according to Muthu

[37], the apparel and footwear supply chain is very complex due to the wide spectrum of the products. Given this manufacturing strategy and complex features of the entire supply chain, the supply chain of the industry is highly decentralized and globalized, and most apparel and footwear companies rely on outsourcing and subcontracting services of its production [17]. The footwear giants (i.e., Nike Inc., Adidas AG) are good examples of dependence on outsourcing and subcontracting because they do not even own one factory to make the sample product. For example, Nike has most of the sneaker factories in Southeast Asia, while most of the accessories and components factories in East Asia. The high decentralization and globalization are the recipe for the low price of the footwear products because such sourcing strategy has created a highly competitive manufacturing environment, and thereby, the cost could be reduced dramatically. The increased labor cost in the mature production market (i.e., China and Vietnam) has also encouraged the footwear companies to source in the countries with lower cost (i.e., Myanmar, Bangladesh). All these factors fast response market strategies, diversified products, and components and low-cost sourcing plans drove the footwear companies to decentralize their supply chain in almost all continents of the world.

3.4 Challenges Throughout Supply Chain

Given the nature of the footwear decentralized supply chain and the risks existing in the production processes, there are several issues that the industry is facing throughout the supply chain: (1) social and environmental concerns; (2) transparency and traceability demands; (3) counterfeit products and quality.

3.4.1 Social/Environmental Concerns

According to Boström and Karlsson [3], the apparel and footwear industry poses social threats, given the complexity in the supply chain side of the industry. Forced labor, child labor, and other occupational health issues could be the potential social problems identified throughout the extended and decentralized supply chain. For example, it will pose serious threats to the health of the relevant workers who are involved in these processes due to the improper use of hazardous chemicals without personal protection equipment, the inappropriate store, and disposal of these chemicals. The recent investigation regarding forced labor issues identified in Chinese suppliers has jeopardized most brands' reputation and images. Rana Plaza collapse in Bangladesh is another example, which has called most brands and consumers' attention to sustainability [10]. The industry also causes environmental problems because of the resources used and pollutants and effluents associated with the production processes. For example, the emissions from the production process of the vulcanized shoes could have a tremendous negative impact on the air quality around the workshop if there were no treatments for these perilous emissions. The industry also poses environmental threats due to other issues such as energy consumption, intensive water use from the tanning processes, and improper chemical wastes treatment. Besides, the hazardous chemicals associated with the production process could definitely pollute the environment if the disposal of these chemicals was not treated properly. The hazardous chemicals are involved in the production processes for both tier 1 and tier 2 manufacturing units. In terms of tier 2 manufacturing units, Cao et al. [4] indicated that hazardous chemicals such as chromium salts are used in tanning production which converts the animal hides to the leather while the toxical substance, such as dioxins, is one of the most important by-products throughout the whole life cycle of producing PVC, a substitute of the leather material for the shoe manufacturing. These two materials are mostly used as components and materials for manufacturing in the footwear industry. When it comes to the shoe assembling processes (tier 1), Rajnarayan [43] showed us that organic solvents are frequently used in the assembly processes as the adhesive and glues, which are detrimental to workers' heath due to neurotoxicity.

3.4.2 Transparency/Traceability Demands

According to Doorey [11], information disclosure could mitigate the social sustainability impact from the supply chain, address the forced labor issues, and improve labor practices. James and Montgomery [21] also identify the consumer demand for the sharing of information throughout the supply chain from fashion retailers. In the meantime, the apparel and footwear industry is foreseeing a trend to go beyond auditing, monitor beyond the first tier of suppliers, integrate sustainability to core business practices, and bring transparency to the supply chain [23]. As mentioned earlier, the Higg index has been created due to these increasing demands and trends. Several papers and researches also back up the trend and customer needs for information disclosure, highlighting the importance of creating a transparent and traceable supply chain [24]. Thus, the footwear brands and retailers should prioritize establishing a transparent and traceable supply chain when decision-makers are developing short- or long-term strategies.

3.4.3 Counterfeit Products and Quality

Counterfeit products did threaten the manufacturers and retailers of originator products [6]. Most of the footwear brands will allocate tremendous resources to address fake product issues each year. Apart from the counterfeit products, quality assurance is another challenge that most of the brands are facing. Although the product is manufactured in a standardized way, brands and retailers are still facing quality issues due to the decentralized and globalized supply chain. The quality issue of the products would not only lead to the return or recall of the product but also making the customer loose their confidence toward their purchase of the product.

3.5 Supplier Relationship Management

Given these issues and obstacles that the apparel companies are facing, most companies champion the importance of supplier relationship management. One of the most common ways is to build a healthy relationship between their business and global supply chain to drive incremental value identified in strategic sourcing projects. For example, some of the footwear companies have created a supplier relationship management (SRM) system to classify their suppliers into "Platinum," "Gold," "Silver," and "On and Off" based on the performance and business value of these suppliers. Other companies have established a risk system to categorize the suppliers based on their risk in terms of sustainability and quality issues. The better the performance of the suppliers in either SRM or risk system, the more business will be rewarded to these suppliers. Although this seems a smart method to keep the suppliers in the fence of the code of conduct of the companies, there are several additional issues. For example, most of the footwear companies will need to monitor the performance in terms of sustainability and quality through auditing their suppliers on a frequent basis. These monitoring will increase the budget of these companies, given the resources such as workforce and frequent business travel needed. In addition, the opaque supply chain often prevents the auditing team from finding issues such as forced labor and child labor lurking inside the irresponsible suppliers.

4 Blockchain

4.1 What Is Blockchain

You may hear about buzz words such as cryptocurrency, bitcoin, and litecoin even if you are not an investor of virtual currency. If you have heard these buzz words before, you may be familiar with "blockchain"—a technology behind all these virtual currencies.

The idea of the blockchain architecture was first mentioned in Haber and Stornetta [15].

But this concept has been applied by Satoshi Nakamoto until 2008 in one of his paper titled "Bitcoin: A Peer-to-Peer Electronic Cash System." [38]. This concept has been further developed and evolved to the blockchain technology that could be applied in more commercial and economic fields [16]. So, what is the blockchain? According to Drescher [12], blockchain is a digital ledger with blocks of information linked together. Each party on a blockchain has access to all the information, and no single person or entity controls a blockchain. Here is an easier way to explain this definition: The blockchain consists of a chain of blocks. Each block has two parts: header and body. Information such as the time stamp (time of the transaction) and hash (the identity of the block) will be calculated and stored in the header part while all the transaction information will be stored and in the body part. The hash is a function that converts your input (transaction information) into an encrypted output. The beauty of the blockchain is that all the transaction information (previous ones and the current ones) will be hashed altogether as a unique identity of the block—just like the fingerprint—and make any tiny change of data—even change in one block—impossible. Iansiti and Lakhani [19] identified the underlying architecture and characteristics of the blockchain: (1) distributed ledger or database, (2) accessibility within the network, (3) node to node communication, and (4) immutability. According to Drescher [12], each block of the chain structure has the following key characteristics: (1) immutable, (2) time-stamped, (3) append only, (4) secure, and (5) open and transparent. Blockchain technology could create a secure, transparent, authentic, and trustworthy supply chains due to its decentralized system and its features [1].

4.2   Key Terms (IOT/Ledger/Smart Contract/Proof of Work)

Several key terms should be explained to understand the concept of the blockchain fully. These terms include (1) Internet of things (IoT), (2) digital ledger, (3) smart contract, and (4) proof of work (PoW). These terms are not only the fundamental building block of the blockchain but also are the reasons why this technology could create a secure, transparent, authentic, and trustworthy supply chains.

4.2.1   Internet of Things (IoT)

According to [40], IoT refers to the interconnection of smart devices (i.e., digital devices, computing machines) to collect data and facilitate the decision-making processes based on the data collected. Panarello et al. [40] further stated that the blockchain could address the security and privacy issues given the IoT are vulnerable to hacker attacks and data safety threats. In the context of the footwear supply chain, the IoT could connect all the digital devices and computers in the production facilities and collect the data out of this system. The introduction of the blockchain could secure the data is not hampered or hacked due to the characteristics and features of the blockchain. The decision-makers from the manufacturers and retailers could, therefore, trust the quality of the data and make the decision based on the data collected from the IoT system.

4.2.2   Digital Ledger

The essence of the blockchain is to record data or information collected by the IoT system digitally. The recorded information will then be encrypted in the blocks so that stakeholders could track the untampered information from the blocks. This is the reason why the blockchain is called the digital ledger.

4.2.3 Smart Contracts

A smart contract is a computer program (functions and applications) having self-verifying, self-executing, and tamper-resistant properties [34]. According to Karamitsos et al. [22], the contract code will be uploaded and executed without the third parties into the blockchain network, and the blockchain network will assign each contract a unique address. In other words, the mechanism of smart contracts takes transaction—the information we want—as an input, executes the function, and transfers to the output of the events.

4.2.4 Proof of Work

Proof of work (PoW) is another important mechanism to secure the safety and security of the entire blockchain network. Although PoW was introduced by bitcoin to secure the safety of the transaction, this mechanism was used in the blockchain network, which assumes that each peer votes with his "computing power" by solving the proof of work instances and constructing the appropriate blocks [13]. Due to this consensus mechanism, not a single party in the blockchain network could tamper the information in the blocks.

4.3 Blockchain Frameworks and Services

4.3.1 Ethereum Versus Hyperledger Fabric

There are two most popular frameworks for developing and building the blockchain structure: (1) Ethereum and (2) Hyperledger Fabric. According to Prerna [42], Ethereum[2] is designed upon smart contracts to be decentralized and is for mass consumption while Hyperledger[3] leverages blockchain technology for business. Ethereum is a public blockchain network, while Hyperledger is more a private network for most of the business users due to the confidential information shared in the network. For footwear companies, they could develop their blockchain network by adopting both Ethereum and Hyperledger frameworks. The basic information of the suppliers (i.e., address and factory size) could be disclosed to the consumers through the public blockchain network. In the meantime, the confidential information such as the cost, sourcing price, and standard allowed minutes (SAM, etc.) should be delivered in the private application of the blockchain.

---

[2]Ethereum [Online]. Available: https://www.ethereum.org/.

[3]Hyperledger [Online]. Available: https://www.hyperledger.org/.

4.3.2 Services: Platform (AWS/IBM)

Although there are several platforms and services in terms of building the blockchain network, the authors highlighted two of them: Amazon Web Service (AWS) and IBM blockchain platform due to the needs of the most footwear brands and retailers. Both platforms are open-source platform and claimed that they could support the business to build scalable blockchain networks and ledger applications. One of the most significant differences between the two platforms is the application of AWS could be built upon both Hyperledger and Ethereum frameworks, while that of IBM could only be designed by using Hyperledger Fabric frameworks. In addition, there are other nuances between the two platforms. For example, AWS also provides AWS Blockchain Templates which could be used by the developers from the small business to build the blockchain architecture based on their own needs. The IBM platforms target most big corporates and also have created several business cases from the retailer giant's globalized supply chain (i.e., Walmart blockchain). When it comes to the footwear industry, the decision-makers could select the right platforms based on their business needs and internal capabilities. For example, footwear companies like Nike Inc. could select AWS platforms for their small supplier, given the flexibility and the AWS Blockchain Templates provided. At the same time, Nike Inc. could customize the blockchain network for its closely cooperated big suppliers by using the IBM platform since the successful business case from the retailer giant's globalized supply chain has been generalized.

5 The Benefits and Gains from Adoption

To answer the first research question: What can the footwear companies gain from the adoption? This section focused on exploring the benefits and advantages that the blockchain technology could provide to the footwear supply chain and companies.

5.1 TOWS Matrix

Based on current trends, characteristics of the footwear industry, and the power of the blockchain, a threats, opportunity, weakness, and strength (TOWS) matrix (see exhibit 1) was created to provide the reasons why companies should adopt the blockchain technology and what they can gain from the adoption. TOWS matrix is a tool by which strategies can be developed, and decisions can be made based on the identified strengths, weaknesses, opportunities, and threats [14]. As a variant of the strength, weakness, opportunity, and threats (SWOT) analysis, the TOWS matrix combines both internal factors (strengths and weaknesses) and external factors (opportunities and threats), making the decision process more comprehensive and thoroughgoing. Internal factors refer to both the strengths and weaknesses of the footwear companies and suppliers due to the unique features of the footwear industry and decentralized supply chain. External factors reflect the trends of the industry and the demands of the market. Four situations were created based on the combination of the internal and external factors: strengths and opportunities (SO), strengths and threats (ST), weaknesses and opportunities (WO), and weaknesses and threats (WT). The results derived from these situational analyses were condensed to answer the research question: What can the footwear companies and their supply chains gain from the adoption of blockchain technology?

5.2 Threats, Opportunity, Weakness, Strength

5.2.1 Opportunity

Opportunities refer to any external factors that companies and suppliers could tap into when it comes to the adoption of blockchain technology. Based on the literature review, the author identified below opportunities that are most relevant to the blockchain technology in the footwear industry:

1. Opportunity 1 (O1): The advent of digital technologies such as RFID and 3D designing and printing has shifted the traditional apparel and footwear supply chain to a digital era [7]. More and more digital devices will be introduced into manufacturing processes, which could generate tons of amount of data for decision-makers' usage. All these data could be used as input for the entire blockchain network before being processed for the visualization from the stakeholders of the whole value chain.
2. Opportunity 2 (O2): By incorporating the Internet of things (IoT) and radio-frequency identification (RFID), blockchain is likely to affect critical objectives of supply chain management such as cost, quality, risk reduction, sustainability, and flexibility [26]. The combination of the IoT and RFID could strengthen the power of the blockchain and render it to address the issues mentioned earlier in terms of the sustainability, quality, and marketing strategy.
3. Opportunity 3 (O3): The apparel and footwear industry is foreseeing a trend of creating a transparent and traceable supply chain [24]. Such a trend could drive the change of the supply chain and push the industry to explore the innovative technology or methodology to render its supply chain more transparent and traceable.

5.2.2 Threats

Unlike opportunities, threats are the adversary conditions that companies or suppliers should overcome in terms of the adoption of the technology. Most of the treats could be risks due to external factors and market trends. This study identified the following threats and risks footwear companies and suppliers would be facing if they adopted blockchain technology:

1. Threat 1 (T1): Supply chain and reputational risks such as Rana Plaza collapse and forced labor issue could undermine companies' reputations and public images [20]. Such scandals will cost not only the companies' public representations but also corrupt the revenues and profits of the companies because of the potential boycotts from the customers.
2. Threat 2 (T2): The supply chain of the textile industry is highly decentralized and globalized [17]. This characteristic of globalization increases the risk of the suppliers being exposed to the different cultural and legal system and therefore increases the risk of violating the laws and regulations of the sourcing country. The decentralization of the supply chain also doubles the difficulties for the footwear retailers to monitor their suppliers and scramble to keep suppliers in the fence of companies' code of conduct or sustainability commitments.
3. Threat 3 (T3): The viral message from social media has a huge impact on companies' reputations and public images [49]. As mentioned earlier, the viral information from social media could impact target consumers' purchase decisions even with trivial wrongdoings from one of the footwear companies' suppliers.

5.2.3 Weaknesses

Weaknesses are the internal problems that most of the footwear companies and suppliers are facing. Normally, the inherent organizational structure and the characteristics of the industry could be the sources of these problems. According to the literature review, the authors identified the following weaknesses existed internally in the footwear industry:

1. Weakness 1 (W1): Apparel and footwear companies are facing substantial inventory issues due to quick response strategies [5]. Such problems lead to a tremendous amount of material waste generated annually [35]. Most of the waste ended up in the landfill and therefore have a huge impact on our ecosystem.
2. Weakness 2 (W2): Most footwear companies rely on outsourcing and subcontracting services of its production. There is a vendor relationship between apparel and footwear companies and their suppliers [17]. Companies such as Nike Inc. and Adidas AG do not even own one production facility through their entire value chain. This makes the companies have limited leverage over their entire supply chain and therefore lack the ability to monitor their whole supply chain.
3. Weakness 3 (W3): Most apparel and footwear companies require a shorter lead time than traditional manufacturing schedules [45]. This situation would increase issues such as long working hours and forced labor.
4. Weakness 4 (W4): Most of the apparel and footwear products are providing products at a low price to the market [17]. The lower price strategy makes these companies transfer their sourcing plan to lower-income countries and make the supply chain more decentralized.

5.2.4 Strengths

Strengths are some areas in which companies have competitive edges. These competitive edges could differentiate the companies from their competitors in the marketing environment. In the competitive market, more competitive advantages the company has, more profits, and customer loyalty the company could be rewarded. The author identified three key strengths relevant to most footwear giants' visions and values:

1. Strength 1 (S1): Most of the apparel and footwear companies have integrated sustainability into their business operations [46]. The integration could make these companies in line with conscious consumption demands based on the increasing market trends regarding sustainability.
2. Strength 2 (S2): Most apparel and footwear companies have a substantial economic and social impact due to their economic scale and great employment opportunities generated by the companies [17]. As mentioned in the previous part, three footwear giants hired more than 170,000 employees last year globally.
3. Strength 3 (S3): Most of the retailers and brands have committed to mitigating their environmental impact, such as reducing greenhouse gas emissions and recycling textile wastes.

5.3 Gains from the Adoption

Based on internal and external factors listed in the previous part, this study analyzed four situations derived from the combination of these factors: strengths and opportunities (SO), strengths and threats (ST), weaknesses and opportunities (WO), and weaknesses and threats (WT). Under the scenario of SO and ST, the author figured out benefits by analyzing how footwear companies can use their strengths to take advantage of opportunities or avoid potential threats by adopting the blockchain technology. In terms of WO and WT, the author identified the gains by investigating how footwear companies can overcome weaknesses by tapping into the opportunities or avoiding external threats when applying blockchain technology throughout the entire supply chain.

Next, the author listed all the benefits and gains from four perspectives: transparency and traceability, cost, speed (lead time), and mitigation of the social and environmental impact. The benefits and gains of the adoption could provide the fitting solution to the challenges that the footwear industry and supply chain are facing.

5.3.1 Transparency and Traceability

1. S1, S2, S3 and O2, O3: The characteristics of the immutability and chain structure could help to create a more transparent and traceable supply chain [26]. The potential of creating a transparent supply chain is in line with the customers'

increasing demands of previous inaccessible information disclosure throughout the supply chain. This could also address the counterfeit product issues since all the products could be tracked down from the raw material to the final assembly attributed to this new technology.

2. S2, S3 and T1, T3: The transparent and traceable supply chain with supported data could provide a positive image of the company. Although most of the footwear companies have committed to mitigating social and environmental impact, few of them could deliver concrete results to respond to their commitments, especially when it comes to the issue existed throughout companies' decentralized supply chain. After the adoption of the technology, the stakeholders and consumers would be able to quickly get access to previous inaccessible data (information) derived from the supply chain. These verified and hamper-resistant data could convince the stakeholders that companies were taking actions in fulfilling their commitments, and therefore, more investments and business opportunities would be rewarded.

5.3.2   Cost

1. W2 and T1, T2: Blockchain could build up a trusted relationship between suppliers(vendors) and fashion companies and audit-free business model, which reduces the cost of the audits in terms of the sustainability and quality issues of the products and the cost of the middleman (vendors) in the business [26]. As mentioned earlier, most footwear companies champion the importance of supplier relationship management (SRM) depending on the monitoring system of the performance of the suppliers in the quality, sustainability, and business spectrum. However, the maintenance of such an SRM system will increase tons of budgets of the company by using either the internal or external auditing team to catch up with the performance of the suppliers. The adoption of the blockchain would reduce such expense to a minimum or even zero if the old system being successfully replaced.

2. W2, W4 and O1, O2: The mechanism of blockchains such as automation, stream-lined process, and processing speed could lessen the labor requirements and opti-mize the production process, which thereby reduces the production-related costs [18]. In the section of the introduction of the footwear industry, we mentioned the characteristics of labor intensive of the shoe production process, and several social and occupational health issues are associated with the manufacturing processes. Automated production processes could not only save the labor cost due to the reduction of the workers in the production facilities but also lessen the risks associated with the old-fashioned production processes.

5.3.3   Speed (Lead Time)

1. W1, W3 and O1: Blockchain could seamlessly capture data end-to-end through each digitalized production process [7]. These data could provide insights to

optimize the production process and therefore reduce the lead time. For example, the exponentially increased amount of data could be fed into machine learning or optimization models and could improve the performance and accuracy of these models. The decision-makers could provide solutions to improve the working efficiency according to the hint from the outputs of the models.

5.3.4 Mitigation of the Social and Environmental Impact

1. S1 and T1: According to Kshetri [26], blockchain architecture could provide secure, trusted, and transparent data. These data could support the buying team of the footwear companies and brands to make responsible sourcing decisions and thereby refrain from social risks associated with the decentralized and globalized supply chain. The data and sourcing specialist could pick the right suppliers and reward the suppliers with more business opportunities based on the high-quality and hamper-resistant data extracted from the blockchain network. Also, the business team could build up their simulation models based on historical data to evaluate the new suppliers and therefore shorten and facilitate the due diligence processes before working with the new supplier or sourcing in uncharted territories.
2. W1 and T2: Blockchain technology could mitigate production waste in a twofold way. First, the technology could track the data in each production process and identify the hot spots of material usage. Thus, the companies could focus on researching new solutions to tackle the hot spots identified by the technology. Second, the accessible data of the products could facilitate the recycling process due to the recognizable ingredients of the digital tag generated from the blockchain network. This could prevent tons of waste from the products from being landfilled and therefore encourage the usage of recyclable and eco-friendly materials.

6 Strategic Position in the Market

In this section, we attempted to answer the last research question—Where can the blockchain position the fast fashion companies in the competitive market?—in two steps. First, the author tried to dig into the potential of this technology and identify if this technology is disruptive by using the definition of disruptive technology. Second, the author sought to figure out what competitive advantages could bring to the companies and how this technology could differentiate the companies from the competitors if they adopted blockchain technology by analyzing Roger's diffusion of innovation theory.

6.1 Disruptive Technology (Potential of the Blockchain)

6.1.1 Disruptive Technology Definition

According to Danneels [9], disruptive technologies would be those technologies that render established technologies obsolete and change the base of competition by developing the performance metrics along which firms compete. In other words, certain technology could change how the supply chain operates and replace traditional working methods with innovative ones. Also, the entire industry should change the methods on how to evaluate each participant's performance in the industry. Kostoff et al. [25] also stated that disruptive technology could provide exponential improvements in the value of the process/products/services received by the customers. That is to say, disruptive technology could enhance customers' experiences by its enormous improvements and added value to the services. In the context of the supply chain, we could expect the growth of the production efficiency and the development of the capability of the production processes. Kostoff et al. [25] further declared that disruptive technologies typically draw upon many diverse technologies and can be either a new combination of existing technologies or new technologies that cause major technology product paradigm shifts or create entirely new ones.

6.1.2 Why Blockchain Disruptive?

According to Azzi et al. [1], the blockchain network could replace the traditional and centralized enterprise resource planning (ERP) system with a blockchain-based and decentralized network. Unlike the ERP system, the specific architecture of the blockchain would totally change the way how the supply chain deals with their data extracted from the production process. Azzi et al. [1] also claimed that the unique chain structure and proof of work mechanism had lifted chain management to the next level, making the supply chain more transparent, authentic, and trustworthy. For decades, most of the retailers had been struggling to establish and traceable and transparent supply chain and scrambling to collect the data that could reflect real situations throughout their complete supply chains. These are fragments of evidence that blockchain is a disruptive technology that renders established or traditional technologies obsolete.

Besides, Hughes et al. [18] mentioned most of the companies should redesign their business models and organizational structures so that they could benefit from the adoption of blockchain technology. That is to say, blockchain could flip over the supply chain ecosystem and therefore change all the evaluation methods throughout the supply chain, which proved this technology is disruptive ones as it could turn the base of competition by improving the performance metrics. Apart from these two features, blockchain could also affect key supply chains management objectives such as cost, quality, speed, dependability, risk reduction, sustainability, and flexibility [26]. Thus, blockchain is a disruptive technology because this technology does provide exponential improvements in the value of the process/products/services received by the customers based on Kshetri's argument. Finally, this technology evolves from diversified technologies. By incorporating technologies such as RFID, proof of work, smart contracts, and cloud-based solutions, the technology could meet the key supply chain management objectives mentioned above [26]. This feature demonstrated that technology is a disruptive one, as blockchain is the combination of existing technologies and new technologies. These features and traits of blockchain technology extracted from the literature review perfectly satisfy the definition of disruptive technology.

6.1.3 What Disruptiveness Means to Footwear Companies?

Companies could gain from the adoption of this disruptive technology from three perspectives. First, the adoption of this disruptive technology could address the challenges and issues companies are facing, such as inventory issues, perplexing supply chain risks, and social and environmental concerns. As mentioned earlier, the blockchain network could provide high-quality and immutable data extracted from the production processes and facilities throughout the footwear supply chain. The decision-makers could refer to these previously inaccessible data and, in turn, to make the decisions that could address these long-existing issues in the industry. Besides, the adoption of this disruptive technology could support footwear companies to establish the competitive edges in response to the new blockchain era given the blockchain technology could rewrite the rule of the competitive market. The early adopters could create their standards and evaluation methodology of the supply chain and became the standards makers and leaders in this uncharted territory. Finally, the adoption could build a long-term competitive advantage by reducing the risk and saving the cost in the long run, given the benefits the technology could bring to the supply chain and disruptive features of the blockchain technology.

6.2 Diffusion Theories

In this section, the author investigated Roger's diffusion of innovations and the growth model of emerging technology to identify the position of the companies in the competitive market if adopting blockchain technology. The author also referred to the analysis of blockchain from Hughes et al. [18], Azzi et al. [1], and Kshetri [26] to figure out what the potential for the development of this technology is and what competitive edges this technology could bring to the companies.

6.2.1 Rogers' Diffusion of Innovations Theory

Rogers [44] classified the adopters based on the innovativeness, including inno-vators, early adopters, early majority, late majority, and laggards. The differences among these adopters are associated with the timing of the adoption of the new technology and the acceptance level of embracing the latest technology. Based on Rogers' theory, only one-sixth of the adopters have been categorized into innovators and early adopters. Rogers [44] also claimed that the innovators and early adopters are more likely to hold leadership in the social system. Since the application of the blockchain in the apparel and footwear world is in its "fledgling stage," the footwear pioneers could be regarded as innovators or early adopters if they started adopting technology at this point. The leadership position gained from the early adoption of the technology could bring these companies the competitive edges in the market if the trend of the transparent and traceable supply chain was inevitable. The first practice of the adoption could establish the novel standards, innovative working methods, and new evaluation metrics of the performance for the supply chain, which would make the majority competitors in the market to follow the leaders in the latter stage of the implementation of this technology in the industry.

6.2.2 Growth Model

The growth model theory could support the argument that technology could bring a leadership position to the companies if they adopted the technology at this point. According to Porter et al. [41], there are four phases for technology growth and development: (1) emerging, (2) rapid growth, (3) maturity, and (4) decline. Once the technology emerged, there is a period of rapid growth, followed by an inflection point and slower growth as the product enters a period of maturity. Like the organic entity, the technology will be declining after it reached the maturity period. Thus, before reaching the inflection point, the technology still has enormous potential for the development in the period of rapid growth. Based on Hughes et al. [18], Azzi et al. [1], and Kshetri [26], blockchain is still in the phase of rapid growth but not yet hit the inflection point. For example, several case studies indicated most of the companies in other industries are still piloting the blockchain projects in their supply chains. The technology is still a fresh concept for the footwear industry. Besides, several challenges and barriers need to be addressed before blockchain technology can be fully adopted in the supply chain ecosystem. Thus, there is considerable potential for developing this technology at this moment. The pioneers of the adoption could enjoy the bonus dividend due to the exponential development of the technology used in the context of the supply chain.

6.3 Position of the Company

Based on the theories abovementioned, the disruptive blockchain could make footwear companies become the innovators and early adopters at this moment. Such roles in the competitive market could bring the leadership position to these pioneer companies and make them gain competitive edges before this technology hits the inflection point. These competing advantages could render these companies to establish the novel standards, innovative working methods, and new evaluation metrics of the performance for the supply chain and distinguish them from the competitors in the footwear industry.

7 Application Area and Cases Analysis

7.1 Application Area

Based on the literature review, blockchain has a variety of application areas in the industrial, financial, and commercial world. This section would focus on the application area related to the supply chain due to the scope of this paper. According to Hughes et al. [18] and Tian [48], blockchain could be applied in the transportation and logistics management, inventory management, supplier management, raw material or ingredients identification, and transaction/payment in the supply chain. These areas are highly relevant to the businesses of the footwear supply chain. The blockchain technology has already been used frequently in some areas, while it is still in the theoretical stage for other areas. For example, the payment system for the transaction has been developed due to the popularity of the virtual cryptocurrencies. This practice could be extended to all the transactions that happened throughout the supply chain. In addition, IBM Food Trust—a data-sharing platform/network among all the participants in the food industry—has created a successful case about inventory management and identification of the hot spots for food waste by implementing the blockchain technology. However, applications, such as raw material or ingredients identification, are still in the pilot phases as there are several hurdles that companies need to overcome (i.e., equipment replacement, workforce training, etc.).

7.2 Cases Analysis

We identified three cases that are related to the supply chain of retailing companies. Also, we noted that functions realized in these cases could also address the issues and problems that the footwear companies are facing mentioned earlier.

7.2.1 Alibaba

Chinese online retail giant Alibaba collaborated with AusPost, Blackmores, and PwC to investigate the application of blockchain to fight against food fraud [27] As mentioned earlier, most of the footwear companies have committed to recycling the wastes generated from their products or production processes. However, inaccurate material information creates obstacles for the recycling processes, and most of the waste ended up in the landfill instead of being recycled as promised by these companies. This case shows a successful example to address the issue by tracking the raw material for the footwear products and therefore overcome the challenges from the recycling process. This case also shows the potential of addressing another counterfeits issue in the footwear industry since the consumers could track the information of the supply chain and prevent from buying fake footwear products.

7.2.2 Walmart

Walmart has worked with IBM to track food safety by incorporating the blockchain technology. The company also used the technology in authenticating not only the package but also the customers and couriers in its logistics sector [26]. The solution used in the logistics area created an excellent example of how to track every production step from the production of the components to the finished products manufacturing and the shipping information from the component manufacturing mills to the assembly production facilities. This practice could prevent social risks such as undeclare outsourcing, which refers to the products that are produced in the facilities without permission from the footwear retailers and brands. Most undeclared facilities have high risks of social sustainability issues such as child labor, forced labor, unannounced outsourcing, and occupational health and safety issues. The footwear companies are normally struggling in tracking their supplier outsourcing their orders to some sweatshops or even prisons under the current suppliers' monitoring system. This solution could solve this issue and therefore mitigate the social impact accordingly.

7.3 Everledger

Everledger, a London-based start-up, uses the blockchain-based solutions to verify the origin of products [26]. The system could refrain from the certification process for the diamond products. Currently, most of the footwear companies require their suppliers to certify their factories or accept the audits from external organizations due to the quality and sustainability concerns. However, most suppliers are paying tons of money to get such certifications or being audited by third parties, while the footwear brands are reluctant to increase the sourcing prices due to the competitive production market. The cost could undermine the profit of the suppliers and, in turn, exploit the workers' wages to cover the loss from the certifications and audits. This "Everledger" case is an excellent example to showcase the possibility of creating a certificate-free and audit-free for the footwear supply chain. In the meantime, the footwear brands could still monitor their suppliers' performance in terms of quality and sustainability because of the implementation of the new technology.

8   Barriers and Potentials

8.1   Barriers

Although there is a promising future of blockchain being adopted in the footwear supply chain, the early adopters still need to face the barriers and challenges from politics, social, economic, and technical perspectives. We identified the following barriers in this research:

- Complexity of laws, regulations, jurisdictions, and institutions across a global supply chain [26]. The blockchain technology may not be able to get access to every corner of the supply chain given different countries may have distinct laws and regulations in terms of the implementation of the blockchain technology.
- Collaboration among all relevant parties [26]. As mentioned earlier, the footwear brands have limited leverage over most of the small suppliers due to the business quantity. Thus, the footwear companies should figure out how to convince the suppliers to embrace the new technology and disclose all the data from the production processes.
- Implementation of technology-based solutions for suppliers, both in developed and developing countries [26]. The full implementation of the blockchain technology will need to integrate other technology and digital devices and services. However, this will be the obstacle of the suppliers in underdeveloped countries where the suppliers could not easily get access to these digital devices and services.
- High cost for incorporating cloud service and RFID technologies to make the blockchain technology work in the supply chain; and [18].
- Lack of knowledge concerning blockchain at the workplace of the supply chain. In this case, more blockchain technicians or specialists should be involved during the adoption processes.

8.2 Potentials

These challenges are just like a double-edged sword. The solutions behind these challenges could push the development of this blockchain technology to the next level. For example, the collaboration between tech companies and apparel and footwear companies (i.e., Vechain and H&M collaboration) could lessen the technical barriers of the adoption. The technical breakthrough could prompt more companies' adoption and thereby reduce the cost due to the economic scale. More successful use cases and adopters could also push the policy-makers to redesign the policies to facilitate the development of this technology. Thus, there exists a tremendous potential for the adoption of the technology, though the barriers and challenges mentioned above.

9   Conclusion

In conclusion, the authors recommended the footwear companies to adopt the blockchain technology throughout their supply chain at this moment since the benefits and competitive advantages could bring to these pioneers if they adopted the technology. To answer the first sub-research question, the authors fed the TOWS matrix with the inputs extracted from characteristics of the footwear supply chain and the challenges that the footwear industry is facing. The authors then obtained the output by synthesizing and analyzing based on the literature reviews.

According to the analysis of the results from the TOWS matrix, we highlighted what benefits the footwear companies and their supply chains can get from the adoption of the technology. The results were investigated from four perspectives: (1) transparency and traceability, (2) save the cost (eliminating middleman/vendor/verification), (3) speed (lead time), and (4) mitigation of the social and environmental impact. In addition, the authors analyzed where the blockchain will position the company if the implementation of the technology is in the supply chain by reviewing the definitions and characteristics of the disruptive technology and Roger's diffusion of innovations theories. The author concluded that disruptive blockchain could make footwear companies become the innovators and early adopters at this moment. Such roles in the competitive market could bring a leadership position to these companies and make them gain competitive edges before this technology hits the inflection point. Several cases were displayed to prove the potentials and possibilities of this technology could bring to the supply chain of the footwear companies and the possibilities of addressing the issues that have been troubling the industry for decades. Finally, the author mentioned that early adopters would face the barriers and challenges from politics, social, economic, and technical perspectives. But these challenges are just like a double-edged sword. The solutions behind these challenges could push the development of this blockchain technology to the next level. Thus, the technology still holds for promise though the implementation of the technology has several barriers and challenges.

Appendix

Exhibit 1: TOWS analysis

	Strengths: S1: Most fast fashion companies have a huge economic and social impact due to their economic scale and high employment opportunities generated by the companies [17]. S2: Most of the apparel and footwear companies have integrated sustainability into their business strategies [46]. S3: Most of the apparel and footwear companies have committed to mitigate their environmental impact by reducing greenhouse gas emissions and recycling textile wastes	Weaknesses: W1: Apparel and footwear companies are facing substantial inventory issues due to quick response strategies [5] Such problems lead to a tremendous amount of waste generated annually [35]. W2: Most apparel and footwear companies rely on outsourcing and subcontracting services of its production. There is a vendor relationship between these companies and their suppliers [17]. W3: Apparel and footwear companies require shorter lead time than traditional manufacturing schedule [45]. W4: Apparel and footwear products are providing low price to the market [17]
Opportunity: O1: The advent of digital technologies, such as RFID and 3D designing, and printing has shifted the traditional supply chain to a digital era. [7] O2: By incorporating the Internet of things (IoT) and radio-frequency identification (RFID), blockchain is likely to affect key supply chain management objectives such as cost, quality, risk reduction, sustainability, and flexibility [26]. O3: The apparel and footwear industry is foreseeing a trend of creating a transparent and traceable supply chain [24]	S1, S2, S3 and O2, O3: The characteristics of the immutability and chain structure make the supply chain more transparent and traceable [26]	W1, W3 and O1: Blockchain could seamlessly capture data end-to-end through each digitalized production process [7]. These data insights could be used to optimize the production process and therefore reduce the lead time. W2, W4 and O1, O2: The mechanism of blockchains such as automation, streamlined process, and processing speed could lessen the labor requirements and optimize the production process, which thereby reduces the production-related costs [18]

(continued)

(continued)

Threats:	S2, S3 and T1, T3: The	W2 and T1, T2:
T1: Supply chain and reputational risks such as Rana Plaza collapse and forced labor issue could undermine companies' reputations and public images [20]. T2: The supply chain of the apparel and footwear industry is highly decentralized and globalized [17]. T3: The viral message from social media has huge impact on companies' reputations [49]	transparent and traceable supply chain with supported data could provide a positive image of the company S1 and T1: According to Hawlitschek et al. [16], blockchain architecture could provide secured, trusted, and transparent data. These data could support the buying team of the companies to make responsible sourcing decisions	Blockchain could build up a trusted relationship between suppliers(vendors) and retailers and audit-free business model, which reduces the cost of the audits in terms of the sustainability and quality issues of the products and the cost of the middleman (vendors) in the business [16] W1 and T2: The blockchain technology could mitigate product waste from two perspectives. First, the technology could save the materials and reduce wastes during the production process. Second, the traceable data of the products could facilitate the recycling process due to the identifiable ingredients of the digital tag generated from blockchain technology

References

1. Azzi R, Chamoun RK, Sokhn M (2019) The power of a blockchain-based supply chain. Comput Ind Eng 135:582–592
2. Bhagwat VM (2006) RFID technology for textile supply chain management. Asian Text J 15(2):33–36
3. Boström M, Karlsson M (2013) Responsible procurement, complex product chains and the integration of vertical and horizontal governance. Environ Policy Govern 23(6):381–394
4. Cao H, Wool RP, Bonanno P, Dan Q, Kramer J, Lipschitz S (2014) Development and evaluation of apparel and footwear made from renewable bio-based materials. Int J Fashion Des Technol Educ 7(1):21–30
5. Caro F, Martínez-de-Albéniz V (2010) The impact of quick response in inventory-based competition. Manufact Serv Oper Manage 12(3):409–429
6. Carpenter JM., Edwards KE (2013) U.S. consumer attitudes toward counterfeit fashion products. J Textile Apparel Technol Manage (JTATM) 8(1):1–16
7. Crawford C (2019) The smart supply chain: a digital revolution. AATCC Rev 19(3):38–45
8. Credence Research (2019) Global footwear market—Growth, future prospects and competitive analysis, 2018–2026. Retrieved from https://www.credenceresearch.com/report/footwear-market
9. Danneels E (2004) Disruptive technology reconsidered: a critique and research agenda. J Prod Innov Manage 21(4):246–258

10. Donaghey J, Reinecke J (2018) When Industrial democracy meets corporate social responsibility—A comparison of the Bangladesh accord and alliance as responses to the Rana Plaza disaster. Br J Ind Relat 56(1):14–42
11. Doorey DJ (2011) The transparent supply chain: from resistance to implementation at Nike and Levi-Strauss. J Bus Ethics 103(4):587–603
12. Drescher D (2017) Blockchain basics (vol 276). Apress, Berkeley, CA
13. Gervais A Karame GO, Wüst K, Glykantzis V, Ritzdorf H, Capkun S (2016) On the security and performance of proof of work blockchains. In: Proceedings of the 2016 ACM SIGSAC conference on computer and communications security, pp 3–16
14. Gottfried O, De Clercq D, Blair E, Weng X, Wang C (2018) SWOT-AHP-TOWS analysis of private investment behavior in the Chinese biogas sector. J Clean Prod 184:632–647
15. Haber S, Stornetta WS (1990) How to time-stamp a digital document. In: Conference on the theory and application of cryptography. Springer, Berlin, Heidelberg, pp 437–455
16. Hawlitschek F, Notheisen B, Teubner T (2018) The limits of trust-free systems: A literature review on blockchain technology and trust in the sharing economy. Electron Commer Res Appl 29:50–63
17. Hines T (2007) Globalization: global markets and global supplies. In: Fashion marketing. Routledge, pp 25–50
18. Hughes L, Dwivedi YK, Misra SK, Rana NP, Raghavan V, Akella V (2019) Blockchain research, practice and policy: applications, benefits, limitations, emerging research themes and research agenda. Int J Inf Manage 49:114–129
19. Iansiti M, Lakhani KR (2017) The truth about blockchain. Harvard Bus Rev 95(1):118–127
20. Jacobs BW, Singhal VR (2017) The effect of the Rana Plaza disaster on shareholder wealth of retailers: implications for sourcing strategies and supply chain governance. J Oper Manage 49:52–66
21. James AM, Montgomery B (2017) Engaging the fashion consumer in a transparent business model. Int J Fashion Des Technol Educ 10(3):287–299
22. Karamitsos I, Papadaki M, Al Barghuthi NB (2018) Design of the blockchain smart contract: a use case for real estate. J Inf Secur 9(3):177–190
23. Khurana K, Ricchetti M (2016) Two decades of sustainable supply chain management in the fashion business, an appraisal. J Fashion Mark Manage
24. Köksal D, Strähle J, Müller M, Freise M (2017) Social sustainable supply chain management in the textile and apparel industry—A literature review. Sustainability 9(1):100
25. Kostoff RN, Boylan R, Simons GR (2004) Disruptive technology roadmaps. Technol Forecast Soc Change 71(1–2):141–159
26. Kshetri N (2018) Blockchain's roles in meeting key supply chain management objectives. In J Inf Manag 39. Retrieved from https://doi.org/10.1016/j.ijinfomgt.2017.12.005
27. Kshetri N, Loukoianova E (2019) Blockchain adoption in supply chain networks in Asia. IT Professional 21(1):11–15
28. Kwok SK, Wu KK (2009) RFID-based intra-supply chain in textile industry. Ind Manage Data Syst 109(9):1166–1178
29. Laufer WS (2003) Social accountability and corporate greenwashing. J Bus Ethics 43(3):253–261
30. Li CN, Zhao Q (2012) RFID technology in garment production line management application. Adv Mater Res 542(543):344–348
31. Li L (2013) Technology designed to combat fakes in the global supply chain. Bus Horiz 56(2):167–177
32. Lou X, Cao H (2019) A comparison between consumer and industry perspectives on sustainable practices throughout the apparel product lifecycle. In J Fashion Des Technol Educ 12(2):149–157
33. McAfee A, Dessain V, Sjöman A (2004) Zara: IT for fast fashion. Harvard Business School
34. Mohanta BK, Panda SS, Jena D (2018) An overview of smart contract and use cases in blockchain technology. In: 2018 9th International conference on computing, communication and networking technologies (ICCCNT). IEEE, pp 1–4

35. Morgan LR, Birtwistle G (2009) An investigation of young fashion consumers' disposal habits. Int J Consum Stud 33(2):190–198
36. Muthu SS (2013) The environmental impact of footwear and footwear materials. In: Handbook of footwear design and manufacture. Woodhead Publishing, pp 266–279
37. Muthu SS (2014) The textile supply chain and its environmental impact. In SS Muthu (ed) Assessing the environmental impact of textiles and the clothing supply chain, Woodhead, Cambridge, pp 1–31
38. Nakamoto S (2008) Bitcoin: a peer-to-peer electronic cash system. bitcoin. org. https://bitcoin.org/bitcoin.pdf. (accessed: 24.02.2020)
39. Nash J (2019) Exploring how social media platforms influence fashion consumer decisions in the UK retail sector. J Fashion Mark Manage 23(1):82–103
40. Panarello A, Tapas N, Merlino G, Longo F, Puliafito A (2018) Blockchain and iot integration: a systematic survey. Sensors 18(8):2575
41. Porter AL, Cunningham W, Banks J (2011) Forecasting and management of technology, second edition. Technology forecasting. Wiley, Hoboken, NJ, pp 15–39
42. Prerna (2019) Hyperledger versus ethereum—Which blockchain platform will benefit your business? Retrieved from https://www.edureka.co/blog/hyperledger-vs-ethereum/#keydiffer ences
43. Rajnarayan R (2005) Child labor in footwear industry Child labor in footwear industry: possible occupational health hazardous occupational health hazards. Ind J Occup Environ Med 9(1)
44. Rogers EM (2003) Diffusion of innovations 5th edn. Free Press, New York
45. Seo MJ, Kim M, Lee K-H (2016) Supply chain management strategies for small fast fashion firms: the case of the Dongdaemun Fashion District in South Korea. Int J Fashion Des Technol Educ 9(1):51–61
46. Shen B (2014) Sustainable fashion supply chain: lessons from H&M. Sustainability 6(9):6236–6249
47. Subic A, Shabani B, Hedayati M, Crossin E (2012) Capability framework for sustainable manufacturing of sports apparel and footwear. Sustainability 4(9):2127–2145
48. Tian F (2016) An agri-food supply chain traceability system for China based on RFID & blockchain technology. In: 2016 13th International conference on service systems and service management (ICSSSM). IEEE, pp 1–6
49. Vo TT, Xiao X, Ho SY (2019) How does corporate social responsibility engagement influence word of mouth on twitter? Evidence from the airline industry. J Bus Ethics 157(2):525–542
50. White GRT (2017) Future applications of blockchain in business and management: a delphi study. Strat Change 26(5):439–451. Retrieved from http://dx.doi.org/10.1002/jsc.2144
51. Wikipedia Contributors (2019) Radio-frequency identification. In: Wikipedia, the free encyclopedia. Retrieved from https://en.wikipedia.org/w/index.php?title=Radio-frequency_identi fication&oldid=912774629

Designer Activism Strategies for Sustainable Leather Product Designs

V. Nithyaprakash, S. Niveathitha, and V. Shanmugapriya

Abstract In this chapter, we discuss the leather product usage in ancient times and contemporary age, the properties of leather and the trends of the leather processing. Later, the environmental impact of leather processing is elaborated along with the breakdown of the assessment tools. To sustain the leather demand and face the challenges posed by it, the agenda proposed by the research conglomerates and affiliated authorities, designer activism strategies for promoting recyclability of leather and measures to advocate second life for virgin leather products, alternative leather material initiatives and new technology options were investigated through case study analysis. This investigation identifies the possible affirmations concerning the direct role of designer activism in promoting sustainable leather product designs. This study concludes circular design options for leather processing using vegetable tanning methods and promotes the use of biodegradable leather material alternatives.

Keywords Sustainability · Design · Leather · Leather alternatives · Environment · Recyclability

1 Introduction

Leather is one of the materials that has been used continuously ever since Iron Age. Leather product usage could be traced to the ancient time periods. The reason that makes leather as one of the most sought-after materials is attributed to its maker-friendly properties. Leather's preferred material characteristics from the perspective of users and product designers range from being malleable and flexible to being tear

V. Nithyaprakash (✉) · S. Niveathitha · V. Shanmugapriya
Department of Fashion Technology, Bannari Amman Institute of Technology, Coimbatore, India
e-mail: nithyaprakashv@bitsathy.ac.in

S. Niveathitha
e-mail: Niveathitha@bitsathy.ac.in

V. Shanmugapriya
e-mail: shanmugapriya@bitsathy.ac.in

S. S. Muthu (ed.), *Leather and Footwear Sustainability*, Textile Science and Clothing Technology, https://doi.org/10.1007/978-981-15-6296-9_4

resistant. Leather's compatibility with mechanical surface treatments and chemical treatments makes it a versatile substance for large-scale industrial manufacturing. It is processed from the originally biodegradable skins of slaughtered animals using chemicals, and in the process it is converted into an inorganic material that slows down the rate of degradation as well as the amount of degradation. Along the conversion process of leather, toxic substances are left over in the effluents whose impact on the environment remains a global challenge to be combated. As leather is highly valued and treated with esteem respect by consumers, to sustain its stature and demand a multipronged approach and environment-friendly processing strategies are required in the current scenario.

2 Leather Product Usage in History and Contemporary Age

2.1 Leather Product Usage in History

The oldest leather product usage recorded in history is almost 5500 years old and identified as a shoe [1]. It was discovered in an American cave by the archaeologists. In fact, the shoe also bears enormous similarities with the make and style of contemporary European shoe [1]. This speaks volumes about the leather design characteristics and its semiotic attributes evolved since then. The earliest usage of leather in Iron Age corresponds to skin garments traced to the peat bogs of Jutland (200 BC–500 AD) [2]. Leather in garments could be traced to as old as ninth century (Bible AD 845). Leather clothing was preferred for its warmth. Hence, it was heavy, less flexible and difficult to wash [3]. Capes of Denmark indicate the presence of skilled craftspeople who made these leather garments [4]. Leather dress variants in the form of hoods and cloak were discovered in one of the cemeteries at Harford [5]. Leather over garments like caftans and coats was lined with fur and sheepskin [6]. Meanwhile, leather's use in foundation garments such as bikinis is also identified among the female athletes of ancient Rome [7]. Archaeological evidence of bikinis indicates the date to be as old as first century AD [8]. Further evidences of leather foundation garments were traced to the mosaic Villa Romana del Casale in Italy where more elaborate applications such as breast bands, briefs, acrobats and other entertainers were identified [7]. Archaeological evidence of gloves made of fur was identified in medieval graves of Unterhaching near Munich, Germany [9]. In another such excavation in the Saint-Denis Basilica, leather belt with silk embroidery and silver buckles was identified [10].

2.2 Leather Product Usage in Modern Era and Contemporary Age

Technology revolution and research developments in science during the early phases of modern era helped to diversify the leather process [11]. The shiny black colour shoe leather, patent leather, was invented in 1819 as a result of one such development [12]. Prior to the mechanization of tanning in the 1860s, leather usage was traced to products like costume and clothing, working equipment bags, military tools, household objects, bookbinding, toys, saddles, shoes and boots [11]. Industrialization of leather processing in the late nineteenth century [13] and decentralization of fashion in the early twentieth century expanded the usage of leather in many products as far as highly personalized leather gifts, niche shoes, interior upholsteries, automotive accessories and iconic leather handbags. First, leather product among French leather goods to be accorded luxury status is a wallet made from a single piece of leather without any stitches, which earned itself the name "Sans Couture" [14] in 1898. This leather wallet later fetched a silver medal at the universal exhibition in 1900 for its innovative features [14]. The newly ordained leather's luxury status paved the way for its increased preference in highly personalized accessories. By the 1930s, many leather accessories for both men and women gathered momentum as it witnessed the discovery of new evening bag styles including the clutch bag [15]. 1955 marked the issue of beautiful crafted leather gifts and presentables [16].

Soon leather occupied a central stage in the global luxury fashion value stream. At the moment, leather material usage could be traced to extended fashionable leather products like Chanel's patchwork stockings [17] and Kanye West's joggers [18], designer automotive interiors, designer upholsteries, computer accessory covers, stationary wraps and prosthetics.

Thus, the material options for leather products have expanded like never before and helped evolve the leather material perceptions even more. Product life cycles in fashion have shortened, and consumers are motivated to purchase new items every season. The fast-changing fashion cycles are encouraged through planned and aesthetic obsolesce [19]. In these circumstances, they do not form any personal relationship with them [20]. Further fashion creates symbolic boundaries between what is fashion and what is not fashion [21]. The designers take credit for infusing these symbolic values in fashion goods. All the agents involved in creation and promotion of fashion area are susceptible to cultural innovations, changes in lifestyle, values and attitude [22]. According to McCracken, fashion values newness contrary to patina that acts as a visual proof of kind of status. In postmodern world, consumers are driven by new tastes and preferences [23]. Fashion is not without fetishism; leather corsets, lingerie and catsuit are among the common clothing fetishes [24]. The emergence of the Italian leather further enhanced the appeal of luxury leather products drawing from its ethnic fine traditional craft [25]. By 2004, leather could be identified in almost every category of fashion starting from haute couture to fast fashion at all price points [24]. Just like other fashion goods, leather also experiences trend cycles.

Apart from fulfilling the performance attributes by virtue of its mechanical properties, the symbolic values surrounding the leather material and its craft also evolved along with the progressive applications. In fact, leather has become a synonym for fine craft, aesthetics and classic quality, especially among the fashion accessories. It further escalated the value propositions of leather products never mind even if it happened to be a staple utility product or an exclusive high-end luxury fashion product. In the current scenario as per the European Union source, 13% of finished leather is used for automotive industries, 17% of finished leather is used for upholsteries, 19% of finished leather is used for handbags, 8% is used for apparel making, 41% is used for footwear and 2% is used for glove making [26]. The global luxury leather goods market research report for 2020–2026 considers leather a luxury good meant to be extravagant and treasured [27]. A classic example for being most treasured material in USA is shell cordovan, and a name of rich Burgundy or dark rose raw material consumed in making exclusive leather shoe has been the ubiquitous choice for US Presidents from Ronald Reagan to George W Bush [28]. The leather product market is expected to engross worth US $629.65 billion by 2025 [29] that projects a requirement of 430 million cows annually by 2025 to make handbags, belts and shoes.

3 Leather Properties

3.1 Leather Properties from the Perspective of Product Design Characteristics

The desired leather characteristics streamline two aspects, namely 1. mechanical and physical properties, and elasticity and fracture toughness and 2. tactile property of softness [30]. The mechanical properties are assessed by quantifying the modulus of the material; however, the tactile property of leather stems from psycho-epistemic logic of the cultural prototypes constructed since the ancient period instead of engineering property. The ability of leather to deflect during handle or stretch under tensile force actions and recover to its original shape constitutes a supreme characteristic property [31]. It is suitable for a whole lot of traditional leather craft techniques like stamping, embossing, carving, etching, embroidery and engraving. Each traditional technique imparts unique aesthetic effects on the leather in such a way that the legacy surrounds the place of craft and the end product made using that technique. One such classic example of the legendary craft is Moroccan leather embroidered Berber bags. The iconic artisan embellished leather bag comes with different patterns. The quality and finish of the artisan-made leather bags bear significance not only to its fine soft malleable material nature but also to the craftsmanship practices survived until today [32]. In fact, both fine craft and Moroccan leather's superior quality endowed the name "maroquinerie" for fine leather goods.

In fact, from the perspective of product design characteristics, the leather is finished accordingly, to meet the end product functions and desired design attributes. Leather possesses the flexibility to be finished both on its grain side and on its flesh side [33]. Leather is capable of coloured, split, conditioned, polished and paintable appropriate to the end use requirements. Hence, leather comes in various types, namely: full-grain leather—where the skin is presented as such replete with animal's scars and stains, corrected grain—leather surface sanded for uniformity, hot-stuffed—conditioned with grease, split leather—where the skin is sliced into thinner layers for use in gloves and garments and suede—split leather sanded on both sides [34]. Leather can be tanned and finished to be any of the following types [35]: i. patent leather—the hide is finished and lacquered to give a glossy shiny surface, ii. glaze—the leather is finished to provide a hard polished surface, iii. natural grain surface—the leather is finished to retain its natural grain and iv. embossed—the leather surface is pressed with a picturesque art on its surface [35].

3.2 Leather Properties by Source Types

In the global leather product platform, cattle, cow, goat and sheep form the major livestock species from which the raw material leather is extracted [33]. Leather from cow is generally addressed as hides, and leather from goat, sheep and pigs is classified as skins. Cow leather, the most common type of leather, is renowned for its strength, stiffness and durability. According to an estimate, 66% of the leather products are made from cow leather [36]. It is this stiffness and strength which render it suitable for pyrography processes of varying temperatures. Cowhide is processed and finished to yield different surface effects such as patent leather, embossed leather and split leather. Nine percentage of the leather products are made from goat and sheep leathers [37]. Unlike the cow leather, goat leather is more soft and supple, an intrinsic leather property that it is allegiance to the presence of lanolin in it [37]. Goatskin leather is water resistant contributing to easy wash care maintenance. Sheepskins and goatskins finished on grain side are used in soft leather products like handbags. Meanwhile, the goatskin finished on the flesh side yields velvet like nap with good tear strength and tongue strength, thus making it a preferred choice for exclusive fashion garments. Moreover, sheepskin is compatible with several types of leather treatments like sanding, suede, grain correction, full-grain treatment, hot-stuffed and split. Fur skin is a general name given to products made from the skins of animal species like rabbits, wildcats, deer, fox, etc [33]. They are used in caps, hats, gloves and garments. Suede leather surface is brushed to produce a soft velvet-like texture [35] considered a luxury in fashion owing to its soft pliable drape. Unlike the other surface-treated leather types, suede effect is produced from cowhides, pigskins, goatskins and sheepskins.

4 Leather Processing Trends

Leather is one among the three majorly used material groups along with fur and skin ever since prehistoric period [38] for its product design characteristic properties and maker-friendly engineering properties. But it is not a readily available material from nature. It is obtained from the animal skins which are subjected to highly toxic treatments in order to convert the skin into a most sought-after and maker-friendly raw material.

Animal skin peeled from the slaughtered animal's carcass needs immediate treatment to prevent it from decay. The process of converting the skin into a flexible and malleable material is referred as tanning. Processes such as salting also aid in persevering it temporarily. This temporarily preserved leather offers the potential to be processed either as parchment, oil- or fat-cured leather, tanned leather or fur [39]. If the hair on the skin is retained and processed to hold strongly on the skin, then it is converted into a fur material. For the other three products, parchment, tanned leather and oil-cured leather, the skin needs a lime treatment for removing the unstructured proteins present on the flesh side of the skin [40]. The processed skin thus yielded is categorized as pelt. Further, the pelt could be used for manufacturing parchment or transparent leather for drum linings [41]. For producing tanned leather, the pelts are treated with tanning agents [41]. Tanning methods range from vegetable tanning, oil tanning and chrome tanning. The pelts are treated with oil at warm temperatures to facilitate oxidation, which in turn produce a soft form of leather referred to chamois leather [42]. The pelts treated by vegetable tanning or chrome tanning produce a more versatile substrate with improved stability and properties. The leather substrate thus produced by vegetable tanning and chrome tanning is compatible with subsequent processes such as splitting, dyeing and finishing [41]. Texture imparting finishes are applied only at the last step when the leather maintains a dry state. In other words, leather has been allocated for a product and performing of leather has begun. The surface of the tanned leather is treated with pigments and dyes to yield desirable surface characteristics that enhance the aesthetic appearance of the leather substrate [43].

Industrial tanning process constitutes 85–90% of the leather quantity produced across the globe [44]. Industrial tanning uses chromium VI which is classified as toxic cancer causing allergen. The remaining 10–15% of global leather produced employs oak bark tanning and vegetable tanning [45]. Leather tanned with oak bark yields a softer variant of the leather suitable for producing white clothing [46]. For producing leather gloves, the tanned leather is treated with alum oil and other materials as combination or formalin [46]. In oak bark tanning, the tanned hides are allowed to be hanged on ropes in open sheds for several days and involve processes such as paring or shaving to level thickness, colouring and treatment with oils and greases to produce attractive surface finishes [47]. Grain surface is treated with waxes, proteins such as blood and egg albumins to produce attractive surface finishes. This method originated in San Segundo, Avila, Spain [47]. Tanning process know-how was passed to the subsequent generations orally until seventeenth century. Aftermath, King Louis

XIV of France initiated documentation of scientific process of leather tanning [47]. Late nineteenth century marked the demand for many kinds of heavier duty leather catering to the evolving mechanization of industries [48]. For example, belting leather for driving machines, leather for textile looms, saddler leather, harness leather and shoes ushered the requirement for flexible and large volume tanning methods [48]. Vegetable-tanned leather suffered from irregular properties and uneven characteristics [48]. This necessitated faster and large-scale tanning methods like chromium tanning to produce sturdier industrial leather products. The newly evolved large-scale tanning comprised an expanded set of operations like beam house operations, tanning, post-tanning, dyeing and finishing [48]. Beam house operations further incorporated several stages of processing like soaking, liming, fleshing and splitting to clean the skin, remove the adipose tissue, remove hair and adapt the leather thickness to a desired value [47]. Tanning includes deliming, bating and pickling. The objective of tanning is to partially degrade the skin structure to facilitate penetration and subsequent fixing of chemicals, adjust pH and stabilize the structure of collagen by adding tanning agents. Industrialized tanning uses inorganic products or minerals consisting of chromium salts, aluminium, iron, titanium, etc [47]. On the contrary, vegetable tanning only used natural vegetable extracts but required more time. Organic compounds like quinines, aldehydes, sulphochlorinated paraffin and multiple resins are also used for tanning [49]. Another alternative method of tanning is enzyme tanning which is less pollutive compared to the chrome tanning. Potential alternative evolved subsequent to the abolishment of chrome salts and leather formaldehydes appears to be active cross-linking (ACL) agents, as it claims the finished leather to be toxic free [50]. This is attributed to the ability of ACL to cross-link the functional groups and chemically stabilize the fibrous protein collagen in leather structure without leaving behind traces of chrome in the finished product [50].

Once leather has been tanned, post-tanning processes involve: shaving, neutralization, retanning, drying, fat liquoring, summying and drying. Dyes and surface treatments are added to lend itself suitable for current fashion trends [35]. Dyeing refers to group of operations generically represented as colouration. From chemistry perspective, dyes are classified as natural and synthetic dyes. In the earlier days, leather was dyed with plants with naturally occurring materials that includes indigo, saffron, green span or verdigris [51]. Today, 70% of the leather is dyed with plastic dyes and aniline dyes [52]. In order to guarantee the dyeing quality, dyestuff parameters such as fastness, penetration power, matching capability, degree of opacity and method of use are held accountable. So, the desired synthetic dyes are acid dyes, direct dyes, basic dyes, metal complex dyes and reactive dyes [53]. Pigments form an alternative method of colouration. Contrary to dyes, these substances are insoluble and applied in aqueous or organic dispersion [54]. Unlike the dyes where the dye molecule is absorbed into the morphological structure of leather, the pigments are deposited in the surface pores and hold on them. Hence, pigments cover the leather surface defects.

Finishing comprises the various surface treatments meted out on the leather to yield final surface texture, character and appearance. Process like trimming and ironing helps to remove the peg marks and holes [55]. Buffing with emery paper produces more even leather but leaves behind lot of dust, so dedusting with air blasts improves the leather life by removing the impurities [56]. Buffing and embossing were the two mechanical treatments applied in finishing [57]. Wet finishing techniques such as spraying, padding and roller coating are used [57]. Glazable finishes, protein binders, thermoplastic finishes and nitrocellulose finishes are preferred surface coating techniques to achieve new homogenous surface textures. Furthermore, depending on the amount of pigment contained in the finish, more classifications came into practice [58]. Types of such finishes are full aniline finish; completely transparent without any type of pigment, semi-aniline finish; finish with certain covering power by moderately adding pigment and dyestuffs, pigmented finish; and greater covering effect by adding high quantities of covering pigments [58].

4.1 Impact of Leather Processing

According to UNIDO report, out of 1000 kg of rawhides processed only 250–300 kg of leather is produced and leaves behind 600 kg of solid waste [59]. Moreover, in this conventional transformation process from hides to final leather, 452 kg of chemical substances are added to 1000 kg of wet salted hides and around 380 kg of the added chemicals go as part of waste [60]. Chrome used in tanning ends up as carcinogenic compounds (chromium VI) in finished articles, thus posing health hazards for the end user. Leather processing inflicts disastrous twin effects on both human health and environment. And the industrial waste produced by leather during its processing is classified under Class 1—solid waste, highly toxic [61].

The environmental impact of the leather processing and the consequent transport is measured via life cycle assessment tool that indexes a numerical value calculated by assessing the hazardous impact on the environment produced per unit conversion of raw materials as per the stated process. It is presented as Higg Index. The Higg Index of conventionally processed leather (chrome tanning method) hovers around a value of 161 [62], whereas the alternative synthetic leather produces a Higg Index of 59, almost just above one-third of the conventional method [62]. Faux leather and faux suede are the synthetic alternatives to animal hides and skins. Faux leather and suede can be used on almost all products made of natural leather [35], but the inherent difference in property is they are non-porous and very hot compared to the natural leather. But large numbers of colours are possible with synthetic alternatives. However, the synthetic alternatives have a polyurethane film coating which again has its own share of toxic load on the environment. On the contrary, Pete Lankford design director for Earth Keepers and Timberland Boot in 2011 claimed "A pair of leather boots that last twice as long as a synthetic alternative, will end up with half the environmental impact in the long run" [63]. The leather processing industries initiating sustainable measures also co-asserts the strategic decision-making functions pertaining to the business process of leather products, thus covering all and sundry players as a whole in the leather product trade. Hence, life cycle assessment encompasses and multiplies the effects of sustainable measures, a deterrent for reducing the toxic load of leather processing industries.

4.2 Breakdown of the Assessment Methodologies

The necessity to weigh the toxic impact generated at each stage and regulate the process stream gathered momentum when studies identified chromium (VI) as carcinogenic relevant to nose and nasal sinus infections [64]. Thus, imperatives, perspectives and rationales fostering sustainability measure emphasis on how to apportion and account for the industrial wastes generated at every phase of leather process stream got initiated [65]. Subsequent developments in the sustainable front lead by the Leather Working Group established protocols to monitor the environmental impact caused by leather processing [66]. Life cycle analysis provides a factual analysis of products during its entire life cycle in terms of sustainability. It assesses the environmental impacts of product or services from cradle to grave. The life cycle of leather comprises four major stages: 1. hide processing stage alias tanning stage, 2. leather material conversion stage, 3. product distribution and retailing stage and 4. post-consumer stage [67]. Meanwhile, the waste generated at the product distribution and retailing stage comprising returned goods, unsold goods and seized counterfeit goods is more clean and valuable [67]. Life cycle assessment as per ISO 14044 comprises four main phases, namely goal and scope, inventory analysis, impact assessment and interpretation. Goal and scope defines the product and its life cycle in addition to the description of system boundaries [68]. The guidelines for ISO 14044 were framed with reference to the elemental definitions coined in ISO DIN 14067 for quantifying the carbon footprints at each stage of processing [69]. Inventory analysis observes all environmental inputs and outputs associated with product or service by monitoring the raw materials used, energy consumed, volume of pollutants emitted to the atmosphere and quantity of effluent streams let out [68], whereas the impact assessment phase classifies the environmental impacts, evaluates on terms of what is important or essential to the leather manufacturing firm and translates them to environmental themes. A classic example among them is effects of global warming on human health laid down by the analysis and recommendations of UNIDO 2012 report [70]. The chosen theme is presented as an element of corporate social responsibility drawing reference to the audience they cater to. In the last phase, interpretation, ISO 14044 standards describe a number of checks to test whether conclusions are adequately supported by data and procedures complied by the corporate organization [68]. There is direct correlation between increase in anthropogenic greenhouse gases (methane CH_4), nitrogen oxide (N_2O) and carbon dioxide (CO_2) and increase in global average temperature. Leather industry activities use hazardous chemical substances that generate a major portion of CO_2 with a direct impact on global warming [70]. Leather processing activities have a direct impact on global warming through CO_2 emissions [71]. Currently, there is no universal methodology or a general accepted agreement on calculating the environmental impact of leather processing.

Best available techniques (BAT) reference document for tanning of hides and skins provides valuable insights into leather processing [72]. Impact categories are scientific definitions that link the specific substances to specific environmental issues. The major impact categories identified so far as severely affecting the environment and used in leather processing impact assessment are 1. global warming potential (GWP), 2. acidification potential (AP), 3. photochemical ozone creation potential (POCP), 4. ozone layer depletion potential (ODP) and 5. eutrophication potential (EP) [72]. Global warming potential measures the impact of greenhouses on the environment [73], whereas acidification potential assesses the intensity and amount of acid gases released into atmosphere that culminates in acid rain [74]. Meanwhile, photochemical ozone creation potential quantifies the amount of change inflicted on the ozone layer due to gas emissions and the direct degradation of ozone layer is monitored via ozone layer depletion potential [71]. According to epa.gov, CO_2 has a GWP of 1 regardless of time period used as reference. Its concentration on earth will last for thousands of years [75]. As per the Romanian tanning industry case study, finishing phase generates highest pollution in terms of CO_2 followed by dying phase and beam house phase for processing a full-grain assortment of bovine leather that is chrome tanned [76].

The technological process reference materials and resources considered for life cycle assessment of full-grain bovine leather assortment include bovine wet salted hides, bovine pets, wet blue and crust weighed in kilogram unit and the equivalent quantity of natural gas in kilogram units consumed for water and air heating [77]. Electricity in kilowatt hour consumed by tanning machines, Process water quantity in kilogram units, Chemical substances equivalent quantity in kilogram units, waste for recovered from technological process in kilogram units and the waste water resulting from leather processing in kilogram units [77].

5 Strategies for Producing Environment-Friendly Leather

Leather as a brand is highly valued and respected by its consumers all over the world irrespective of the environmental impact it leaves behind [78]. In order to maintain and carry forward its value from the perspective of well-made sustainable piece of leather, all the value chain partners in the leather product network should work together in eliminating the toxic hazards produced as a result of leather processing [78]. Given the hazardous nature of leather wastes generated during the life cycle of leather goods processing and its iconic status, an inclusive design approach as well as a flexible process which could promote better relationships with the environment and the consumer demand forms the need of the hour [79, 80]. So, green peace treaties and eco-friendly awareness have etched sustainability requirement as mandatory for every leather product design, thus making it an essential design ingredient. This inculcates a need for engineering sustainable design constituents. Rather in other words, sustainability shall be an inbuilt function for each of the materials and methods used in converting the design concept into product henceforth. In order to express the luxury fashion's solidarity to sustainable initiatives, the Paris Ethical Fashion Show was organized in 2004 that comprised 20 fashion designers. Later, this effort garnered support and mobilized popularity that went to grew up to 90 designers by 2009 [81]. At the same time, the fashion consumer moved towards individual expression and sought new ways of expressing their identification in the city life [82]. This new found consumer tastes were pioneered by Gen Y consumers who embarked on living their passion over acquiring status symbols [83]. Again non-indulgence in impersonating others provided the impetus to nurture value-driven fashion solutions [84]. The novelty for twenty-first-century customers is defined in terms of personalized products and services [85]. In 2018, the world's leading business event on sustainability in fashion with 1300 international key players was organized [86]. French fashion, the largest shareholder of luxury fashion segment, pronounced its intentions to become the most sustainable fashion capital by 2024 [87]. These efforts have also roped in global organizations like "Leather Working Group" (LWG) to ensure and promote sustainable practices of leather-processing industries [88].

The crucial stage necessitates major design interventions. Further, it creates a scope for rethinking the entire value chain cycle across a new plane. The value chain partners include brands, designers, craftsmen, testing societies and standardization institutes, new technologies, rawhide suppliers, alternative materials, new process routes, tanners, research institutes and technologies for recycling. Among all the progressive efforts taken up by different partners of the value chain, the role of designer is crucial to the promotion of sustainable or circular leather product design stemming from the fact that the motives of the fashion system in terms of sustainable fashion demand and social agenda of fashion revolve around the creative work of the product designer [23]. Several scholars like Manzini, Margolin, Papanek [89, 90] and Thackara [91] do agree social innovations and design as a catalyser for creating visions of possible futures. The role of research institutions also assumes importance as advent of research tools like EDFR futuristic techniques add enterprise to identify and forecast future demands of design [92]. From the sustainable leather product design perspective, the designer's role could range from not only creating a fashionable product, but also his responsibility to use the resource for a longer span of time in the design's life cycle, the ability to reuse waste materials in the design development and predict the scalability of the designs. In overall, the direct role of designer in producing a sustainable leather design encompasses activities right from i. the development of new conceptual framework of design, ii. advocating reuse of waste and second life product strategies, iii. developing key performance indicators for ensuring sustainable practices along the material supply chain, iv. engaging craftsmen, customers and entrepreneurs in co-creation, v. developing new material alternatives and vi. adopting environment-friendly new technologies and process routes.

5.1   Development of New Conceptual Framework for Design and Key Performance Indicators for Ensuring Sustainable Practices

FAMEST, acronym for the project footwear, advanced materials, equipment and software technologies, aims to produce new concepts of fashion, technical, personalized and customized work shoes of high added value [93]. These high-value additions along with the sustainable design needs are incorporated by deploying suitable material development plans, production process and post-consumption processes [94]. This is crucial as 41% of the leather produce is consumed by footwear products [26]. The entire life cycle of the product is monitored through flexible and agile technologies with the sole objective of developing solutions to reuse the waste generated in production and post-consumption [95]. This project is supported by a consortium of twenty-three companies spread across the whole value chain of footwear which includes makers and distributors of leather material, insole and sole components, chemical substances, software firms, equipment and logistic firms [94]. Thus, multi-disciplinary and complementary competencies are entitled through selective choice of the tools used in the footwear manufacturing, futuristic design concepts, new chemical substances, new material options, new functionality components, Tech 4.0 advanced production methods and marketing technologies for 4.0 [94]. The project scope draws from the belief, and true environmental sustainable footwear products begin with creation process that is design [96]. It is at the design stage that the entire footwear value chain and the environmental aspects from selection of raw materials to the end of product life can be envisaged. So, the following sets of design strategies are floated to reduce environmental impact and develop ecological footwear products [94]. The project embarks on a set of ten guidelines, namely "i. improved efficiency of the material used for the product making, ii. improved energy efficiency of the processes through reduction in the number of production process steps engaged or in other words the lesser the number of production processes greater the energy efficiency, iii. design for cleaner production thus less harm to mother earth and atmosphere, iv. design for durability, v. design to optimize functionality to facilitate multiple product functions, functional optimization and product modularity, vi. design to reuse and recycle demonstrating the capability of easy disassembly of the product and simplification of materials for easy recycling, vii. avoid potentially hazardous substances and materials, viii. design to reduce the environmental impact in the use phase, ix. employ environmentally more efficient distribution channels to deal with the objective of less packaging waste, lesser packaging requirements and eliminate unnecessary packaging and x. optimize the end of product life opportunities through reduction of product complexity, easy separate marking of material for channelizing their recyclability and facilitate communication regarding end of product life opportunities to the consumer" [94].

The NYU stern centre for sustainable business in partnership with council of fashion designers of America developed the kerning standard for leather goods and shoe manufacturers. These KPIs are defined with reference to a set of nine topics framed within CFDA guide for sustainable strategies [97]. These metrics can be broadly classified into i. manufacturing process metrics that includes materials, process, design philosophies and other energy resources used in the manufacturing, ii. supply chain metrics for monitoring raw material suppliers, distribution service providers and other logistic service providers and iii. community and social responsibility metrics for managing people-related matters and business strategies. The community and social responsibility metrics represent the first KPI that focuses on establishing pay structure, measuring employee satisfaction, improving minority representation and committing to X% of women workers [97]. Supply chain metrics advocates local purchase policy in case of raw materials advocates stringent audit procedures to ensure the logistic supply chain partners strongly abides by ESG guidelines [97].

Manufacturing process metrics encompasses all the activities right from design development to packaging. It adopts the design philosophy "design for sustainability" with the objective of creating zero waste patterns for cutting, manufacturing waste reusage and recyclability of product at its end of first life [97]. The design scope is defined on the lines identifying circular opportunities for materials used up in the manufacturing. Further, it also pledges to reduce the usage of HVAC equipment and adopt the policy of reducing energy consumption every year [97]. It enforces optimized sustainable process methods with emphasis on reducing environment impact caused by them.

Thus, the Kering Standard for sustainable leather goods and shoe manufacturers encompasses all and sundry means by which a designer could pursue active strategies along the design stream, sourcing stream, production stream, sales and distribution stream, machinery and energy stream, material stream and by-product stream. The affordability of the KPI's deployment lies in the fact that either any combination of them or as a whole or as an individual metric is incorporated by the designers or brands parse or entrepreneurs alike.

5.2 Advocating Reuse of Waste and Second Life Product Strategies

Recycling of the products alone is inadequate to question or change the attitude of fashion buyer, given the semiotic value of leather as a luxury product. For true change, design should focus on sustainable lifestyles instead of mere sustainable products [98]. Therefore, a broader understanding of design is required especially while advocating second life product strategies. The challenge lies in reaching critical mass of designers and consumers to be powerful enough to start change on a large scale [22]. For a more sustainable production system, the general relationship between product, producer and end user should shift towards slower cycles, more local production and much more valued product–person relationship [99]. An exceptional example of resold value being greater than its original retail value could be traced among celebrity-owned Birkin handbags and Kelly handbags [100]. Further, Brundtland's definition of sustainability states the all-encompassing responsibility to fulfil the needs of the society without depleting or reducing the environment's resources. In other words, strategical relook at the market share of the traditional virgin materials used, obtaining materials from sources declared as waste and their scalability for expansive end uses of the contemporary world, needs assertive measures.

5.2.1 Scope of Recycling Concept

The challenge lies in finding appropriate markets for leather recycled products [67]. From the perspective of product value and economic viability, recycled leather product is classified into three categories, i. downcycled leather product, ii. recycled leather product and iii. upcycled leather product. Downcycling refers to the failure to recover leather's original value, and hence, the recycled materials shall end up as feedstocks or supplements for alternate processes such as agricultural engineering [101]. Recycling leather refers to reconstituted leather recovered from leather wastes whose value relies on its purity based on which they are classified into low range (<70% purity), medium range (70–90% purity) and high range (>90% purity) [67]. Upcycling refers to the processes that are applied to extract the embedded chemical compounds from the leather waste [67]. At hide processing stage, anywhere around 70% by wet weight of the hides might go into waste streams [101]. Microbial fermentation of solid wastes helps to recover the waste fluid streams and is reused for pertaining applications. Chemical processing of trimmings and split pieces generated during the conversion stage produces films and adhesives [102]. Soares discovered a new approach to life cycle analysis of footwear products by advocating reintroduction of clean solid waste generated during production by re-engineering the process from design to product modelling [103]. But still each category of recycled leather product requires separate technology and enterprise framework to realize its product value.

5.2.2 Reuse of Wastes Generated in Manufacturing Process

Wastes generated during second stage of leather life cycle such as fabric trimmings and leather cut pieces are assembled into square patterns of 25 cm^3 and later rectangles of 50 cm^3 in the course of converting the clean waste into a high heel boot [104]. However, the polygon pattern shape and size are determined from the volumetric area of the target upcycled product to be manufactured via a mathematical method [104]. DNP acronym Design na Pele is a project initiated at a children shoe-producing factory situated at Rio Grande do Sul, Brazil. This company produces around 15 thousand shoes per day. The project aims to reuse the solid waste materials generated in the production phase, and code names the process as reverse logistics [105]. As usual, the design development starts with traditional factoring of trends and seasons. Along the design development process, the volume of leather waste in the form of smaller pieces is also studied upon as a factor. So, the design development team is also vested with the task of building a unique model that shall consume the leather scraps generated in the production floor which is otherwise discarded as landfills. The more the amount of small wastes generated in the process goes as waste, the higher gets the end product price. Thus, reuse of small leather scraps and large leather scraps produced during manufacturing begins to be planned right from the footwear design development and specification stage itself [105]. In this new approach, the designers are informed about the nature of the waste generated in the process and the amount available. Using these inputs, a new conceptual shoe design is developed determining which waste type, colour and quantity in what proportions are deployed. However, the varnish type, opaqueness and texture are factored according to the design type and ultimately, no single leather product produced by this method has the same character and texture [105]. Every product is unique by way of the composition and colour usage. After the approval of the model, the waste scraps are reintroduced into the production process. The larger waste scraps and the smaller waste scraps are used separately. The smaller leather scraps are used for decorative articles such as petals of a flower or strips of ribbons, and the disadvantage is that it consumes more time and cost. The waste savings in the form of reusage accounted for 15% reduction in solid waste disposal, approximately 21 kg of leather waste [105]. It belongs to the category of green product design that aims to build completely from recycled materials. In this cradle-to-cradle life cycle planning process design for disassembly, choose only decomposable materials that are rapidly renewable—bamboo, cotton, natural rubber, cork, etc., are emphasized [106].

5.2.3 Reusability of Design Information for Leather Goods

Design information pertains to all and sundry involved in design process starting from design brief, design details, design process, design methodology and leather product-making process [107]. Design process starts with understanding the product attributes by analysing the design brief or looking forward to solve a design problem. The design requirements are understood through brainstorming the brief and identifying the key design elements. Design research process is about being artistic and open ended. Thus, the whole design process is guided by the inspirational references unlike the scientific research process where the objectives are framed to solve the defined problem or to take you closer to the problem.

Perceived design details of the leather products are those availed from the leather products offered on sale, forecast reports and consumer preferences. Design details per se comprise the whole picture beginning with the conceptual part explaining the manner in which the design elements are presented along the design construct and its supposed purport. Further, the product-making phase incorporates all the design details with the help of craft and technology on the material leather. The material per se encompasses all the constituent parts of the product structure. Meanwhile, the visual composition of the leather product design is constructed by employing the elements and principles of design upon the material in a conceptual framework. It is the conceptual framework that defines the attributes of the design as per the trend. So, an unsold leather product in the sales and distribution stream might require design reconciliation in order to render it marketable as per the new designed value propositions and product concept. Design reconciliation may start anywhere from reorganization of visual design composition by shredding the product components and refabricating the whole design to just rearrangement of visual appearance through reorganization of visual hierarchy. Achieving new visual hierarchy through minor tasks like surface treatments will be the effective process route in terms of time consumed and cost involved, whereas refabricating of design concept shall entail higher expenses. Technology, skill and energy, all these are incurred during the redesigning process.

Kate Fletcher's project "local wisdom" took an inventive step in proposing that sustainability can emerge from a wealth of simple interactions with people. The interactions were recorded as a story shared by the people which in turn opened up insights relevant to the people's reflections on the experience about the product [108]. By any means, these interactions might not fit the objectives of commercial opportunities but rather open up a new front especially for discussing the everyday practice of product usage. In fact, it indirectly reflects on the culture each and every person has pursued diligently. Just as in the local wisdom project where reflective practices of dress culture provided the means to frame cultural understanding of the clothing as critical to garment's sustainability [108], it is essential to frame cultural understanding of the leather products to determine the reuse of design information. Drawing from the initial findings of "local wisdom project", a questionnaire was developed to comprehend the cultural understandings of leather product usage among the southern metropolitan cities of India. And it was circulated among 500 leather product customers who are identified as regular users of leather products.

1. Which age group you belong to?
A. 17–21, B. 22–26, C. 27–31, D. 31–40 yrs.
2. Do you have the habit of carrying a leather article or wearing a leather jacket?

   YES/NO

3. If yes, how often you don a leather jacket or any leather carried accessory such as handbag or pouch in a week?
A. Twice a week, B. No matter how many times a week leather article is an indispensable part of my outerwear. C. Only when I do outdoor travelling. D. Only once or twice.
4. What sort of bonding does the leather product give you like _____?
A. I wish to carry a leather article or don a leather jacket as it gives me a status symbol.
B. Leather product adds on to the overall outlook and elegance.
C. For etiquette reasons to fit on to the occasion.
D. A leather product gives me a personal attachment and makes me feel intuitively more accessible.

5. During leather product selection and purchase are you interested in
   _____.
A. Avant-garde design concept of the product, its presentation.
B. Brand values, legendary leather design details.
C. Design, leather raw material sources and its claim of how delicately it was made using artisans (craft).
D. Only physical appearance of the product; latest look, celebrity endorsed product, latest fashion, etc.
6. Would you prefer to appreciate innovations in leather product silhouettes wrought by recycled and re-engineered materials?

   YES/NO

7. If your mother/father owned a leather product of genuine leather and good quality, would like to _____.
A. Refurbish (renovate and add value) with new style features and functions (change the visceral quality and consequently behavioural experience also).
B. Retain its renowned traditional form and original design concept by giving preservative treatments and mending (retain the visceral quality and more emotional).
C. Give preservative treatments and convert it into modern silhouette appropriate for today's usage as its original design has the potential for getting repopularizing (change the visceral quality and behavioural quality—defining new behaviour).
D. Though its original design did not have the potential for repolarization (coming back into fashion), yet would like to convert into suitable product and reuse due to its very good quality and value of leather (change visceral quality, change behaviour but appreciate the value of leather).

Forty percentage of the surveyed respondents belonged to the age group of 17–21 years old. And in every age category, there were nearly about 20% of the respondents. Among the responses to leather product usage in a week, 84.5% accounted for leather product usage as an indispensable part of outerwear. This conveys a mass preference for leather product usage. With leather scoring high on being considered an important and essential life style product underscores the significance of its cultural value. However, the emotional bonding with leather products reflecting on their cultural understanding could be confirmed from the subsequent response where only 54% of the respondents indicated that leather products give them a personal attachment and make them feel more accessible. Later, 67% of the respondents supported brand values and legendary leather design values. And almost 68% of the respondents preferred refurbishing the leather product with new style features and functions. The preference for refurbishing undermines the possibility of reusing the material in two different ways, 1. adding new value additions on existing design details and 2. extending the cultural value with modified functions and design details. By either way, design information is reused but with necessary changes. Hence, cost components, technical know-how and technology all are incurred in the redesigning phase. Actually, it is reused by redesign.

5.3 Co-design

Co-design aims at interlinking the designers, craftsmen, customers and entrepreneurs to create leather products of unique value by accommodating the environmental, cultural and social principles of sustainability. In the co-design framework, the working supply chain partners are classified into two categories, i. stake holders and ii. duty holders. The supply chain partners who align with the demand positions or the pull positions are grouped as stakeholders, and the supply chain partners those who take the role of fulfilling the demand and satisfying the customers are deemed duty holders [109]. This discussion on co-design shall feature the brand initiatives and activism strategies undertaken by the designers. Tuscany a word eponymous for genuine leather design still pioneers in the production of vegetable tanned leather product [109]. Every year, "Craft the Leather International workshop" brings budding designers to observe and learn the value of leather material through interaction with leather craftsmen of Tuscany. It is a perceived endeavour to package "arti-sanality" approach in producing innovative leather goods, the cultural component in sustainability [109].

At Hidesign in India, the leather design requirements are transferred to the product by understanding and gaining insight about the vegetable tanned leather properties, detailing, product aesthetics and conceptualizing for whom they are producing for [110]. Through this interaction between designers and craftsmen, the fashion trends are communicated to the craftsmen so that they can enhance the design and in turn the designers learn from the craftsmen how they translate the design requirements and aesthetics in the product. Similarly, eco-craft also evolved through the interaction with the craftsmen of Jaipur who produced handmade sandals. Jaipur craftsmen were spurred on by the great Mahatma Gandhi's Swadeshi Movement emphasizing make your own chappals and cloth as a symbol of independence [111]. And India's growth in leather exports is attributed to the strong link between traditional craftsmanship and innovations the designers are able to bring in [112]. In another case, the art of making fish leather old to several centuries and indispensable part of Nordic culture was revived. The difference between fish leather and other biological leather materials in the case of brand "Mayu" identifies with the zero waste process methodology followed by skilled Indian craftsmen and upcycled leather produced from fish skin a by-product of food supply chain [113].

Stella of Stella Soomlais is a pioneer in manufacturing handbags in a sustainable and innovative way. The design strategies are developed to minimize waste such that the cutting leftover materials are converted into small accessories like purse, wrist band and key chains with the help of expert craftsmen. The brand's innovative cutting pattern management plans the second life product patterns on the perimeters of the first life product patterns in order to close the loop [114]. Overall flat design methodology is followed with emphasis on functionality and efficacy. Rather in other words, this strategy provides an interaction design platform for both designers and craftsmen to co-create leather designs. Around 2.4% wastage has been reported one of the lowest across the industry as of now, but they further endeavour to bring the waste down to zero percentage [114].

In the perspective of Luna Mazzolini, leather is a by-product of food industry; if not processed into a useful material, it becomes just waste threatening to pollute the world. Her design ideology revolves around goal of using only hand stitching techniques and screwable fasteners in making the leather product [115]. So that, the entire product can be dismantled and reconverted into another product after the end of the life of the original product [115]. Innovative double-row hand stitching was applied which added on to the flexibility of disassembling the layers. Thus, handmade craft was once again used do the fore in redesigning the end of life leather products. Other discarded scrap materials were recycled into bonded leather which ended up as linings for new product development [115]. She also promoted the use of bio-based polyurethane coatings to extend lifespan of leather products further. The choice of bio-based polyurethane not only eliminated the fossil fuel petroleum sources but also provided better durability and water resistance. This coating exhibited the property of ageing gracefully upon exposure to sunlight. The treated areas shone brighter and brighter upon prolonged exposure to light [115].

5.4 Developing New Material Alternatives

5.4.1 Biofacturing

As the name suggests the product, a garment or an accessory is synthesized with the help of bacteria from sugary tea solution, a unit-level stage directly. It bypasses all the intermittent stages of material development. The merging of biological organisms and human-made manufacture will cause a radical shift as we seek to grow products, rather than manufacture them, from cells or mycelium. In other words, grow made products. The fermentation between the bacteria and sugary tea solution leaves behind a sheet layer-like substance on the top which is taken out once it is 2 cm thick and washed in cold soapy water. The thick layer-like substance, when dry, is suitable for cutting and conventional sewing. This new innovative method of production is being pioneered by Designer Suzanne Lee and seen as a cradle-to-cradle approach with no waste and direct synthesis from the raw material in a span of few days. Luxury brand Gucci is pioneering a new eco-friendly and sustainable alternative. This material is already 20% of Gucci's ready to wear collection, and about 40% RTW collections use this material [116]. Another example is microbe fermented fashion where the bacterium Acetobacter is used to convert wine into vinegar resulting in a by-product, a skinny rubbery soft skin-like substance. This substance is scaled up into a garment form without any seam requirement. As the material produced by the action of a bacterial culture on the alcohol is cellulose, it is naturally degradable in the environment.

In another variant of biofacturing, manipulated grown structures and products were directly produced [117]. Designer David Benjamin identified two different types of genetically modified bacteria. He mixed them in a large Petri dish with nutrients and cultivated their growth [118]. Along its growing phase, the bacteria were allowed to interact with compatible fungal structures. Thus by growth, flat sheets of material with distinct rigid and flexible regions were formed [118]. It produced a natural pattern texture and was lightweight. It was used as envelopes in boats, airbuses and buildings. In one other pavilion architecture, a pavilion geometry was created using an algorithm that aligns a continuous silk filament across the scaffold [119]. These scaffolds were fed with groups of silkworms which filled the gaps between the scaffold frames. An outcome of digital and biological fabrication was pioneered by MIT media laboratory [119]. Similarly, young tree branches of willow oak tree were pruned and allowed to grow on the former frames. At intermittent nodal locations, the tree branches were grafted along with the former so that it grows into a single solid piece [120]. Once the graft reaches the required shape through continuous growth, it is harvested as table or chair determined by the base former shape. In a similar initiative, Designer Jen Keane grew rhaeticus bacteria on interwoven yarn membranes in the shape of shoe mould [121]. The bacteria grow around the bamboo yarn and form a impregnated surface.

5.4.2 New Vegan Leather Alternatives

The first ever vegan fashion week was held at Los Angeles in 2019. It was a conscious effort to cut down on fur and exotic leather trade besides exploring the technologies and manufacturing means for producing and branding leather alternatives. Parallelly, sustainable fashion weeks were also held along the similar lines emphasizing the virtues real, sustainability and recyclability alike. Continuous efforts paid off with the finds like coffee, grass leather, pineapple leather, mushroom leather, fungi leather, coffee leather, crop waste leather, apple leather hogging the limelight of fashion shows. One such pioneering effort was by Sebastian Thies, a sixth-generation shoemaker who helped his brand to creating sneakers from real, sustainable, recycled and fully vegan materials. Another such brand initiative saw Humour.noir a luxury brand succeeding in producing and marketing animal-free, sustainable and exclusive accessories [122]. The brand prides itself on its handbags especially which are handcrafted by Italian craftsman using natural vegan materials like pineapple leaf-based leather, Cartina: a composite of recycled paper and cork. Another value addition to the handbag is its 24-carat gold-plated accessories that is free from both lead and nickel. On top of all these innovative measures is a bio-leather alternative, a product conceived and developed by modern meadows. Modern meadows advocate biofabrication process, where animal-free, recombinant collagen is produced by means of DNA editing tools that assist to engineer special collagen-producing yeast cells [123]. The DNA editing tool-assisted yeast cells produce collagen equivalent to that of cow's leather. The biotechnology kit further organizes the collagen to recapitulate it into fully grown biological collagen structure characteristic of that identified in an animal hide. Later, this collagen is forwarded to tanneries to undergo the remaining process in leather production. This technique of producing bio-leather draws from inspiration from 3D bioprinting technique adopted by Organovo that makes skin models of liver and kidney organs. This bio-fabricated material is also available in liquid form that provides endless opportunities to spray or pour and moulded into a 3D form without any joining operations such as sewing [124]. This material is commercially sold in the name of "Zoa". Zoa can be produced in any required density, thus capable of being moulded and shaped; besides, it is also compatible with other materials in forming composites [125]. Further, it can also take many textures. In 2017, Zoa graphic shirt was launched by modern meadows which is now partnering with world class luxury brands for new product developments [126].

Meanwhile, brands like Bolt Threads identified the prospects of growing leather directly from mushroom mycelium which even capitalized with the support rendered by renowned fashion designers like Stella McCartney. This work gained momentum when British furniture maker Sebastian Cox collaborated with Designer Ninela Ivanova to investigate mushroom mycelium's potential in growing contemporary furniture out of moulded wooden strips [127]. At the beginning of the process, the mushroom that feeds and grows upon the base of wood or corn or straw, etc., is wetted. Once it is damp, the substance is put into a bag and allowed to pasteurize. Pasteurization kills bad bacteria and propels the growing phase. Then, the mushroom colonies are allowed to develop which usually takes around two weeks of time. Later, the grown up substrate is broken and compressed into desired shape and size. Desired dye colours are added to finish the leather alternative. It is commercially known as Mylo. Actually, the base used for growing mushroom is a post-consumer waste comprising anything from corn cobs, wood chips, straw, etc [128]. Brand ZVNDER that makes mushroom leather fashion accessories has also collaborated with Nat-2 to make vegan shoes out of mushroom leather and recycled plastic water bottles [128]. Mylo is a breakthrough material that looks, feels and behaves like handcrafted leather [30].

Carrying forward the trend, the brand Vegea introduced leather alternative made from by-products of wine industry that had expanded horizons of usage ranging from fashion to automotive and transportation industries. Brands like nuuwaï, Poétique Paris and Happy Genie initiated the usage of apple fibre leather in fashion accessory designs. Apple fibre leather was made from leftover apple wastes. Frumat is a leather alternative prepared from 50% leftover apple waste and 50% polyurethane [129]. It is available in variety textures and surface treatments such as laser printing and embossing that can be afforded on its surface. Even more environment-friendly material was developed by Tree Tribe and Elpis Studio by soaking, drying fallen teak tree leaves and pressing them into sheets. These sheets are capable of being mended with cotton to yield soft interior structures [130]. Another leather alternative from fruits is pineapple leather that is made from composite mixture of pineapple harvest waste, PLA and petroleum-based resins [131]. PLA is linear aliphatic thermoplastic polyester made from plant-based matter. PLA is made by extracting the starch from plants and converting it into a fermentable sugar, such as glucose, by enzymatic hydrolysis. Fermentation of the plant sugars produces lactic acid. When water is removed, a cyclic intermediate dimer—lactide—is produced [132]. The pineapple fibres are extracted through decortication process which are washed, dried and degummed later. These degummed fibre produce cotton like flexibility of virtue of which it is converted into a felt like sheets. These sheets are later converted into leather alternative through a subsequent process along with resins [132]. The Pinatex leather alternative is composed of is currently composed of 80% pineapple leaf fibre (PALF) and 20% polylactic acid (PLA) fibres and is 100% biodegradable [30n]; 85–90% of the Pinatex leather alternative is biodegradable except the remaining 10–15% that forms the coating material which is not biodegradable. The dyes are used for colouring the Pinatex material as per the GOTS standard. The resins used on them are compliant with the standards of the Apparel and Footwear International RSL Management (AFIRM), an organization that serves to advance the global management of restricted substances [133, 134]. It was developed by a leather industry consultant Carmen Hijosa. Pinatex now works with 43 high-end labels and brands that include players like Hugo Boss and H&M [30]. Between the efforts of the designers and these leather alternatives to see through their successful transition into product forms lies the compatibility of their mechanical properties vis-a-vis viscoelastic properties.

A vegan leather alternative that transcended these requirements is Desserto a cactus-based leather alternative produced from nopal cactus leaves [135]. Other forms of vegetal leather include the one produced by coating cloth with raw rubber and then smoking it to make a durable, waterproof material. This technology had been used for years by rubber tappers to make waterproof bags out of old sugar sacks [136]. The products made from vegetal leather feature mainly assorted bags and purses, cosmetic cases, cool briefcases, duffel-style travel bag, etc. Luxury brand Givenchy has introduced an eco-friendly vegetal leather velvet belt [137]. This piece features a wide-to-narrow style for a balanced play on functionality and femininity to achieve a high-fashion look.

Among all the bio-leather alternatives, these three materials, Zoa, Mylo and Pinatex, are seen as potential leather alternatives and commercially feasible for bulk production. All the three materials Zoa, Mylo and Pinatex seek to eliminate the most environmentally impactful phase of leather, the rearing and slaughter of livestock for its raw material [30]. While Mylo and Zoa are able to eliminate the industrial chromium tanning process by choosing natural tanning methods, Pinatex does not require this stage as it is finished with alternate materials [30]. All the three materials require comparatively shorter periods of growth. Mycelium mushroom the raw material for Mylo grows in a matter of weeks [138], and Pinatex raw material "pineapple" requires eighteen to nineteen months for harvest [132]. In case of Zoa for developing collagen from yeast, it takes around two weeks [127]. In fact, these material properties are far better than PU-coated artificial leather which require a minimum of two weeks' time for their production [132]. Conceptually, the possibility of creating a finished material equivalent to natural leather from the hides in terms of composition and structure has been evolved [30]. They offer the scope of incurring less waste material; rather in other words, the material can be grown to required specification. These materials can be handled by artisans and industrial technology alike.

5.4.3 Natural Textile Composite Alternatives

Designer Tamara Orjola developed a composite from pine needles and bioplastic binders. It produced high-quality look accompanied by a fibrous feel [139]. Technically, it was durable, hard and tough capable of fulfilling functions like load bearing, insulation, decorative and styling functions [117]. It was 100% biodegradable. Normally millions of pine needles go to waste. The pine needles were crushed and soaked. Later, the soaked material was steamed and carded. Subsequent to carding, binding and pressing operations were carried out to finish the material [139]. It was found suitable for applications such as carpet, paper and furniture. Designer Tjeerd Veenhoven developed palm leather from areca palm tree leaves and wood [140]. The dry leaves and brittle leaves of the areca palm tree were dipped in a biological softening solution and processed into substrates [117]. It offered resistance to tear; at the same time, it was flexible and strong, it provided a natural leather like feel and it was also capable of fulfilling functions like protection, decoration, comfort and identity. It held immense potential to be used as a replacement for leather, plastic and rubber materials [140]. Footwear and upholsteries started experimenting with palm leather for new product developments. A manufacturing firm Sedacor JPSCORK Group based out of Portugal developed a smooth, soft and warm material from cork [141]. The cork was steamed and boiled to impart elasticity to it. Later, the substrate was cut into thin sheets. These sheets were capable of performing functions like protection, comfort and decoration; hence, it found itself used in applications such as apparels, upholsteries and footwear [117]. Technically, it was durable, water resistant and stain resistant. Studio Sarmite based out of Latvia developed pine skins from the bark of the pine trees [117]. The pine tree bark was peeled as soon as the tree was cut and blended with natural ingredients and processed into flexible tenacious substrates [142], and it was soft, resistant to water and tear. Because of its tensile properties, it ended up in construction of apparel products. Another firm Dekodur GmbH & Co of Germany produced bark cloth [143]. The bark cloth is a substrate made from a composition of fig tree bark, high-tech phenol and amino plate papers [117]. It offered UV protection and found to be resistant to scratch, water and heat. Hence, it was preferred for furniture, wall panels, indoor coverings and ceiling panels. The fig tree bark was hammered into soft material, pressed with high-tech phenol and amino plates and processed under high pressure [143]. Designer Buro Belén developed a biodegradable wooden substrate from mulberry tree bark [144]. The bark was peeled as skin and soaked in solutions for softening up. The softened material was beaten to spread the fibrous stuff in the form of sheets [117]. Aesthetically, it was matter and soft and produced a warm feeling besides being scratch resistant and flexible. It was used in interior decorative textile applications [144].

5.4.4 Advantage of New Material Properties

The material experience as stated by Karana defines it as "the grasp of material properties the designer could interpret while they used them in product designs"

[145]. Later, the material interpretations in the form of sensitivity to the five senses, their meanings and emotions bonded to them were consolidated [146]. Such consolidation of information pertaining to these novel materials provides a platform to compare them vis-a-vis with the traditional leather properties. And it also furthered the classification of new materials. Among all of them, especially the DIY materials produced by way of individual efforts or small group collective efforts with designer intervention are unique on their own. The unique properties in terms of aesthetics or engineering attributes are self-honed by the designer which also provides the royalty privileges until and unless it is mass customized. In due course of time, there might be modified versions and further developed versions of them [147]. These materials are also becoming popular for other reasons such as sustainability attributes of the material, novel unique material expressions, one of kind appeal, their exclusivity in the market due to their limited availability and in overall avant-garde categorization of products made out of it [147, 148].

For exploring the perception of the aesthetic properties and their conceptual attributes, [149] classifies the materials on the basis of the source from which they are developed. The material source categories are identified as vegetable sources, animal sources, manmade sources, etc. Materials developed out of vegetable sources naturally refer to the intrinsic variations in texture otherwise known as imperfections. These imperfections are sensitized by touch arising out of uneven surface roughness or smoothness that is attributed to their non-homogenous nature. However, the consumer's strong predisposition about vegetable material source is their sustainable properties. The combination of natural visual appearance and sustainability properties has started to influence the user's affective attitude according to Mangier and Schoormans [150]. In fact, the above-discussed surface unevenness has become an active stimulus to user's cognition for it elicits various types of emotions [151]. The conception of natural materials as a healthy environmental choice holds the beliefs in ransom intuitively [79]. The most positive emotion types reported about these vegetable materials are [151] gratification followed by animation and enjoyment, while negative emotion types recorded are discontent, uncertainty and aversion. According to Wabi Sabi concept [152], imperfection is often embraced as it communicates the intrinsic unevenness of the natural materials. Natural materials lend uniqueness about it by drawing from the phases of their development and existential nature on mother earth [153]. Events and track record of its development phases have always held important in Western culture owing to its cultural and anthropological associations with the local place and its people. It is this that develops into semiotics, an indispensable attribute of every design product, thus paving the way for the development of positive emotional bond with both the material and the product [149].

The surface characteristics of these vegetable source materials and biopolymers attribute their grain texture to their morphological make-up. The intrinsic natural random variations and non-homogenous character yield uneven patterns on the surface. These surface patterns and grain make-up remain decipherable and obvious. Irrespective of the inherent grain rendering brought about by their morphological make-up, obvious differences arise as a result of variation in size and difference in clustering density of the grains on the surface [154]. These attributes indicate the hallmark characteristic of natural materials; in other words, they represent the pedigree traits of natural materials. Rather, it paves the way for expressing the naturalness [154]. It is the visual modality that dominates the experience in the beginning of the user–product relationship [155]. Among all the design elements, pattern and texture score the foremost on the sensorial experience scale. These material traits regarding naturalness have the potential to form long enduring relationships [147], which might end up in sustainable design obtained through emotional endurance [156]. Research about "natural" materials [153] predicted that the natural colours and patterns which show commonalities with basic aesthetic features of nature are considered to be more durable and long-lasting. Further, these surfaces are capable of being treated with different coating substances that change the tactile characteristics of the surface as well as its visual textural outlook. Beeswax, carnauba wax, latex, shellac, pine rosin and ghee are the different types of coating substances deployed for finishing both vegetable source materials. Overall, the interpretive features such as natural and handcrafted outlook attributed to its inherent visual pattern and texture along with the perceptive characteristics of the sensorial experience related to its tactile properties define the experience for these vegetable source materials which is capable of taking on the animal source materials.

Sustainable materials from the animal sources such as vegetable tanned leather and other forms of skin and exoskeleton composites as already discussed owe their tactile characteristics like softness and flexibility to the presence of lanolin in the skin. Meanwhile on the visual texture front of animal source leather and skin and exoskeleton-based materials, a plethora of colouring and texturing finishes are available to render innumerable number of visual effects. The craft techniques evolved over continuously since its usage in the ancient time to what we see today. It is this amazing property of leather to converge with very many treatments that mark their versatility. The softness, suppleness, elasticity, toughness, patina and durability together index the leather as the super-luxury material. And leather surface can have matched to any synthetic or natural surface characteristic [120]. It is important to recognize that leather substitutes can offer completely consistent surface appearance and advantages in cut-component yields. It is also important to understand that some substitute materials perform better than leather in one or selected properties. In addition, these developments in materials science will produce ever-improving products [120]. While the material morphology bound characteristics such as visual texture and pattern, tactile traits are irreplaceable considering their unique value and propositions they are capable of. So, the factors to be considered while comparing leather with its potential alternative shall be comfort, durability and recyclability [120]. Due to the difference in their mechanical properties, the techniques and technologies used for construction and assembly of fashion accessories differ from the conventional craft and industrialized techniques used for leather. In addition to the material benchmark properties discussed above, its role in the society needs to be complemented for adoption [117]. The experiential studies conducted in Italy yielded the following results. The new materials shared affective connotations such as respect, kindness of empathy type, worship of aspiration type and surprised of animation type [117]. All these connotations fall under pleasant emotion category.

5.5 Adopting New Technologies

The average handbag takes 75 steps on a factory line, while the Traveller Duffle takes around 139 steps [157]. On the contrary, XYZBAG an Italian brand advocates digital craftsmanship: it combines traditional design methods and additive manufacturing technologies to produce unique handbags. By this, 3D technology means it has made possible to manufacture the handbag straight away at few work stations in quick time. Further, it adds a wide range of flexibility in personalizing the handbag designs at the design preview stage. So, designs can be directly shown to the customers and based on the requirement they could be produced. Thus, the dimension of post-distribution unsold bags in retail stores can be weaned away [158] as it eliminates the need for holding inventory because of remarkable lead time affected by additive manufacturing. At XYZBAG not only 3D printing, traditional Italian craftsmanship is also used in combination. By virtue of cloud-based CAD software, design options based on 3 master products (DADA, GRETEL and TRADA) are shown to the potential customers. Customers have the privilege of deciding on the design with recommendations for change in details. This paves the way for customer-driven design approach and production only on demand. The benefits of additive manufacturing or 3D printing are personalized designs in terms of specification and creative attitude. Rather, it incorporates both sustainability and co-creation, the future of fashion industry. The role of customer in design interaction and co-creation adds a new dimension to the evolving fashion industry [158]. Another Designer Lana Hopkins pioneers in the production of 3D-printed handbags [159]. The sports brand Nike launched a 3D-printed sports kit bag Rebento Duffel on the eve of FIFA World cup 2014 [159].

6   Conclusion

There has been several significant research developments and design process innovations imbuing faith in sustainable goals and circular design. Well waste generation in design directly translates to failure of the design. The major problems in the design and development of leather products are the impact on environment it leaves behind along the process of raw material to product form conversion and identifying appropriate second life for leather products at its end of first life. According to Burall, Author of cradle-to-cradle design approach, the designer is entitled to know the end of the products right from the product design phase itself and develop methods on how to begin another cycle [160]. So, circularity is the future that requires the coordination of designer activism strategies. Our study of the literature and case studies pertaining to designer activism strategies revealed the following enterprising avenues, namely; i. production of genuine leather products from by-product hides sources using vegetable tanning methods and fine craftsmanship and ii. production of fashion accessories from biodegradable and recyclable leather alternatives. Hence, understanding and appreciating the role of designers, who do not want to be confined to only the synthetic leather alternatives and the highly coated industrial leather, are important as they influence fashion and its demand [65]. From a design-thinking perspective, all the efforts seem to be directed towards producing designs on the lines of circularity. Available literature supports the claim of genuine leather products using vegetable tanning was found to be effective in cutting down the environment and hazardous impacts. Not only does such genuine leather preserve its symbolic luxury image along the fashion domain but also it preserves the legendary craftsmanship of artisans. In a way, it fulfils the basic premise of the sustainable tenet that "*Sustainable development is development that meets the needs of the present without compromising the ability of future generations to meet their own needs*" [161]. Engineering circular design product from leather-based biodegradable and recyclable substrates has earned its new material experience and exclusivity in the fashion domain already. This endeavour holds promise for meeting the burgeoning demand for fashion accessories. Overall, the study helped to conclude the scope and potential growth opportunities for such designer activism strategies but could not ascertain the actual energy demands and environment impact which if carried in near future shall provide a more detailed idea for further progressive developments in circular design.

References

1. Peng W, Qin Z, Zhang W (2009) Research on the reusability of design information for leather goods. In: Proceedings of the 2009 international symposium on web information systems and applications, Nanchang, pp 429–433
2. Bijma-Aerts AT et al (2004) Dating bog bodies by means of 14C-AMS. J Archaeol Sci
3. Kania K (2010) Kleidung im Mittelalter, Materialien—Konstruktion—Nahtechnik. Ein Handbuch. Bohlau Verlag, Koln-Weimer-Wien, p 432
4. Mannering U, Gleba M, Bloch-Hansen M (2012) Denmark. In: Gleba M, Mannering U (eds) Textiles and textile production in Europe from prehistory to AD 400. Ancient textile series, vol 11, pp 91–121
5. Penn K (2000) Excavations on the Norwich Southern Bypass, 1989–91, part—II: the anglo saxon cemetery at Harford farm, Caisor St. Edmund, Norfolk. East Anglican Archeol 92:137
6. Soares BO (2015) Da Moda ao Resíduo e do Resíduo à Moda, Um contributo sustentável no uso da pele residual na Indústria do calçado. Dissertação de mestrado, Universidade do Minho, Portugal
7. Cleland L, Davies G, Llewellyn-Jones L (2007) Greek and Roman dress from A to Z. Routledge Taylor and Francis Group, London and New York, p 240
8. Croom AT (2002) Roman clothing and fashion. Tempus, Charleston, p 160
9. Nowak-Bock B, Von Looz G (2013) Mit siede und Pelz ins Grab: Die Textilien aus den fruhmittelalterlichen grabern von Unterhaching (lkr. Munchen). In: Banck-Burgess J, Nubold C (eds) The north European symposium for archeological textiles XI (NESAT XI). Verlag Marie Leidorf, Rahden/Westf, pp 173–180

10. Rast-Eicher A (2010) Garments for a queen—first results of the new investigation on the merowingian grave of Queen Arnegundis. In: Strand A et al (eds) North European symposium for archaeological textiles X. Oxbow, Oxford, pp 208–210
11. https://www.blackstockleather.com/history-of-the-leather-tanning-industry. Last accessed 4 Jan 2020
12. https://www.en.m.wikipedia.org/PatentLeather. Last accessed 4 Jan 2020
13. Ellsworth LF (1972) Craft to national industry in the nineteenth century: a case study of the transformation of New York state tanning industry. J Econ Hist 1(1):399–402
14. https://www.avellano.com. Last accessed 3 Jan 2020
15. https://www.vintagedancer.com/. 1930s purses, handbags and evening bags. Last accessed 3 Jan 2020
16. https://www.Le-tanneur.com. Last accessed 4 Jan 2020
17. https://www.lyst.com. Last accessed 7 Jan 2020
18. https://www.thedailybeast.com. Last accessed 7 Jan 2020
19. Burns B (2010) Re-evaluating obsolescence and planning for it. In: Cooper T (ed) Longer lasting products: alternatives to the throwaway society. MPG Books Group, UK, pp 39–61
20. Fletcher K, Grose L (2012) Fashion & sustainability—design for change. Laurence King Publishing, London
21. Thronquist C (2018) The fashion condition: rethinking fashion from its everyday practices. J Des Creat Process Fash Ind 289–310
22. McCracken G (1988) Culture and consumption. Indiana University Press, Bloomington, p 80
23. Kawamura Y (2018) Fashionology: an introduction to fashion studies, 2nd edn. Bloomsbury Publishers, London, p 90
24. Steele V (ed) (2005) Encyclopedia of clothing and fashion, vol 2. Thomson Gale, Farmington Hills, pp 80, 345, 354
25. Ruggeri A (2019) 'Made on earth' why is Italian style so seductive? Available at: www.bbc.com. Last accessed 4 Jan 2020
26. https://www.ec.europa.eu/growth/sectors/fashion/leather/eu-industry_en. Last accessed 3 Dec 2019
27. The global luxury leather goods market research report for 2020–2026 (2020) Available at: https://www.marketwatch.com. Last accessed 21 Mar 2020
28. https://www.gentlemansgazette.com. Last accessed 5 Jan 2020
29. Leather goods market size worth $629.65 billion by 2025 CAGR 5.4%. Available at: https://www.grandviewresearch.com/pressrelease/global-leather-goods-market. Last accessed 4 Jan 2020
30. Ashby M, Johnson K (2014) Materials and design the art and science of material selection in product design, 3rd edn. Elsevier Science & Technology Books
31. Qua FJS (2019) a qualitative study on sustainable materials for design through a comparative review of leather and its modern alternatives. Masters degree project, Massachusetts Institute of Technology
32. https://www.artsandculture.google.com. Last accessed 15 Jan 2020
33. Nofal RM (2019) Seam performance of nano-titanium treated microfiber polyester fabrics. J Stud Search Specif Educ 5(1), 207–237
34. https://www.octaneseating.com. Last accessed 11 Jan 2020
35. Baugh G (2011) The fashion designer's textile directory. The Quarto Publications, London, pp 120, 122, 124
36. Future trends in the world leather and leather products industry and trade (2010) United Nations Industrial Development Organization (UNIDO)
37. https://www.highonleather.com. Last accessed 12 Jan 2020
38. Harris S, Veldmeijer AJ (eds) (2014) Why leather? The material and cultural dimensions of leather. Sidestone Press, Leiden
39. Gromer K, Russ-Poppa G, Saliari K (2017) Products of animal skin from antiquity to the medieval period. Ann Naturhist Mus Wien Ser A 119:69–93

40. Moog GE (2005) Der Gerber. Handbuhc fur die Lederherstellung. Verlag Eugen Ulmer, Stuttgart, p 199
41. Fasol T (1954) Was ist leder? Eine Technologie des Leders. Franckh'sche Verlagshandlung, Stuttgart, p 98
42. An overview of tanning hides—both ancient and modern methods. Available at: https://www. native-art-in-canada.com. Last accessed 7 Jan 2020
43. Finishing techniques for leather (2018) Essays, UK. Available at: https://www.ukessays.com/ essays/chemistry/finishingtechniques. Last accessed 6 Jan 2020
44. Ecopell: chrome in leather. Available at: https://www.table-tanned-leather.com. Last accessed 8 Jan 2020
45. https://www.leather-dictionary.com. Last accessed 12 Jan 2020
46. Norman IE, Slee Smith PI (2016) Making or tanning of leather, Chap IV. In: Leather craft manual. Crosby Lockwood & Sons
47. Manual for oxazolidine tanned leather. Available at: https://www.ec.eurpoa.eu/projects. Last accessed 13 Jan 2020
48. Church RA (1971) The British leather industry and Foreign competition 1870–1910. Econ Hist Rev New Ser 24(4), 543–570
49. Covington AD (2001) Theory and mechanism of tanning: present thinking and future implications for industry. J Soc Leather Technol Chem 85(1), 24–34
50. Setchi R, Howlett RJ, Liu Y, Theobald P (eds) (2016) Sustainable design and manufacturing. Springer Publications, UK, pp 151, 154
51. Management Association, Information Resources (eds) (2017) Fashion and textiles: break-throughs in research and practice, p 446
52. https://www.stahl.com. Last accessed 14 Jan 2020
53. https://www.chemicalbook.com. Last accessed 10 Jan 2020
54. Dyes: application and evaluation (2006) Available at: https://www.onlinelibrary.wiley.com. Last accessed 10 Jan 2020
55. Sammarco & Umberto (2007) Flaying and trimming. Available at: https://www.intracen.org. Last accessed 12 Jan 2020
56. Tanning and leather finishing (2011) Available at: https://www.iloencyclopaedia.org. Last accessed 13 Jan 2020
57. History of tanning (2018) Available at: https://www.leatherworksforbeginners.com. Last accessed 12 Jan 2020
58. https://www.all-about-leather.co.uk. Last accessed 14 Jan 2020
59. Leather Carbon Footprint Review of the European Standard EN 16887:2077 (2017) United Nations Industrial Development Organization (UNIDO)
60. Buljian J et al (1997) Mass balance in leather balancing. In: IULTCS congress
61. Ferreira MJ (2012) Contribuições para a Gestão de resíduos de Couro Curtido com Crómio da indústria do Calçado. QREN 2007–2013, Portugal
62. Sustainable apparel coalition. Higg materials sustainability index (MSI) methodology. Higg MSI (2018) Available at: https://msi.higg.org/uploads/msi.higg.org/sac-textpage-sec tion-files/22/file/MSLMethodology_7-31-18.pdf. Last accessed 27 Dec 2019
63. Redwood M (2018) Leather naturally—a white paper view of the sustainability of responsibly made leather. Available at: https://leathernaturally.org. Last accessed 23 Dec 2019
64. Gibb HJ et al (2000) Lung cancer among workers in chromium chemical production. Am J Ind Med 38:115–126
65. Bulijan J, Kral I (2019) The framework for sustainable leather manufacture, 2nd edn. United Nations Industrial Development Organization (UNIDO)
66. Who We Are. Leather Working Group (2019) Available at: https://www.leatherworkinggroup. com/whowe-are. Last accessed 27 Dec 2019
67. Pringle T, Barwood M, Rahimifard S (2016) The challenges in achieving a circular economy within leather recycling. Procedia CIRP 48:544–549
68. https://www.pre-sustainability.com. Last accessed 12 Jan 2020

69. ISO 14067: 2018 greenhouse gases-carbon footprint of products—requirements and guidelines for quantification (2018) Available at: https://www.iso.org. Last accessed 27 Dec 2019
70. Christner J (2014) Sustainability in the leather supply chain. TFL
71. Albu LA, Deselnicu V (2016) The 6th international conference on advanced materials and systems, Bucharest. ISSN: 2068-0783
72. Michael B et al (2013) Best available techniques (BAT) reference document for the tanning of hides and skins: industrial emissions directive 2010/75/EU. Available at: Semanticscholar.org. Last accessed 7 Feb 2020
73. Borduas N, Donahue NM (2018) The natural atmosphere. In: Green chemistry; an inclusive approach. Available at: https://doi.org/10.1016/B978-0-12-809270-5.00006-6. Last accessed 7 Feb 2020
74. Chowdhury ZUM et al (2018) Environmental lifecycle assessment of leather processing industry: a case study of Bangladesh. J Soc Leather Technol Chem 102, 18–26
75. Understanding global warming potentials (2017) Available at: https://www.epa.gov. Last accessed 12 Jan 2020
76. Adrian SG et al (2016) Applications of life cycle assessment to leather industry—an overview and a case study. In: ICAMS 2016—6th international conference on advanced materials and systems, Bucharest
77. Kirchain RE Jr, Gregory JR, Olivetti EA (2017) Environmental life-cycle assessment. Nat Mater 16:693–697
78. Redwood M (2008) The challenges of the leather industry. J Soc Leather Technol Chem
79. Joshi Y, Rahman Z (2015) Factors affecting green purchase behaviour and future research directions. Int Strateg Manag Rev 3(1–2):128–143
80. Manzini E, Vezzoli C (2008) O desenvolvimento de produtos sustentáveis: os requisites ambientais dos produtos industriais. EDUSP/Ed. Universidade de São Paulo, São Paulo
81. Ethical fashion hits big time with German takeover (2010) The Independent. Available at: http://www.independent.co.uk/life-style/fashion/ethical-fashion-hits-big-time-wit hgerman-takeover
82. Christoper B (2003) Fashion. Oxford University Press, Oxford
83. https://www.susannaives.com/wordpress/2012/03/frenchfashion1910. Last accessed 10 Jan 2018
84. Nithyaprakash V (2018) Fashion choices of Gen Y in retrospective. Lat Trends Text Fash Des 1(4). LTTFD. MS.ID.000116
85. Niinimaki K, Koskinen I (2011) I love this dress, it makes feel beautiful! Emphatic knowledge in sustainable design. Des J 14(2):165–186
86. Copenhagen Fashion Summit (2018) Turned words into action. Global fashion agenda. Available at: https://www.globalfashionagenda.com/copenhagen-fashion-summit-2018-tur ned-words-into-action/. Last accessed 15 Apr 2019
87. De Ceglia G (2018) Helsinki fashion week, absolutely green. Vogue Italia. Available at: https://www.vogue.it/en/vogue-talents/news/2018/08/01/helsinki-fashion-week-2018-100-sustainable-ospite-tiziano-guardini/. Last accessed 17 Apr 2019
88. https://www.fossilgroup.com/going-global-commitment-sustainable-design. Last accessed 2 Feb 2020
89. Papanek V (1984) Design for the real world. Thames and Hudson, London
90. Papanek V (1985) Design for the real world: human ecology and social change. Academy Chicago Publisher, Chicago
91. Thackara J (2005) In the bubble: designing in a complex world. MIT Press, Boston
92. Fashion design guidelines for exporting leather goods to Japan the creation of Thai handicraft identity in the avant-garde postmodern style (2019) Inst Cult Arts J 20(2), 77–89
93. Famest: a new generation of products (2019) Available at: https://www.worldfootwear.com. Last accessed 3 Feb 2020
94. https://www.famest.ctcp.pt. Last accessed 4 Feb 2020

95. Soares D et al (2017) FAMEST footwear, advanced materials, equipments and software project. Available at: https://www.researchgate.net. Last accessed 3 Feb 2020
96. FAMEST: it started the mobilization project to stimulate the national footwear (2017) Available at: https://www.ccg.pt. Last accessed 3 Feb 2020
97. https://cfda.com/resources/detail. Last accessed 8 Mar 2020
98. Ceschin F, Gaziulusoy I (2016) Evolution of design for sustainability: from product design to design for system innovations and transitions. Des Stud 47:118–163
99. Kongprasert N, Butde S (2016) A methodology for leather goods design through emotional design approach. Int J Fash Des Technol Educ
100. https://www.vogue.co.uk. Last accessed 14 January 2020
101. Amir S, Benlboukht F, Cancian N, Winterton P, Hafidi M (2008) Physico-chemical analysis of tannery solid waste and structural characterization of its isolated humic acids after composting. J Hazard Mater 160:448–455
102. Dixit S, Yadav A, Dwivedi PD, Das M (2015) Toxic hazards of leather industry and technologies to combat threat: a review. J Clean Prod 87:39–49
103. Soares BO et al (2015) Modular design: contribution to sustainable development within ecodesign and upcycling. In: 3rd international conference, Viana do Castelo
104. Oliveira M, Soares B, Broega C (2016) Sustainable design applied to residual animal leather from the footwear industry. In: 3rd international fashion and design congress, CIMODE 2016, Portugal
105. Ashton EG (2018) Analysis of footwear development from the design perspective—reduction in solid waste generated. Strateg Des Res J 11(1). https://doi.org/10.4013/sdrj.2018.111.01
106. https://www.crowdspring.com. Last accessed 3 Feb 2020
107. Lu L, Peng W-L (2012) Research on design information management system for leather goods. Phys Procedia 24(Part C), 2151–2158
108. Fletcher K (2014) Sustainable fashion and textiles: design journeys, 2nd edn. Routledge Taylor and Francis Group, London, pp 166, 169
109. Catalogue, Craft the Leather International Workshop (2015) Consorzio Vera Pelle Italiana Conciata al Vegetale Piazza Spalletti Stellato, Italy. Available at: www.pellealvegetale.it. Last accessed 5 Feb 2020
110. Schaber F et al (2013) INDIA matters, leather from production to co-design, consilience and innovation in design. International Association of Societies of Design Research, Tokyo, pp 5733–5740
111. Jain-Neubauer J (2000) Feet and footwear in Indian culture. Bata Shoe Museum, Toronto
112. Romano S (2012) New India designscape, triennale design museum. Maurizio Corraini, Mantova
113. https://in.makers.yahoo.com/from-waste-to-luxury-mayura-davda-shah-is-upcycling-fish-leather-into-art-030017964. Last accessed 6 Mar 2020
114. https://www.stellasoomlais.com. Last accessed 5 Feb 2020
115. https://www.leatherbagstage.com. Last accessed 7 Mar 2020
116. Lee S (2012) The textile reader. In: Wang XQ, Jin ZM, Zhang AD (eds) Innovation of transformation of textile industry moving from material to finished good suppliers. Advanced materials research, vols 482–484. Transtech Publications, Switzerland, pp 2551–2554. https://doi.org/10.4028/www.scientific.net/AMR482-484.2551
117. Ford D (2018) Growing a durable yet delicate natural fiber composite textile. Master of science thesis, Faculty of Industrial Design Engineering, Delft University of Technology, Netherlands
118. Emerging voices: David Benjamin of the living (2014) Architectural leagues' annual emerging voices awards. Available at: https://www.archdailly.com. Last accessed 28 Mar 2020
119. Keating S (2016) From bacteria to buildings: additive manufacturing outside the box. PhD thesis
120. https://www.fastcompany.com. Last accessed 28 Mar 2020
121. https://www.jenkeane.com. Last accessed 29 Mar 2020
122. https://www.humournoir.com. Last accessed 28 Mar 2020
123. https://www.modernmeadow.com. Last accessed 7 Mar 2020

124. https://www.leathermag.com/features/featurematerial-gains-6713680. Last accessed 8 Mar 2020
125. Modern Meadow (2019) About us. Available at: https://www.modernmeadow.com/about-us. Last accessed 5 Mar 2020
126. Modern Meadow launches Zoa, the first ever biofabricated leather material brand (2017) Modern Meadow. Available at: http://www.modernmeadow.com. Last accessed 5 Mar 2020
127. https://www.eclectictrends.com/mushroom-mycelium. Last accessed 8 Mar 2020
128. Madaria D (2018) Product design history, December 1 2018, I search paper: mushroom leather
129. https://www.premierevision.com/en/news/smart-creation/frumat-apples-skins. Last accessed 8 Mar 2020
130. https://www.plantbasednews.org/opinion/10-new-innovative-vegan-alternatives-to-leather. Last accessed 8 Mar 2020
131. https://www.fastcompany.com/3059190/this-gorgeous-sustainable-leather-is-made-from-pineapple-waste. Last accessed 8 Mar 2020
132. https://danandmez.com/blog/pinatex. Last accessed 9 Mar 2020
133. Ananas Anam, Ltd. Pinatex branding guidelines. Last accessed 5 Mar 2020
134. AFIRM Group I reduce the use of harmful substances (2019) AFIRM Group. Available at: https://www.afirm-group.com. Last accessed 6 Mar 2020
135. https://www.mymodernmet.com. Last accessed 6 Mar 2020
136. Vegetal leather. Available at: http://environment.yale.edu/publication-series/documents/downloads/0-9/98ginu.pdf. Last accessed 10 Mar 2020
137. Vegetal leather. Available at: http://www.lanecrawford.com/product/givenchy/-show-velour-vegetal-leather-belt/_/LYK134/product.lc. Last accessed 11 Mar 2020
138. Bolt Threads (2019) Available at: https://boltthreads.com. Last accessed 10 Mar 2020
139. Tamara Orjola makes furniture and textiles using pine needles (2016) Available at: https://www.dezeen.com. Last accessed 25 Mar 2020
140. https://www.Tjeerdveenhoven.com. Last accessed 26 Mar 2020
141. https://www.jpscorkgroup.com. Last accessed 28 Mar 2020
142. Pineskins (2018) Material district, Rotterdam, Netherlands. Available at: https://www.materialdistrict.com. Last accessed 29 Mar 2020
143. https://www.dekodur.com. Last accessed 28 Mar 2020
144. https://www.signenorgaard.com. Last accessed 28 Mar 2020
145. Karana E (2009) Meanings of materials
146. Karana E, Barati B, Rognoli V, Der Laan V, Zeeuw A (2015) Material driven design (MDD): a method to design for material experiences
147. Rognoli V et al (2015) DIY materials. Mater Des 86:692–702
148. Ayala-Garcia C, Rognoli V (2017) The new aesthetic of DIY-materials. Des J 20(sup1):S375–S389
149. Schifferstein R, Wastiels L (2014) Sensing materials: exploring the building blocks for experiential design. In: Karana E et al (eds) Materials experience: fundamentals of materials and design. Elsevier, Oxford, pp 15–24
150. Bahrudin FI, Aurisicchio M (2018) The appraisals of sustainable materials
151. Desmet PM, Hekkert P (2007) Framework of product experience. Int J Des 1(1):57–66
152. Salvia G et al (2010) The value of imperfection in sustainable design. In: LeNS conference, Bengaluru
153. Karana E (2012) Characterization of 'natural' and 'high-quality' materials to improve perception of bio-plastics. J Clean Prod 37:316
154. Rognoli V, Karana E (2014) Toward a new materials aesthetic based on imperfection and graceful aging. In: Materials experience. Butterworth-Heinemann, pp 145–154
155. Fenko A, Schifferstein HNJ, Hekkert P (2010) Shifts in sensory dominance between various stages of user product interaction. J Appl Ergon 41(1):34–40. https://doi.org/10.1016/j.apergo.2009.03.007
156. Mugge R et al (2009) Emotional bonding with personalized products. J Eng Des 20(5):467–476

157. https://www.willleathergoods.com/blogs/news/the-traveler-duffle-deconstructed.   Last accessed 11 Mar 2020
158. https://www.3dnatives.com/en/xyzbag-3d-printed-handbags-020820194. Last accessed 11 Mar 2020
159. https://www.3dprint.com/186347/3d-printed-handbags. Last accessed 12 Mar 2020
160. Burall P (1991) Green design. Design Council, London
161. Jarvie EM (1987) Brundtland report, World Commission on environment and development. Available at: https://www.brittanica.com. Last accessed 8 Mar 2020

Bacterial Cellulose—A Sustainable Alternative Material for Footwear and Leather Products

R. Rathinamoorthy and T. Kiruba

Abstract Issues related to sustainability are an inevitable factor which always associated with leather products as the raw material is associated with animal slaughtering. At the same time, research agencies predicted that the footwear industry is expected to grow 371.8 billion USD in 2020 with a CAGR of 5.5%. Out of different raw materials used in the industry, leather products are occupying a significant market share as premium goods. The major pollution by the footwear and other leather products not only comes from the disposal but also from the manufacturing stages like machine usage, energy confirmations, chemicals, etc. It is estimated that approximately 50.2 m^2 of land and 25,000 L of water required to develop leather boots. On average, the production of single boot emits 30 lb of carbon-di-oxide to the environment. The existing leather alternative materials like polyurethane, synthetic textiles, and rubber will take roughly 50 years to decompose totally. Material selection is one of the important solutions for sustainability-related issues and to reduce the negative environmental impacts. Bacterial cellulose is one such material that attracted the footwear industry due to its special properties like unique structure, biodegradability, mechanical strength, and high crystallinity. The chapter discusses various research works performed on leather alternative materials and specifically details the potential nature of bacterial cellulose. The production method, factors influencing the production, material properties, and application scopes will be analyzed with specific concern on the leather and footwear industry. The chapter also details the advantages of bacterial cellulose over other alternative material in terms of wearer comfort, durability, disposability, biodegradability, and cost factors.

Keywords Footwear industry · Sustainability issues · Bacterial cellulose · Production · Properties · Leather alternative

R. Rathinamoorthy (✉) · T. Kiruba
Department of Fashion Technology, PSG College of Technology, Coimbatore 641004, India
e-mail: r.rathinamoorthy@gmail.com

© The Editor(s) (if applicable) and The Author(s), under exclusive license to Springer Nature Singapore Pte Ltd. 2020
S. S. Muthu (ed.), *Leather and Footwear Sustainability*, Textile Science and Clothing Technology, https://doi.org/10.1007/978-981-15-6296-9_5

1   Introduction

Animal leather is used in apparel, accessories, furniture, and upholsteries due to its unique characteristics like higher esthetic, rich look along with its flexibility and durability. Out of the total leather produced, approximately 25% of leather produced is used in the footwear industry. The remaining percentage majorly occupied by clothing, furniture and upholstery application. The main objective of the leather industry is to process the animal skin or animal hides into useful material for several applications. In this process, the skin or hide is subjected to several chemical and physical treatments subsequently and this process is called tanning. Due to the excessive usage of water and chemicals, leather industries tend to pollute the environment in terms of bad smell, organic wastes, and leather processes effluents. The level of pollution caused by each industry is different from each other based on the types of skins used in the process and their required process sequence for the hides and skins processed. Hence, the waste produced from each industry also defers significantly based on the process adopted. In general, this process causes air, water pollution, solid organic waste disposal, and a higher amount of bad odor and toxic gas generation.

The leather processing industries are identified as one of the highly polluting industries due to the use of a lot of toxic chemicals used in the tanning process and disposal of wastes. The effluent from the leather tanneries contains a higher quantity of toxic chemicals like chromium slats and metal ions which ultimately increases the biological oxygen demand and makes the water the most polluted one. One of the other issues with the leather processing and tanning industry is the generation of unpleasant odor to the environment due to the decomposition of protein waste along with sulfides and ammonia used in the tanning process [1]. Reports indicated that approximately 800 kg of solid wastes are developed during a ton of rawhide processing along with the emission of harmful chemical substances like chromium salts, sulfuric acid, sulfides, ammonium salts, and vegetable tanning materials. Further, most of the industries discharge the unprocessed tanning effluent into the soil which affects the soil productivity at a major level along with significant environmental pollution. Figure 1 represents the activities involved in the conversion of rawhide into leather [2].

The first section of the leather manufacturing industry is to handle the animal skins to prepare the rawhide. This section is generally known as beam house. In this process, the animal skin is soaked in water and processed for fleshing and unhairing of skin. Finally, the skin is treated with lime and other chemicals to dissolve hair. The second step in the leather manufacturing is tanning process in which the animal skin is processed for the removal of proteinaceous matter and stabilized. For this purpose, tanning chemicals like chromium salts, zirconium, etc. were used. As the process is performed in the acidic medium, the wastewater discharged from the process is highly acidic. The last step of the leather production is finishing activity where the tanned skin is re-tanned and then finished as per the requirement. Further at this stage, the tanned skin is treated with dye for coloring or bleached and lubricants applied based on the final endues requirements. During this skin or hide to leather

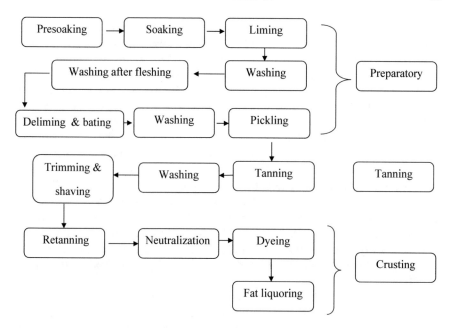

Fig. 1 Process flow of rawhide to leather conversion

conversion process, the different forms of pollutions/wastes generated as listed in Fig. 2.

Fig. 2 Types of waste generated from the leather processing industries

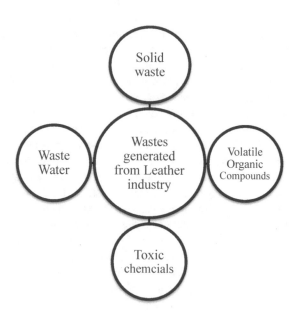

In leather production activity, the degraded hide, hair from the first stage of the skin process and trimmings are the major solid wastes developed. Though the leather is biodegradable in nature, the treatment process involved during the tanning process converts this waste leather as the non-biodegradable one at the disposal. The tanning process is one of the most toxic processes in the leather processing sequence and it majorly contributes the wastewater pollution. Due to the treatment nature, the effluent released from the tanning process is highly acidic and so the pH of the effluence causes higher chemical oxygen demand (COD). Also, the added chemicals like sulfates and chloride during the second phase of the leather making develop higher total dissolved solids (TDS) in the wastewater [3]. Researchers also identified that the usage of sodium sulfide in the tanning process not only causes pollution in terms of TDS but also reduces the effluent treatment efficacy during the post-tanning process. Sometimes, the poor processing of leather with chrome results in higher environmental pollution by leaching in the subsequent processing stages. The post-tanning activity causes higher metal pollution in the wastewater. These chemical sediments in the water bodies adversely affect the ecological balance in the aquatic life [4].

In the case of volatile organic compounds (VOC), the leather industry is well known for its emissions like ammonia, hydrogen sulfide, volatile hydrocarbons, amines, and aldehydes. The major source of pollution is the manufacturing process and tannery plant effluents. During the leather manufacturing first stage, beaming room, the ammonia is majorly emitted from the process like de-liming, unhairing, or drying processes. Further, the liming and subsequent washing processes are responsible for the release of sulfides. In the leather buffing process, the chromium is emitted as a result of the handling of chromic sulfate powder or directly from the tanning process, in which the chromate is reduced as chromium on treatment. Finally, the effluent obtained from the tanning process is highly acidic as mentioned earlier. Out of the treatment or storage of effluent, a higher amount of hydrogen sulfide is released as a major pollutant [5]. The other important pollution source in the leather conversion process is toxic chemicals used in the manufacturing stages. The previous researcher performed a detailed analysis of the various chemicals used in the leather industries and their toxic nature as listed in Table 1 [4].

Though the regulatory bodies have implemented serious regulations on using harmful chemicals in the leather industry, the usage of chemicals is still in practice. The most commonly used plasticizers in the leather industry are phthalates. This chemical has a major influence on the reproductive systems of animals and human and so the usage of the chemical is restricted to 0.5% on the final product. During the leather final finishing process, two different chemicals used, namely nonylphenol and biocides. Whereas the nonylphenol is more of non-biodegradable and so the usage on the final product is restricted to 0.1% [6], the biocides are mainly the water-based chemical used to restrict the microbial growth on the surface of the leather. The usage of these chemicals also restricted due to its irritant nature.

Another important chemical used in the coalescence, plasticizing, leveling, and wetting and as a swelling agent is N-Methyl pyrrolidone. This finish provides a high-performance finish to the leather; at the same time, it is identified as a reproductive

Table 1 Toxicity of chemicals used in the leather industry [4]

S. no.	Name	Uses	Target organs
1.	Benzyl butyl phthalate	Used in process for the production of a microporous artificial leather coating/water vapor-permeable sheet materials	Eyes, lungs, liver, reproductive system
2.	Bis(2-ethylhexyl) phthalate (DEHP)	Used as plasticizers in the processing of shoe soles, and artificial leather manufacturing	Liver and testes
3.	Dibutyl phthalate (DBP)	Used as a phthalate plasticizer in the artificial leather industry	Eyes, lungs, gastrointestinal tract, testes
4.	Anthracene	Used as a tanning agent (for leather)	Kidney, liver, fat, and carcinogen
5.	Short-chain chlorinated paraffin's (PBT)	Additive for the leather treatment (renders smoothness to leather), leather clothing and belts and as a leather oiling agent	Liver, kidney, thyroid, and carcinogen
6.	Sodium dichromate	The principal raw material used in the production of chrome-tanning materials for the leather industry like chrome-tanning salts	Blood, kidneys, heart, lungs, eyes, and carcinogen
7.	Cobalt dichloride	Used in leather dyeing and finishing as well found in tanned leather	Lungs, liver, kidney, heart, skin
8.	Nonylphenol	Used in finishing	Blood, lungs, eyes, skin, CNS, kidneys and low biodegradability
9.	Methyl isothiazolinone	Biocide, microbiological protection	Skin, eyes, and carcinogen
10.	N-methyl pyrrolidone	Coalescence, plasticizers, wetting agent	Eyes, kidney, lymphatic system, liver, lung, testes
11.	Formaldehyde Heavy metals arsenic	Leather finishing	Eyes, lungs, and carcinogen Liver, kidneys, skin, lungs, lymphatic system and carcinogen
12.	Chromium	Used for dyeing	Kidney, CNS, hematopoietic system
13.	Organotin compounds (dibutyl tin)	As a catalyst	The gastrointestinal tract, liver, and carcinogen

(continued)

Table 1 (continued)

S. no.	Name	Uses	Target organs
14.	Azo dyes (orange II)	Used for dyeing	Blood, liver, testes, and carcinogen

toxin by the California Office of Environmental Health Hazard Assessment [7]. Similarly, formaldehyde is used as a cross-linking agent in the top coating process of the leather; however, due to its carcinogenic nature, the usage is banned by regulatory bodies. The other most important chemical pollutions from the leather industry are the inorganic pigments, synthetic, and azo-based dyestuffs used in the coloration process. The toxic heavy metals like lead chromate and cadmium are commonly used in the leather industry due to its efficient color and superior fastness properties [8]. The results of previous research works reported that 1 ton of the wet salted hide can yield 200 kg of useful leather and this amount is approximately 20% of the total raw material processed. In this process, more than 600 kg of solid waste created this is approximately 60% of the total raw material [9]. These wastes are disposed into the environment without processing further. In this process, nearly 50 m^3 of water is generated per ton on rawhide. Research data from FAO mentioned that in a year, approximately 11 million tons of rawhide is processed, and by results, 8.5 million tons of solid waste produced [10].

Higg index, a popular indexing method of environmental impact caused in the apparel production process, measured the leather industry pollution to 159 points. It is very high when compared to the high index of 44 for polyester, and for cotton, it is 98. The highest value of this index represents the contribution of the particular manufacturing industry toward global warming and water pollution [11]. Out of all the process stages mentioned in Fig. 1, the process stage two, known as the tanning process, causes 90% of the pollution in leather manufacturing. It produces higher carcinogenic chemical discharge like lead, chromium, etc. [12]. Out of all these issues in leather production, the material is still one of the best choices for luxury products due to its elegant feel, touch, and esthetical look. The application not only limited to the clothing, footwear industry, and residential furniture but also in the marine, aviation, and automotive industry in recent times to create a luxury [13]. Other than the environmental impact caused by the various processing activities of the leather industry the usage of leather also linked with animal cruelty. The People for the Ethical Treatment of Animals (PETA) indicate that the increased sales of leather goods may increase the sufferings of these animals during the process of transportation. Further to add, the association claims that though the leather produced from the domestic animals, sometimes the exotic animals are also haunted for its skin [14].

2 Synthetic and Bioleather Alternatives

As an alternative to animal leather, there are several materials used in commercial products. These alternative materials are highly abundant and cost-effective than animal leather. In general, polyvinyl chloride (PVC) and polyurethane are the most commonly used synthetic materials due to its higher temperature stability, chemical compatibility and availability [15]. Another most explored synthetic material as a substitute for animal leather is the thermoplastic polymer, polypropylene. One of the main advantages of the polymer is resistant to the environment and non-biodegradable. At the same time, the effect of sunlight and UV radiation has a significant influence on the properties of the polypropylene. Even though the synthetic alternative materials are free from animal cruelty, it poses serious environmental pollution during the disposal.

In order to reduce the environmental impact, researchers' surface grafted the acrylate-functionalized soybean oil onto the gat leather with the help of UV radiation. In their study, they analyzed the grafting efficiency and their thermal stability on the goat leather. The findings revealed that soybean oil grafting is effective in leather finishing and an alternative to the existing toxic chemicals [16]. Nam and Lee developed a multilayered cellulosic structure using nonwoven cellulosic fiber mat, denim fabric, and hemp fabric and evaluated its performance against the performance of two-layered calf and pigskins leather, which is commonly used in the footwear industry. Their results were promising that the developed composite material had very good thermal equilibrium inside the footwear and also possessed superior strength and comfort properties [17]. Similarly, a patent published in the earlier decade illustrates the possibilities of producing eco-friendly leather by using polymerizing cellulosic fiber along with epoxidized and acrylated triglycerides and vinyl monomers. The developed composite is cast using molds as per the requirements [18]. Cao et al. developed en eco-friendly leather from chicken feather composite material as a substitute for animal leather. They developed footwear with feather fiber-reinforced composite for footwear applications. But at the same time, the researcher did not analyze the properties of the developed product in terms of thermal comfort and tensile strength, which is essential for a footwear application [19].

In recent times, there are several research works performed around the world in order to find an eco-friendly substitute for leather. Pinatex is one of the commercialized leather alternative materials manufactured from waste pineapple leaves. The extracted fibers are degummed, washed, and allowed to dry. The dried fibers were converted into a nonwoven mat and then passed to a special finishing process which makes the mat to look like leather. As the material is totally natural and no synthetic chemicals used in manufacturing, the alternative leather is claimed by the manufacture as eco-friendly material [20]. Mycelium textile is another leather alternative material researched by Stella McCartney. This type of mushrooms creates a very small microthread network during the cultivation. The researcher grows mycelium mushroom in the agricultural waste by controlling or engineering the environmental conditions so that it forms a structurally connected mat by itself. The

materials dyed, processed, and used for the final product development. The process of converting the mushroom to a leather material is a green process and does not have any environmental impact like synthetic alternatives materials [21].

Similarly, the leading French footwear brand has launched a sustainable leather material as an alternative to animal leather from the corn. They developed a material that looks like leather made of canvas material coated with corn husk, which is a food waste mixed with polyurethane. The brand claims that the material is 63% biodegradable [22]. Zoa is a brand name that developed biofabricated leather alternative material out of the collagen obtained from yeast. The manufacturer claims that the material is very sustainable and causes no impact on the environment. Further to add, the material can be molded into any shapes, color, and texture as required. The brand released its first prototype in 2017 and yet to launch its commercial vegan leather in the forthcoming year [23].

3 Bacterial Cellulose—A Potential Substitute

Bacterial cellulose is a kind of pure cellulose than the plan origin cellulose. It is one of the sustainable biomaterials which recently attracted may researchers from various disciplines like medicine, food, chemistry, textile, fashion, environmental and other engineering areas. The main advantage of the bacterial cellulose over the plant cellulose is its pure structural form which is free from the impurities like hemicelluloses, pectin, and lignin as present in the plant cellulose. The second and very unique nature of bacterial cellulose material's self-assembled formation produces a nanostructured three-dimensional fiber network during the formation. The micromorphology generates more interfebrile space in the fiber matrix and the nanofiber shape produces a higher surface area compared to normal plant cellulose [24]. There are several bacterial species are identified as capable of synthesizing cellulose called *Sarcina*, *Agrobacterium*, *Rhizobium*, and *Acetobacter* also called *Gluconacetobacter*. Among the genera of identified species *Acetobacter xylinus*, an acetic acid-based bacterium is noted as capable of producing cellulose in the larger quantity for commercial application. Even though the bacterial cellulose is identified a century before, the research on the bacterial cellulose gained momentum in the recent time [24]. It is identified that cellulose production in the bacterial strains is an action of the protective mechanism of the strains from the environment, fungus, and various yeasts. As the bacteria segregate the cellulose as a byproduct of its growth, the cellulose fibrils are nearly 100 times smaller than the available pant cellulose. The properties of the developed cellulose depend upon the culture medium and the carbon source used for the production of the cellulose [25].

The cellulose production using bacteria is mainly depending upon the actual nitrogen and carbon contents in the source medium used for the production [26]. The first and foremost important nutrient for bacterial cellulose production is a carbon source, which helps in cell metabolism and cell growth of the bacteria type. Sugars and derivatives of the sugars are the main sources used for the production. Commonly sucrose, glucose, and fructose are used for the production of bacterial cellulose. Similar to the carbon source, nitrogen sources help in the cell growth of the bacteria. The most commonly used nitrogen sources in the bacterial cellulose production are ammonium sulfate, peptone, glycin, yeast extract, and casein hydrolysate. The cellulose synthesis is a function of oxygen supply and pressure. The researchers noted that the cellulose production was a very negligible amount under the nitrogen, and at the same time, a relatively higher production noted when the set up is exposed to the atmospheric air. Hence, the cellulose formation at the top of the beaker will be higher than the lower part of the layer which is in contact with the liquid [26]. During the cellulose production, the culture media can be used in two methods for the cellulose production,

1. Static culture—The culture media, after inoculation, kept it undisturbed for a specific period of time and so the layer of cellulose structure forms on the top of the media.
2. Agitated culture—The media kept in a shaker and continuously agitated for the whole period of time. Hence, the cellulose formed in the shape of irregular pellicles on the surface.

Out of the abovementioned two methods, the most widely adopted method for quicker production of bacterial cellulose is the agitated method. However, the irregular growth and various fiber diameters of nanofibers formed in this method in this process. For many proposed commercial applications like wound dressing, textile and fashion material, masks for cosmetic application, etc., the cellulose must be in the form of the continuous sheet. Hence, based on the application requirements, static culture method capable of producing the industry expected outputs. There are many research works performed in the static culture method and the cellulosic films developed in the nonwoven sheet form [27–29]. The appearance of the bacterial cellulose at different stages of production is provided in Fig. 3. Figure 3a shows the tea extract with bacterial cellulose inoculated at day one with a kombucha SCOBY. Figure 3b shows the fermentation setup and Fig. 3c the fifth day after the fermentation process, the formation of immature cellulose film, on the surface of the static culture medium. Figure 3d indicates the fermented cellulose layer after 14–17 days of the fermentation process. The fully grown cellulose sheet is of highly thick sheet and consists of a lot of bacterial and yeast content along with a very high quantity of water content.

4 Carbon Sources for Bacterial Cellulose

As the industrial-scale mass production required simple and cost-effective sources, several researchers tried with different alternative sources for the bacterial cellulose production either along with the standard HS medium [30] or without the HS medium [31]. However, the recent trend in the research is more focused on the utilization of various wastes material as a carbon source material. Table 2 represents various

Fig. 3 Various stages of bacterial cellulose production

research works performed on bacterial cellulose production with different wastes as carbon or nutrient source.

Similarly, the few researchers also developed bioreactors to increase the cellulose yield from the existing medium and bacterial strains. Hornung et al. developed a bioreactor to feed the glucose and oxygen directly into the cells to develop the bacterial cellulose yield in the beaker [60]. Another research work developed a rotating bioreactor to reduce the production time of the bacterial cellulose [61]. An aerosol spray bioreactor was built to spray the glucose and oxygen directly into the living bacteria and air interface for economic production of the bacterial cellulose [60]. In other methods, a plastic composite supports were included in the fermentation medium, in order to increase the cellulose fermentation. The incorporation of the plastic composite support in the culture medium had a significant increase in production. However, they reported that the material properties of the developed bacterial cellulose are different while producing in the presence of plastic composite support [62]. Shah et al. developed surface-modified reactors by using agar coating at the bottom of the reactor. They had dissolved the agar and basal medium in the distilled water and coated in the bioreactor along with a control medium. In results, maximum production was noted in the developed bioreactor when compared to the control reactor [63]. A similar research performed with silicone rubber membrane

Table 2 Different carbon sources used for bacterial cellulose production

S. no.	Bacteria	Source	Additional nutrients	Productivity	References
1.	*Gluconacetobacter xylinum* ATCC 23768	Scum of sugarcane jaggery	Nil	2.51 g/L	[32]
2.	*Acetobacter xylinus* 23769	Hot water extract	Nil	0.15 g/L	[33]
3.	*Acetobacter xylinum* BPR2001	Molasses	Corn steep liquor	5.30 g/L	[34]
4.	*Acetobacter xylinum* ATCC 10245	Sugar cane molasses	Amino acids, vitamins, minerals, and carbohydrates	223% as compared to 100% in HS medium	[35]
5.	*Acetobacter xylinum* NRRL B-42	Grape pomace	Nil	6.7 g/L	[34]
6.	*Acetobacter xylinum* NRRL B-42	Corn steep liquor	Nil	6.7 g/L	[34]
7.	*Gluconacetobacter xylinum* ATCC 23768	Blackstrap and brewery molasses	Nil	3.05 g/L	[36]
8.	*Gluconacetobacter xylinus* NRRL B-42	Grape bagasse	Diammonium phosphate and corn steep liquor	8.0 g/L	[37]
9.	*Acetobacter xylinus* ATCC 23770	Enzymatic hydrolysate of wheat straw	Components used in HS medium	8.3 g/L	[38]
10.	*Acetobacter xylinum* 0416 MARDI	Extracted dates syrup	Components used in HS medium	5.8 g/L	[39]
11.	*Gluconacetobacter xylinus* BCRC 12334	Orange peel fluid and orange peel hydrolysate	Acetate buffer, peptone, and yeast extract	3.40 g/L	[40]
12.	*Acetobacter xylinum*	Sugarcane juice	Nil	11 g/L	[41]

(continued)

Table 2 (continued)

S. no.	Bacteria	Source	Additional nutrients	Productivity	References
13.	*Acetobacter xylinum*	Coconut water and sugar palm juice	2 g ammonium sulfate $(NH_4)_2SO_4$ and 0.5 mL glacial acetic acid		[42]
14.	*Gluconacetobacter* Sp	Grape	Cane sugar (5%)	7.47 g/L	[43]
15.	*Acetobacter xylinum*	Rice bark	Glucose, yeast extract, peptone, Na_2HPO_4, citric acid	–	[44]
16.	*Acetobacter xylinum* DSMZ2004	Poor quality apple residues in combination with glycerol	Apple glucose equivalents, glycerol, ammonium sulfate and citric acid	8.6 g/L	[45]
17.	*Gluconacetobacter xylinus* CH001	Discarded waste durian shell	Nil	2.67 g/L	[46]
18.	*Gluconacetobacter xylinus* BCRC 12334	Thin stillage (TS) wastewater	TS and HS medium were mixed with different portions to prepare TS-HS medium	6.26 g/L	[47]
19.	*Gluconacetobacter xylinus* ATCC^R 10788^{TM}	Makgeolli sludge filtrate	Other components are the same as of HS medium	1.6 g/L	[48]
20.	*Gluconacetobacter xylinus* ATCC 23770	Waste fiber sludge sulfate and sulfite fiber sludges	Yeast extract and tryptone	11 g/L	[49]
21.	*Acetobacter.xylinum* ATCC 10245	A water-soluble fraction from pulping waste liquor	Other components are the same as of HS medium	High yield	[50]
22.	*Gluconacetobacter xylinus* ATCC 700178	Carob and haricot bean medium	Nil	3.2 g/L	[51]
23.	*Gluconacetobacter xylinus* CH001	Waste yeast biomass	Nil	2.9 g/L	[52]

(continued)

Table 2 (continued)

S. no.	Bacteria	Source	Additional nutrients	Productivity	References
24.	*Gluconacetobacter xylinum* BC-11	Wastewater after pullulan polysaccharide fermentation	Nil	1.177 g/L	[53]
25.	*Gluconacetobacter xylinus* CH001	Lipid fermentation wastewater	Nil	0.659 g/L	[54]
26.	*Acetobacter xylinum* KJ1	Saccharogenic liquid food wastes	Nil	0.46 g cellulose/g-reducing sugar	[55]
27.	*Gluconacetobacter xylinus*	Waste dyed cotton fabrics hydrolysate	Peptone and yeast extract	12.8 g/L	[56]
28.	*Gluconacetobacter xylinus* DSM 46604	Glycerol from biodiesel industry	Yeast extract, ammonium sulfate, potassium hydrogen orthophosphate, and magnesium sulfate	2.87 g/L	[57]
29.	*Acetobacter xylinum*	Rice waste water	Glycerol and chitosan	–	[58]
30.	*Acetobacter xylinum*	Rice bran	Components used in HS medium	–	[59]

as a submerged material in the medium in different shapes like flat, tube, balloon in the shape of a cylinder. In their result, it is mentioned that the introduction of flat membrane increased the production of ten folds [64]. Wu and Li came up with different types of bioreactors by including a stirred tank, conventional airlift, and modified airlift with a rectangular wire-mesh draft tube, for the industrial-scale large production of bacterial cellulose [65].

5   Production Process

Tea is one of the commonly identified media to grow bacterial cellulose by fermentation. The fermentation is generally performed to develop a drink commonly known as "kombucha." The drink consumed worldwide due to its higher health benefits over the controversial name as unsafe medicinal tea [66]. The beverage is simply produced from the fermentation of tea leave infusions or decoction with the help of symbiotic association of bacteria and yeasts, popularly known as SCOBY [67]. Upon the fermentation, it develops a sweet, acidic beverage for the food lovers and it also creates a new SCOBY on the top of the fermenting vessel which is of our interest. The newly formed SCOBY is an aid for the next batch of tea preparation in common. Reiss details the standard procedure for the preparation of kombucha through his research and experience [68]. The details of the process are listed in Fig. 4. The general production mechanism of the cellulose from the *Acetobacter xylinum* is extensively documented by various researchers [69–71]. In general, the cellulose is produced from the glucose through various intermediate products material like glucose-6-phosphate, glucose-l-phosphate, and uridine-5-diphosphate glucose.

In Fig. 4, the procedure is explained to prepare a fermented tea beverage for the food lovers. However, the cellulosic layer formed during the fermentation process at the top of the vessel. The SCOBY consists of millions of bacterial strains of *Acetobacter* species along with various types of yeasts. The bacteria and yeasts were layered with the help of the cellulosic matrix in the form of a gel. During the fermentation process, the tea infusion used in the process acts as a resource to the bacterial culture to grow and multiply and helps in the production of cellulose. Several research works performed in developing bacterial cellulose sheets using tea extract as a medium. Various potential applications of bacterial cellulose proposed by the different researchers are provided in Fig. 5.

The characteristics of the bacterial cellulose are another important aspect for their wide applications. Several researchers analyzed the different properties of the cellulose in different forms namely loose fiber from the agitated medium and cellulose matt from the static culture method. The main advantage is its superior physical and mechanical properties than the plant cellulose due to its purity. The usual degree of polymerization is 4000–10,000 anhydrous glucose units compared to the plant cellulose. The crystallinity range of the bacterial cellulose is observed 80–90% compared to plant cellulose [72]. Most of the studies represented the higher water holding capacity and tensile strength properties of the bacterial cellulose sheets developed.

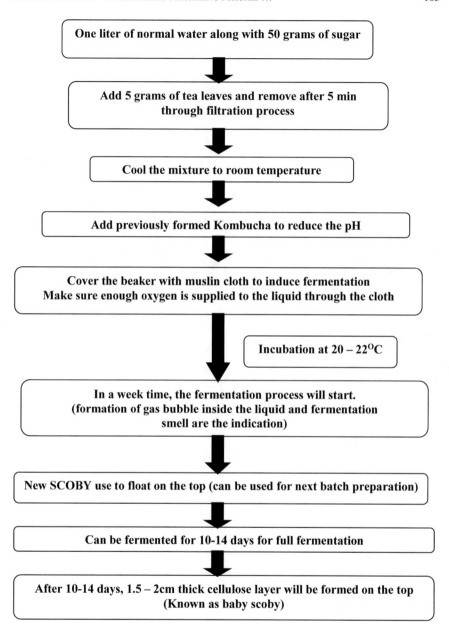

Fig. 4 Steps involved in the production of bacterial cellulose using tea infusion under static culture method

Fig. 5 Application areas of
bacterial cellulose

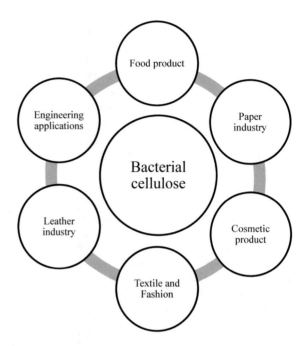

However, the following properties were noted in specific to the bacterial cellulose
[72].

- Higher water holding capacity
- Purity
- High crystallinity
- Biodegradability
- Hydrophilicity
- Mechanical strength
- Porosity
- Nanostructure
- High flexibility
- Easy moldability.

6 Bacterial Cellulose as a Leather Alternative

Bacterial cellulose produced in the form of sheet and tried for its potential application
in various fields like polymer industry, cosmetic products, wound dressing materials
fashion industry [73]. However, due to its leather-like tactile feel, look, and properties
recently, it became an alternative material for animal leather in the footwear industry
[74]. The soft and stretchable nature of the bacterial cellulose along with its look
made it a successful alternative material in the leather and footwear industry. Designer

Suzanne Lee is the person who prolonged the application of bacterial cellulose to the textile and footwear sector. She is the founder of design consultancy, "BioCouture," which makes bio-based fabrics for fashion from living bacterial species. The main objective of their research consultancy is to develop non-hydrocarbon-based feedstock for the fashion industry and her innovation is appreciated by Time magazine in 2010 [75]. Based on her research finding, later then, there are several research works performed on the bacterial cellulose for the fashion applications [76, 77]. However, in her blog, Suzanne Lee mentioned that the bacterial cellulose is just another sustainable material to save our precious natural resource but at the same time it is not a replacement or alternative for leather or cotton-like textile material [78].

Recently, Fernandes et al. tried to utilize the bacterial cellulose in the footwear manufacturing due to its excellent properties like high crystallinity, high Young's modulus, and high water holding capacity with moldability. In spite of good properties, the bacterial cellulose membrane lost its flexibility upon drying. Hence, the researcher attempted to increase the flexibility by treating the bacterial cellulose with different plasticizers and developed bacterial cellulose composites for footwear applications. The researchers emulsified the acrylated epoxidized soybean oil with polyethylene glycol, polydimethylsiloxane, and perfluorocarbon-based polymers. The mixture is then allowed to diffuse into the bacterial cellulose nanofiber matt by the exhaustion process. The results were promising that developed bacterial cellulose polymer composite had higher flexibility and hydrophobicity than the native bacterial cellulose. The characterization results confirmed the diffusion of polymer resin into the matrix of the cellulose. However, the resultant product had poor thermal stability only up to 200 °C, higher water contact angle value, and lower tensile strength. The researcher proposed the developed bacterial cellulose composite for the upper part of the shoes [79].

Ghalachyan developed bacterial cellulose for the application of fashion accessories like leather bags and shoes as a leather alternative. They developed bacterial cellulose from the tea, sugar, and vinegar along with the commercially available kombucha SCOBY culture. Upon fermentation on multiple containers, the researcher developed gel-like bacterial structures. They dried it and used it as a sustainable alternate or substitute material for the leather. After the development, the researcher conducted a subjective analysis with different subjects on the bacterial cellulose material like the smell, feel, texture, overall liking of the material, etc. The findings of the study reported that the participants identified various parameters like; they mentioned the material is see-through, the color is tanned and looks like leather; the texture is wrinkled and unique. However, one of the major attributes the participants reported unpleasant odor or smell on the material [80]. On the extension of the study, the researcher analyzed the overall acceptability of the bacterial cellulose material based on various attributes and the results from the participants noted as the material cannot be used for the clothing application as it is see-through and rough texture and brittle. But at the same time, the leather-like look motivated the participants to accept it as different accessories like bags, hats, and shoes [80].

Though the application of bacterial cellulose is widespread in various technical fields, the textile and leather industry applications are very limited and less explored.

The usage of bacterial cellulose will reduce the overconsumption of non-renewable fiber sources and environmental pollution. Based on this concept, few researchers came with an idea that bacterial cellulose can be used as a leather substitute in bags and clutches. The researcher developed tea-based bacterial cellulose and its leather-like reddish tone and surface texture of the developed mat are the added advantage for apparel accessories. They developed a saddle bag with laser cutting designs [81]. Studies performed in Keio University suggest the bacterial cellulose as a sustainable alternative material for the fashion. In their research, they analyzed the possibility of using it as a 3D printed mold as a part of zero-waste garment development. They reported that this method will eliminate the textile waste at the design stage. Further, they had studied the biodegradation ability of the material [82]. Bacterial cellulose was produced from *Acetobacter xylinus* bacteria in HS medium under the static cultivation method. The developed bacterial cellulose was treated with NaOH and bleached to get rid of un-uniform coloration. The bleached fabric is treated with 4% glycerol to increase flexibility. The developed bacterial cellulose was used as an alternative of leather ottomans, and to compare its properties, the commercial leather ottomans were purchased and analyzed. The researcher reported that the sewing part was easy in the case of bacterial cellulose compared to the original leather and imitation leather. However, the bacterial cellulose material remained with surface wrinkles and a paper-like look. It was reported that the final whiteness of the developed bacterial cellulose is lesser than the white noted immediately after bleaching. Among the selected materials, original leather noted to be whiter and smooth in surface with a typical leather smell. The developed bacterial cellulose possessed no smell and it is the thinnest of all with higher surface texture [83]. Lee et al. reported that the *Komagataeibacter* can produce cellulose pellicles either alone or along with the other bacteria and yeast aerobically. It is also noted that the developed cellulose pellicles can be produced with the desired thickness, along with the leather-like flexible material. The author also suggested that this leather-like material can be used as a substitute in the leather and shoe industry but they have also mentioned that it is not optimized so far concerning the footwear industry [73] (Table 3).

Yim et al. analyzed the effect of tea type as a nitrogen source and the bacterial cellulose development and its surface properties along with sugar as a carbon source. They also evaluated the effect of carbon and nitrogen source on the yield percentage and thickness of the bacterial cellulose. In characterization, they had reported that green tea as nitrogen source and sucrose as carbon source produced the highest yield and with a smooth surface. They evaluated the mechanical properties like thickness, strength and concluded that bacterial cellulose looks like leather and it has higher tensile strength than the top grain leather of similar thickness [29]. Nam and Lee developed bacterial cellulose based on Suzanne Lee's biofabric concept. The researchers developed synthetic-free nonwoven from the biomaterials. They developed shoes using the bacterial cellulose instead of animal/synthetic leather-based on the concept of Scarlett and Rhett, a famous character from the American novel. They also presented that the material is self-colored and so no dyeing used to replicate the leather look [85]. Researchers also developed leather material from

Table 3 Benefits of the implementation of BC in the footwear industry [84]

S. no.	Issues with leather	Benefits of bacterial cellulose
1.	The manufacturing process and maintaining stock is not an environmental friendly process as it releases greenhouse gas	Bio-based production method which is more sustainable
2.	The production process uses harmful chemicals like chromium, chromium III. etc.	No chemicals used in the production process
	Disposal management is very critical and mostly landfilled. Which causes higher environmental pollution	Bacterial cellulose is biodegradable and creates no pollution during the disposal
	Sometimes causes allergy to the user	Skin-friendly and no allergic reactions
	Possibilities of defects and rejections are high	As it is industrially manufactured, the defects can be avoided and produced as such per the requirement (size, thickness, etc.)
	Cost of the animal hide is very higher	Comparatively cheaper

kombucha bacterial culture using different carbon source material other than tea. In research, the commercial sugar was replaced with culled and rotted sweet potatoes are to serve as the sugar source for cellulose production. They have performed the production with different concentrations of sweet potato. During the development, the major issue noted in the cellulose layer is uniformity and quality [86].

With specific consideration to the medical applications of footwears, a research report mentioned that it is expected that to grow 552 million diabetic patients in 2030 from 366 million in 2020 [87]. The major issue with diabetic patients is foot ulcers. Researchers reported that the functionalization of bacterial cellulose material may be a suitable alternative material for the medial footwear sector. The mechanical or chemical characterization along with the proper control of cultivation parameters will yield tailor-made bacterial cellulose for diabetic foot ulcer patients [88]. Other researchers developed bacterial cellulose-based composite material with the help of poly(fluorophenol) and laccase. During the bacterial cellulose production phase, the laccase is efficiently incorporated into the bacterial cellulose nonwoven. Further functionalization with fluorophenol resulted in hydrophobic surfaces on the bacterial cellulose along with improved wetting contact for oils. The researchers also evaluated its efficiency against washing and other mechanical properties after the functionalization and suggested the product as a suitable material for shoes and bags as an alternative material [89].

Many research works indicated that the higher water absorption capacity as an added advantage of the bacterial cellulose based on the specific application. However, when the bacterial cellulose is used in the products like shoes, leather jacket apparel, etc., the water absorption and holding capacity of the material is critical. In this regard, few research works were performed to reduce the moisture handling properties of the bacterial cellulose. Researchers modified the bacterial cellulose using a fabric softener and hydrophobic finishing agent immediately after the fermentation process. In this analysis, they found that the application of softener followed by hydrophobic finish provided better results without affecting the other characteristics of the bacterial cellulose than the hydrophobic finishing followed by the softener treatment method. The results were significant based on water and oil contact angle test and the researchers suggested the applications as home and interior textiles [90]. There are some research works performed on the bacterial cellulose produced from the coconut water as a substitute for leather products like shoes and bags.

In the late nineties, the first bacterial cellulose from the coconut was produced and used as a leather alternative material in the shoe manufacturing process. The idea was developed by Dr. Anselmo S. Cabigan and he researched with his student Amparo Arambulo of St Paul University, QC, Philippines. They developed shoes from the bacterial cellulose developed from the coconut water famously known as nata-de-coco. They developed, dried it to the consistency of leather and developed the product and their shoes passed the DOST test for strength and durability surpassing animal and artificial leather [91]. An Indian start-up company started making leather-like products from coconut water. This is a similar process like kombucha but the raw material is coconut water instead of tea. The startup was founded by Zuzana Gombosva, a Slovakian designer and researcher, and Susmith Suseelan, an engineer from India. They learned the process from the materials library and initiated the process as like kombucha fermentation. The coconut water is inoculated with bacterial strain and allowed to ferment for two weeks. After the defined time, the bacterial jelly is formed on the surface of the broth by the full growth of the bacteria. They named the product as "Malai." After the production, they blend the material with various natural fibers like banana, hemp, etc. to obtain various esthetic looks and textures. In the total manufacturing sequence, there are no synthetic materials used and so the founders claim their products like shoes and bags manufactured out of the bacterial cellulose composites are totally environment-friendly and biodegradable. They claim that even though it is manufactured on a large scale, the heat and energy produced in the manufacturing process are very less [92].

Other researchers from the private patenting office had initiated the possibilities of developing bio-based leather alternative material as an eco-friendly imitative. They had developed baby shoes from the bacterial cellulose sheet developed from kombucha after the dyeing process [93]. Payne reported that they had developed cellulose form kombucha for the making of footwear. They have treated the developed cellulosic fabric with wax to improve its hydrophobic nature without affecting the tensile properties and comfortability of the material. Out of their product, they have developed footwear using the bacterial cellulose as an alternative material for leather. In their research, they outlined that the bacterial cellulose can be developed in tailor-made shapes and so it can be used as zero-waste manufacturing [94]. Table 4 represents the different products developed using bacterial cellulose by various researchers as a substitute for leather products.

Table 4 Bacterial cellulose products produced from different sources as a leather alternative

S. no.	Product	Materials used	References
1.		Kombucha culture in tea medium	[95]
2.		Kombucha culture in tea medium	[80, 81]

(continued)

Table 4 (continued)

S. no.	Product	Materials used	References
3.		Kombucha culture in tea medium	[82]
4.		*Acetobacter xylinum* in HS medium	[83]

(continued)

Table 4 (continued)

S. no.	Product	Materials used	References
5.		Kombucha culture, tea medium	[84]
6.		Kombucha culture, tea medium	[29]

(continued)

Table 4 (continued)

S. no.	Product	Materials used	References
7.		Kombucha culture in tea medium	[85]
8.		Acetobacter aceti bacteria in nata (coconut)as medium	[91]

(continued)

Table 4 (continued)

S. no.	Product	Materials used	References
9.		*Acetobacter bacteri*, coconut water, and wastes as medium	[92]
10.		Kombucha culture in tea medium	[94]

7 Challenges and Future Prospective

Bacterial cellulose is currently being developed at the laboratory level by various researchers and none of the research works performed in the previous decade was commercialized so far. There are so many technical and application issues noted with the bacterial cellulose in the commercialization point of view, either in the leather industry or in the apparel industry. From the previous research analysis, the following points were noted as major challenges in commercial implementation.

- One of the most identified properties of the bacterial cellulose is higher water absorption capacity and higher wettability properties. These properties were advantageous for specific applications like wound dressing, etc. However, this nature of the bacterial cellulose is not suitable for many applications like clothing or leather alternative. To overcome the problem, several surface modifications process was performed by researchers [96].
- Economical viability of developing bacterial cellulose is still under discussion as the synthetic media used for the production is very costly. Many researchers identified various sustainable materials for the production of bacterial cellulose as an alternative carbon source and measured the yield percentage.
- Researchers identified that bacterial cellulose films mainly suffer due to moisture loss behavior. The dried bacterial cellulose film's structure collapses during the moisture loss and loses its texture, tensile, and some other comfort and esthetic properties.
- The quality and repeatability of the cellulose production were uncontrollable during the fermentation process and no works performed on process standardization.
- However, the recent urge for sustainable alternative materials for the leather and footwear sector encouraged many researchers to work by focusing on the abovementioned demerits. The solution to the identified problems will help the industry to commercialize the bacterial cellulose in large-scale production with required physical characteristics.

8 Summary

Bacterial cellulose is one of the biomaterials which are totally sustainable in terms of its manufacturing process and also by its nature. Due to the exiting issues in the leather and footwear industry, it seeks a suitable and viable substitution in the near future. Though the feel and look of the bacterial cellulose are similar to the leather, the material characteristics differ technically based on the manufacturing environment and nutrient sources used for the production process. In the commercialization point, many researchers attempted to improve the productivity and characteristics of the material. The research works are still in progress to optimize the production parameter to get the standard manufacturing method. Similarly, no successful attempts made on the selection of alternative feedstock material such as various industrial wastes. All data were in the infant stage and no large-scale production attempted. Once these kinds of barriers have prevailed over, the cellulose material developed from the biosources will be next-generation leather as a substitute to the animal and imitation leathers exists in the market.

References

1. Kanagaraj J, Velappan KC, Babu NK, Sadulla S (2006) Solid wastes generation in the leather industry and its utilization for cleaner environment. Chem Inform 37(49). https://doi.org/10.1002/chin.200649273
2. Doble M, Kumar A (2005) Tannery effluent. In: Biotreatment of industrial effluents, pp 133–143. https://doi.org/10.1016/b978-075067838-4/50013-0
3. Thanikaivelan P, Rao JR, Nair BU (2000) Development of a leather processing method in narrow pH profile: part 1. Standardisation of dehairing process. J Soc Leather Technol Chem 84(6):276–284
4. Dixit S, Yadav A, Dwivedi PD, Das M (2015) Toxic hazards of leather industry and technologies to combat threat: a review. J Clean Prod 87:39–49
5. Sivaram NM, Barik D (2019) Toxic waste from leather industries. In: Energy from toxic organic waste for heat and power generation, pp 55–67. https://doi.org/10.1016/b978-0-08-102528-4.00005-5
6. EU (2003) Directive 2003/53/EC of the European Parliament & of the Council of 18 June 2003 amending for the 26th time council directive 76/769/EEC relating to restrictions on the marketing & use of certain dangerous substances & preparations (nonyl phenol, nonyl phenol ethoxylate & cement)
7. OEHHA (2001) Chemical listed effective June 15, 2001 as known to the state to cause reproductive toxicity: N-methylpyrrolidone. Office of Environmental Health Hazard Assessment, California Environmental Protection Agency
8. Khanna SK, Das M (1991) Toxicity, carcinogenic and clinico-epidemiological studies on dyes and dye-intermediates. J Sci Ind Res 50, 965–974
9. Taylor MM, Cabeza LF, Dimaio GL, Brown EM, Marmer WN, Carrio R, Celma PJ, Cot J (1998) Processing of leather waste: pilot scale studies on chrome shavings. Part I. Isolation and characterization of protein products and separation of chrome cake. JAL CA 93(3), 61
10. Verheijen LAHM, Wiersema D, Hulshoff Pol LW (1996) Management of waste from animal product processing. In: Tanneries J, De Wit (eds) International Agriculture Centre, Wageningen, The Netherlands. http://www.fao.org/3/X6114E/x6114e05.htm. Accessed on 27 Feb 2020
11. Fibre Briefing: Leather (2018) https://www.commonobjective.co/article/fibre-briefing-leather. Accessed on 27 Feb 2020
12. Human Rights Watch (2012) Toxic tanneries: the health repercussions of Bangladesh's Hazaribagh leather
13. History of Leather (n.d.) Retrieved from https://www.mooreandgiles.com/leather/resources/history/. Accessed on 27 Feb 2020
14. People for the Ethical Treatment of Animals (2019) Animals used for clothing/leather industry. Retrieved from https://www.peta.org/issues/animals-used-for-clothing/leather-industry/. Accessed on 27 Feb 2020
15. Medical Design Briefs (2015) PVC vs. polyurethane: a tubing comparison. Retrieved from https://www.medicaldesignbriefs.com/component/content/article/mdb/features/articles/21705. Accessed on 27 Feb 2020
16. Nunez FU, Santiago EV, Lopez SH (2008) Structural, thermal and morphological characterization of UV-graft polymerization of acrylated-epoxidized soybean oil onto goat leather. Chem Chem Technol 2:191–197

17. Nam C, Lee Y-A (2019) Multilayered cellulosic material as a leather alternative in the footwear industry. Cloth Text Res J 37(1):20–34
18. Cao H, Wool R, Sidoriak E, Dan Q (2013) Evaluating mechanical properties of environmentally friendly leather substitute (eco-leather). In: Proceedings of the international textile and apparel association (ITAA) annual conference proceedings, New Orleans, 15–18 Oct 2013
19. Cao H, Wool RRP, Bonanno P, Dan Q, Kramer J, Lipschitz S (2014) Development and evaluation of apparel and footwear made from renewable bio-based materials. Int J Fash Des Technol Educ 7:21–30
20. Pinatex (2017) https://www.ananas-anam.com/about-us/. Accessed on 27 Feb 2020
21. Mylo (2017) https://boltthreads.com/technology/mylo/. Accessed on 27 Feb 2020
22. Flinn A (2019) BRB, I need these chic French sneakers that are made from corn. https://www.wellandgood.com/good-looks/veja-campo-vegan-sneaker/. Accessed on 27 Feb 2020
23. Transforming the Material World (2018) http://www.modernmeadow.com/. Accessed on 27 Feb 2020
24. Li J, Wei X, Wang Q, Chen J, Chang G, Kong L, Liu Y (2012) Homogeneous isolation of nanocellulose from sugarcane bagasse by high pressure homogenization. Carbohydr Polym 90(4):1609–1613
25. Kaewnopparat S, Sansernluk K, Faroongsarng D (2008) Behavior of freezable bound water in the bacterial cellulose produced by *Acetobacter xylinum*: an approach using thermoporosimetry. Am Assoc Pharm Sci 9:701–707
26. Schramm M, Hestrin S (1954) Factors affecting production of cellulose at the air, liquid interface of a culture of *Acetobacter xylinum*. J Gen Microbiol 11:123–129
27. Chan CK, Shin J, Jiang SXK (2018) Development of tailor-shaped bacterial cellulose textile cultivation techniques for zero-waste design. Cloth Text Res J 36:33–44
28. Gayathrya G, Gopalaswamy G (2014) Production and characterisation of microbial cellulosic fibre from *Acetobacter xylinum*. J Fibre Text Res 39(1):93–96
29. Yim SM, Song JE, Kim HR (2017) Production and characterization of bacterial cellulose fabrics by nitrogen sources of tea and carbon sources of sugar. Process Biochem 59:26–36
30. Keshk SM (2014) Vitamin C enhances bacterial cellulose production in *Gluconacetobacter xylinus*. Carbohydr Polym 99:98–100
31. Matsuoka M, Tsuchida T, Matsushita K et al (1996) A synthetic medium for bacterial cellulose production by *Acetobacter xylinum* subsp. sucrofermentans. Biosci Biotechnol Biochem 60:575–579
32. Khattak WA, Khan T, Ul-Islam M, Ullah MW, Khan S, Wahid F, Park JK (2015) Production, characterization and biological features of bacterial cellulose from scum obtained during preparation of sugarcane jaggery (gur). J Food Sci Technol 52:8343–8349
33. Kiziltas EE, Kiziltas A, Gardner DJ (2015) Synthesis of bacterial cellulose using hot water extracted wood sugars. Carbohydr Polym 124:131–138
34. Bae SO, Shoda M (2005) Production of bacterial cellulose by *Acetobacter xylinum* BPR2001 using molasses medium in a jar fermentor. Appl Microbiol Biotechnol 67:45–51
35. Premjet S, Premjet D, Ohtani Y (2007) The effect of ingredients of sugar cane molasses on bacterial cellulose production by *Acetobacter xylinum* ATCC 10245. Sen'iGakkaishi 63:193–199
36. Cerrutti P, Roldan P, Garcia RM, Galvagno MA, Vazquez A, Foresti ML (2016) Production of bacterial nanocellulose from wine industry residues: importance of fermentation time on pellicle characteristics. J Appl Polym Sci 133:43109
37. Khattak WA, Khan T, Ul-Islam M, Wahid F, Park JK (2015) Production, characterization and physico-mechanical properties of bacterial cellulose from industrial wastes. J Polym Environ 23:45–53
38. Vazquez A, Foresti ML, Cerrutti P, Galvagno M (2013) Bacterial cellulose from simple and low cost production media by *Gluconacetobacter xylinus*. J Polym Environ 21:545–554
39. Chen L, Hong F, Yang XX, Han SF (2013) Biotransformation of wheat straw to bacterial cellulose and its mechanism. Bioresour Technol 135:464–468

40. Lotfiman S, Biak A, Radiah D, Ti TB, Kamarudin S, Nikbin S (2016) Influence of date syrup as a carbon source on bacterial cellulose production by *Acetobacter xylinum* 0416. Adv Polym Technol 37:21759. https://doi.org/10.1002/adv.21759
41. Kuo CH, Huang CY, Shieh CJ, Wang HMD, Tseng CY (2017) Hydrolysis of orange peel with cellulase and pectinase to produce bacterial cellulose using *Gluconacetobacter xylinus*. Waste Biomass Valorization 10:1–9
42. Faridah F, Diana S, Helmi H, Sami M, Mudliana M (2013) Effect of sugar concentrations on bacterial cellulose production as cellulose membrane in mixture liquid medium and material properties analysis
43. Usha Rani M, Udayasankar K, Anu Appaiah KA (2011) Properties of bacterial cellulose produced in grape medium by native isolate *Gluconacetobacter* Sp. J Appl Polym Sci 120(5):2497–3117
44. Goelzer FDE, Faria Tische PCS, Vitorino JC, Maria Sierakowski R, Tischer CA (2009) Production and characterization of nanospheres of bacterial cellulose from *Acetobacter xylinum* from processed rice bark. J Mater Sci Eng 29(2):546–551
45. Casarica A, Campeanu G, Moscovici M, Ghiorghita A, Manea V (2013) Improvement of bacterial cellulose production by *Acetobacter xylinum* dsmz-2004 on poor quality horticultural substrates using the Taguchi method for media optimization. Part 1. Cell Chem Prod Technol 47:61–68
46. Luo MT, Zhao C, Huang C, Chen XF, Huang QL, Qi GX, Chen XD (2017) Efficient using durian shell hydrolysate as low-cost substrate for bacterial cellulose production by *Gluconacetobacter xylinus*. Indian J Med Microbiol 57:393–399
47. Wu JM, Liu RH (2013) Cost-effective production of bacterial cellulose in static cultures using distillery wastewater. J Biosci Bioeng 115:284–290
48. Hyun JY, Mahanty B, Kim CG (2014) Utilization of makgeolli sludge filtrate (MSF) as low-cost substrate for bacterial cellulose production by *Gluconacetobacter xylinus*. Appl Biochem Biotechnol 172:3748–3760
49. Cavka A, Guo X, Tang SJ, Winestrand S, Jönsson LJ, Hong F (2013) Production of bacterial cellulose and enzyme from waste fiber sludge. Biotechnol Biofuels 6:25
50. Uraki Y, Morito M, Kishimoto T, Sano Y (2002) Bacterial cellulose production using monosaccharides derived from hemicelluloses in water-soluble fraction of waste liquor from atmospheric acetic acid pulping. Holzforschung 56:341–347
51. Bilgi E, Bayir E, Sendemir-Urkmez A, Hames EE (2016) Optimization of bacterial cellulose production by *Gluconacetobacter xylinus* using carob and haricot bean. Int J Biol Macromol 90:2–10
52. Luo MT, Huang C, Chen XF, Huang QL, Qi GX, Tian LL, Chen XD (2017) Efficient bioconversion from acid hydrolysate of waste oleaginous yeast biomass after microbial oil extraction to bacterial cellulose by *Komagataeibacter xylinus*. Prep Biochem Biotechnol 47:1025–1031
53. Zhao H, Xia J, Wang J, Yan X, Wang C, Lei T, Zhang H (2018) Production of bacterial cellulose using polysaccharide fermentation wastewater as inexpensive nutrient sources. Biotechnol Biotechnol Equip 32:350–356
54. Huang C, Guo HJ, Xiong L, Wang B, Shi SL, Chen XF, Chen XD (2016) Using wastewater after lipid fermentation as substrate for bacterial cellulose production by *Gluconacetobacter xylinus*. Carbohydr Polym 136:198–202
55. Moon SH, Park JM, Chun HY, Kim SJ (2006) Comparisons of physical properties of bacterial celluloses produced in different culture conditions using saccharified food wastes. Biotechnol Bioprocess Eng 11:26
56. Guo X, Chen L, Tang J, Jönsson LJ, Hong FF (2016) Production of bacterial nanocellulose and enzyme from [AMIM] Cl-pretreated waste cotton fabrics: effects of dyes on enzymatic saccharification and nanocellulose production. J Chem Technol Biotechnol 91:1413–1421
57. Adnan A, Nair GR, Lay MC, Swan JE, Umar R (2015) Glycerol as a cheaper carbon source in bacterial cellulose (BC) production by *Gluconacetobacter xylinus* dsm46604 in batch fermentation system. Malays J Anal Sci 19:1131–1136

58. Rohaeti E, Laksono EW, Rakhmawati A (2017) Characterization and the activity of bacterial cellulose prepared from rice waste water by addition with glycerol and chitosan. J Agric Biol Sci 12(8)
59. Narh C, Frimpong C, Mensah A, Wei Q (2018) Rice bran—an alternative nitrogen source for *Acetobacter xylinum* bacterial cellulose synthesis
60. Hornung M, Ludwig M, Schmauder HP (2007) Optimizing the production of bacterial cellulose in surface culture: a novel aerosol bioreactor working on a fed batch principle (part 3). Eng Life Sci 7(1):35–41
61. Yoshino T, Asakura T, Toda K (1996) Cellulose production by *Acetobacter pasteurianus* on silicon membrane. J Ferment Bioeng 81(1):32–36
62. Cheng K-C, Catchmark JM, Demirci A (2009) Enhanced production of bacterial cellulose by using a biofilm reactor and its material property analysis. J Biol Eng 3:12
63. Shah N, Ha JH, Park JK (2010) Effect of reactor surface on production of bacterial cellulose and water soluble oligosaccharides by *Gluconacetobacter hansenii* PJK. Biotechnol Bioprocess Eng 15, 110–118
64. Onodera M, Harashima I, Toda K, Asakura T (2002) Silicone rubber membrane bioreactors for bacterial cellulose production. Biotechnol Bioprocess Eng 7, Article number 289
65. Wu S-C, Li M-H (2015) Production of bacterial cellulose membranes in a modified airlift bioreactor by *Gluconacetobacter xylinus*. J Biosci Bioeng 1–6
66. Hartmann AM, Burleson LE, Holmes AK, Geist CR (2000) Effects of chronic kombucha ingestion on open-field behaviors, longevity, appetitive behaviors, and organs in C57-BL/6 mice: a pilot study. Nutrition 16:755–761
67. Chen C, Liu BY (2000) Changes in major components of tea fungus metabolites during prolonged fermentation. J Appl Microbiol 89:834–839
68. Reiss J (1994) Influence of different sugars on the metabolism of the tea fungus. Z Lebensm Unters Forsch 198:258–261
69. Masaoka S, Ohe T, Sakota N (1993) Production of cellulose from glucose by *Acetobacter xylinum*. J Ferment Bioeng 75(1):18–22
70. Tonouchi N, Tsuchida T, Yoshinaga F, Beppu T, Horinouchi S (1996) Characterization of the biosynthetic pathway of cellulose from glucose and fructose in *Acetobacter xylinum*. Biosci Biotechnol Biochem 60(8):1377–1379. https://doi.org/10.1271/bbb.60.1377
71. Han NS, Robyt JF (1998) The mechanism of *Acetobacter xylinum* cellulose biosynthesis: direction of chain elongation and the role of lipid pyrophosphate intermediates in the cell membrane. Carbohydr Res 313, 125–133
72. Hussain Z, Sajjad W, Khan T, Wahid F (2019) Production of bacterial cellulose from industrial wastes: a review. Cellulose. https://doi.org/10.1007/s10570-019-02307-1
73. Lee KY, Buldum G, Mantalaris A, Bismarck A (2014) More than meets the eye in bacterial cellulose: biosynthesis, bio-processing, and applications in advanced fiber composites. Macromol Biosci 14:10–32
74. Rognoli V, Bianchini M, Maffei S, Karana E (2015) DIY materials. Mater Des 86:692–702
75. https://en.wikipedia.org/wiki/Suzanne_Lee. Accessed on 27 Feb 2020
76. Ng FMC, Wang PW (2016) Natural self-grown fashion from bacterial cellulose: a paradigm shift design approach in fashion creation. Des J 19, 837–855
77. Ng MCF, Wang W (2015) A study of the receptivity to bacterial cellulosic pellicle for fashion. Res J Text Appar 19:65–69
78. Lee S (2018) My green goodie hero. https://www.mygreengoodiebag.com/blog/2018/6/18/suz anne-lee. Accessed on 27 Feb 2020
79. Fernandes M, Gama M, Dourado F, Souto AP (2019) Development of novel bacterial cellulose composites for the textile and shoe industry. Microb Biotechnol 12:650–661
80. Ghalachyan A (2018) Evaluation of consumer perceptions and acceptance of sustainable fashion products made of bacterial cellulose. Graduate theses and dissertations. 16583. https://lib.dr.iastate.edu/etd/16583
81. Ghalachyan A (2017) Made from scratch. A sustainable handbag made of bacterial cellulose grown in fermenting tea. In: International textile and apparel association (ITAA) annual conference proceedings, p 65. https://lib.dr.iastate.edu/itaa_proceedings/2017/design/65

82. Mizuno D, Kawasaki K (2017) Bio fashion design: a study on design strategy for sustainable production line through DIY bio experiment
83. Solatorio N, Chong Liao C (2019) Synthesis of cellulose by *Acetobacter xylinum*: a comparison vegan leather to animal and imitation leather. Honors thesis, University of Wyoming
84. Garcia C, Prieto MA (2018) Bacterial cellulose as a potential bioleather substitute for the footwear industry. Microb Biotechnol 1–4. https://doi.org/10.1111/1751-7915.13306
85. Nam C, Lee Y-A (2016) RETHINK II: kombucha shoes for Scarlett and Rhett. In: International textile and apparel association (ITAA) annual conference proceedings, p 68. https://lib.dr.iastate.edu/itaa_proceedings/2016/design/68
86. Freeman C, Gillon F, James M, French T, Ward J (2016) Production of microbial leather from culled sweet potato sugars via kombucha culture. In: International textile and apparel association (ITAA) annual conference proceedings, p 109. https://lib.dr.iastate.edu/itaa_proceedings/2016/presentations/109
87. http://www.sohealthyproject.eu/results/main-results-library/publications/50-strategic-research-agenda. Accessed on 27 Feb 2020
88. Hu W, Chen S, Yang J, Li Z, Wang H (2014) Functionalized bacterial cellulose derivatives and nanocomposites. Carbohydr Polym 101:1043–1060
89. Song JE, Silva C, Cavaco-Paulo AM, Kim HR (2019) Functionalization of bacterial cellulose nonwoven by poly(fluorophenol) to improve its hydrophobicity and durability. Front Bioeng Biotechnol 7:332. https://doi.org/10.3389/fbioe.2019.00332
90. da Silva FM, Gouveia IC (2015) The role of technology towards a new bacterial-cellulose-based material for fashion design. J Ind Intell Inf 3(2)
91. The Making of Nata de Coco Shoes (2011) Available online: https://avrotor.blogspot.com/search?q=nata+de+coco. Accessed on 23 June 2020
92. Malai (2019) http://made-from-malai.com/1792-2/. Accessed on 27 Feb 2020
93. Rethinking Foot Wear for Better Future (2018) http://www.patent-shoes.com/. Accessed on 27 Feb 2020
94. Payne A (2016) Will we soon be growing our own vegan leather at home? https://www.dailybulletin.com.au/the-conversation/24701-will-we-soon-be-growing-our-own-vegan-leather-at-home. Accessed on 27 Feb 2020
95. Regine (2013) Artists in laboratories, episode 43: Suzanne Lee (Biocouture). https://we-make-money-not-art.com/ail_artists_in_laboratories_ep_32/. Accessed on 23 June 2020
96. Liu H, Gao S-W, Cai J-S, He C-L, Mao J-J, Zhu T-X et al (2016) Recent progress in fabrication and applications of superhydrophobic coating on cellulose based substrates. Materials 9:124

The Chromium Recovery and Reuse from Tanneries: A Case Study According to the Principles of Circular Economy

Evgenios Kokkinos and Anastasios I. Zouboulis

Abstract In tanning procedures, the main and most widely currently used reagent for skin/hide treatment is the trivalent chromium salt (usually the sulfate one). However, the low yield of the reactions, taking place in the respective chromium bath, commonly results in a significant proportion of this metal, passing unused into the corresponding wastes (mostly wastewater, but also to some extension in the solid waste). These wastes are considered as hazardous, and their sustainable management is recommended/enforced by international and national legislative organizations. An attractive approach to address this issue is the application of circular economy principles, where the Cr(III) content of wastewaters can be appropriately recovered and then reused again into the tanning process, as raw material. The chromium-rich tannery wastes, in which Cr(III) recovery has been examined, are mainly the tanning liquor and the resulting sludge from the physicochemical/biological treatment of wastewater stream. Regarding the tanning liquor, the treatment technique that shows the highest efficiency and also applied in the field is precipitation. This method requires the addition of an appropriate alkaline media in order to increase the solution's pH and to precipitate $Cr(OH)_3$; subsequently, this precipitate can be re-dissolved in sulfuric acid, before feeding it back to the tanning bath. On the other hand, the most effective technology for chromium recovery from the tannery sludge is the selective, mainly acidic dissolution of chromium by applying simple hydrometallurgical principles, followed by precipitation and eventually re-dissolution, before reuse.

Keywords Leather production · Tanning · Waste management · Chromium recovery · Circular economy · Wastewater from tanneries · Tannery sludge

E. Kokkinos · A. I. Zouboulis (✉)
Department of Chemistry, Laboratory of Chemical and Environmental Technology, Aristotle University of Thessaloniki, 54124 Thessaloniki, Greece
e-mail: zoubouli@chem.auth.gr

© The Editor(s) (if applicable) and The Author(s), under exclusive license to Springer Nature Singapore Pte Ltd. 2020
S. S. Muthu (ed.), *Leather and Footwear Sustainability*, Textile Science and Clothing Technology, https://doi.org/10.1007/978-981-15-6296-9_6

123

1 Introduction

By the term "tanning" is meant the conversion of a raw skin or hide (by-products of slaughterhouses) into a high-strength commercial material (leather), which can be further used for the manufacturing of a wide range of consumer products. As a profession, it is considered to be among the oldest in the world and also, in many cases, leather was used historically as a mean of trading goods. Nowadays, it is recognized as an industry of high economic importance. The preservation of skin by tanning and the performance of various stages of pre/treatment will result in a final product with specific practical properties, such as stability, appearance, water resistance, temperature resistance and elasticity. In other words, tanning is the fundamental stage during the treatment of skin through the application of a complex chemical series of subsequent reactions and of several physical–mechanical processes, noting also that tanning has significantly changed as technology, especially over the last decades. It is considered also mostly as an empirical and not highly scientific batch procedure, depending on several parameters, but mainly on the quality, preservation and pre-treatment of the used raw materials.

Nevertheless, tanneries have been criticized numerous times, regarding their (negative) environmental footprint/impact, although other simpler alternatives to manage skins as wastes (e.g. by landfilled disposal) can be also problematic. In fact, the tanning industry is considered as an important source of pollution worldwide and the relevant wastes (such as wastewaters, sludge, residues) are under continuous monitoring by the respective environmental control organizations, since tannery wastes consist of a complex mixture of organic and inorganic pollutants, such as oxidation agents, phenols, chromium. The tannery wastewaters can be generally treated by the application of various physicochemical and biological methods, or usually by an appropriate combination of them. However, due to the complexity of different pollutants' content, it becomes crucial to determine as quantitatively as possible their overall environmental impact (and toxicity), and hence, the respective prevention and remediation strategies can be more effectively applied.

Among them, the proper pollution control devices, the minimization and safe disposal of produced wastes, the effective use of chemicals and the prevention of accidents are between the most important aspects that have to be considered by any tannery in order to reduce and control its significant environmental impact. On the other hand, specific interventions aiming at the sustainability improvement of the overall tanning processes should be carefully evaluated as a part of the respective production, e.g., what would be their financial cost? What is their effect on the convenient production processes and how this can be predicted in advance? Proposing a methodology that is capable to reduce both environmental and financial costs is always desirable and more appealing to apply. In addition, the final product quality should not be negatively affected, but remain quite the same, or even can be further improved. Indeed, in some cases by changing certain parameters of a specific treatment stage during the production process is likely to cause negative effects, regarding product quality, in the following one.

Generally, there are main two strategies, followed primarily by the scientific community and later by the enterprises, which can promote sustainability, namely the application of green chemistry [1] and of circular economy principles [2]. Green chemistry can be briefly described as the application of certain fundamental principles, aiming mainly in waste and pollution minimization/reduction by the appropriate modification/replacement of the used toxic chemicals, whereas the circular economy promotes the exploitation for the effective reuse/recycling of by-products (previously considered as wastes), as well as of services, applied mainly in the industrial scale. Therefore, by taking into consideration the aforementioned criteria, i.e., the green chemistry and the circular economy, it seems they can be quite easily adopted in this specific industrial field (tanning). A typical case of applying the principles of circular economy in a tannery production plant is the recovery of chromium from the respective wastewaters and sludge, aiming to effective reuse.

Both liquid and solid tanning wastes usually contain high concentration levels of trivalent chromium, as this is the most widespread applied chemical reagent $[Cr_2(SO_4)_3]$, commonly used during the skin/hide tanning procedure. However, due to the limited efficiency of respective chemical reactions taking place, leading to the limited consumption of this reagent, a significant portion (up to 40%) may end up in the produced wastewater, which subsequently requires the application of specific management/treatment techniques. On the contrary, by recovering chromium from these wastewaters, this may be reused by the same industry, as raw/feed material, rather than applying supplementary safe treatment and disposal methods.

In this chapter, a simplified description of the tanning procedure is presented, along with the respective environmental impact through the respective waste production and management. The main case study is considered to be the recovery of chromium from tanneries in appropriate form to be subsequently reused also by the same industries, following the general framework of circular economy. It is worth noting that all the methodologies proposed by the literature and described in this chapter are intended, in addition to recovery, to the reuse of chromium in tanning procedures, or by other industries (e.g., electroplating), depending on the specific chromium form (speciation). These technologies were developed based on the principles of the circular economy, and therefore, the evaluation criteria were determined accordingly. In addition, a detailed description of speciation, regarding the major pollutant of interest, i.e., the trivalent chromium, in the various stages of leather production and waste production/management operations, is provided for the better understanding of the respective processes.

2 Short Description of the Tanning Procedure

The first differentiation, regarding the applicable tanning process, can be considered to be the source of raw material. Especially, the term "skin" refers to the raw material delivered from sheep and goats, while "hide" is the respective term from cows and buffalos. Although the final product has almost the same specifications and it is

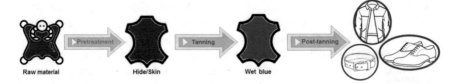

Fig. 1 Basic stages of the overall tanning process

commonly mentioned as leather, despite the different source, the applicable procedures and the respecting requirements may vary. This can be attributed to the different
qualitative characteristics, such as the thickness of the surface layer (epidermis) and
of the subcutaneous tissues. In the following, a detailed list of the fundamental treatment stages (Fig. 1) and steps will be shortly presented, mentioning also that these
are quite likely to vary, depending on the specific raw material (i.e., hide or skin). The
most important between the applied treatment stages is considered to be the tanning
process, and that is the reason why the whole procedure was finally prevailing to be
termed also by this general term.

2.1 Pre-treatment

During the following steps of pre-treatment procedure, the main objective is to
prepare appropriately the skin for the subsequent tanning process in order to achieve
the best possible quality result, i.e., leather with high strength, resistance to external
factors (e.g., humidity, temperature) and elasticity.

- *Trimming and Sorting*: The raw material (feed) of the tannery is likely to include
  some parts of the animal, i.e., not effectively removed during the slaughterhouse
  process, beside the skin/hide (e.g., parts of tail, head, legs), and therefore, these
  should be preliminary removed. Then, the skins can be classified into different
  grades, according to their specific size, weight or quality.
- *Dehydration/Storage*: The removal of moisture from the skin/hide is performed
  by the application of large salt quantities, in order to avoid bacterial degradation during storage. However, if only a short period of storage is required, as an
  alternative procedure, simple cooling may be also applied.
- *Soaking*: Then, the removal of salt is necessary before forwarding the skin/hide
  to the next stages of processing after storage. This procedure is performed by
  washing with water, while removing also other impurities (e.g., blood, dung,
  proteins etc.), and aiming to regain its moisture [3].
- *Liming*: The purpose of this procedure is the removal of hair, flesh, fat (mainly by
  the use of Na_2S) and the fibrous structure open-up (mainly by the use of NaOH,
  Na_2CO_3, $Ca(OH)_2$, etc. and alkaline reagents at the pH range 12–12.5).

- *Fleshing*: The removal of fatty subcutaneous tissue is accomplished commonly by proper mechanical means.
- *De-liming*: The procedure requires the adjustment of pH value in the range 8–8.5 (commonly performed by using NH_4Cl or $(NH_4)_2SO_4$), which allows the collagen fibers' matrix to be smoothly and gradually restored to the initial state.
- *Pickling*: The pH is adjusted to the appropriate value for the following tanning step (commonly by the use of H_2SO_4 and/or HCOOH, by applying quantities proportional to thickness and for the prevention of skin swelling (using NaCl) [4].

2.2 Tanning

Tanning is the second treatment stage, as aforementioned, and the most crucial step during the skin/hide treatment for leather production. In the following, the most widespread applications of tanning methods are enlisted, although it has to be clarified that more than 80% of total leather production is based on chrome tanning:

1. *Chrome Tanning*: The applied reagent is most commonly chromium(III) sulfate (7–10% w/w) and the application pH value ranges from 3.8 to 4. When compared with the other methods, chrome tanning requires less time and addition of chemicals, while delivering better hydrophobic and softening properties of the final product. The semi-treated leather in this case commonly referred as "wet blue." This procedure is mainly applied for the production of leather intended for clothing and for the upper part of shoes.
2. *Vegetable Tanning*: In this method, the applied reagents (physical tannins) are extracted from several plants (e.g., oak, spruce bark, chestnuts, willow bark, valonea, cecidia). Although the process is free from any use of toxic metals, the resulting skin/hide has comparatively reduced heat resistance and color retention. This procedure is mainly used for the production of leather intended for belts, sole and riding equipment. The semi-treated leather referred in this case as "white wet."
3. *Synthetic Tanning*: The used chemical reagents in this case belong to several major categories, such as syntans, resins and polyacrylates (e.g., formaldehyde, glutaraldehyde, phenols, acrylates). Synthetic tanning was developed as an alternative procedure, but it is mostly used in combination with other tanning processes, i.e., either with chrome tanning or with vegetable tanning. The drawbacks of this method are the relative increase of moisture and the heat sensitivity. The semi-treated leather referred in this case as "wet white" and used mainly for leather goods intended for use in the car industry [5].
4. *Combination Tanning*: It is quite common two of the aforementioned treatment methods to be combined, so that the skin can obtain certain specific properties. The already applied combinations are synthetic with chrome or vegetable tanning, the vegetable with subsequent chrome tanning (semi-chrome leather) and the chrome tanning followed by vegetable tanning (chrome re-tanning).

2.3 Post-tanning

At the post-tanning (i.e., the third stage) of the overall treatment process, certain specialized procedures can take place in order to improve the characteristic quality properties of the final product (leather), such as water repellence or resistance, oleophobicity, gas permeability, flame retarding, abrasion and anti-electrostatics. A short description of them is following:

- *Drainage*: The excess moisture is removed in the "wet blue" leather mainly by mechanical means.
- *Trimming*: The previously produced skin/hide is thickened, shaved and leveled by using the appropriate machinery.
- *Re-tanning*: Depending on the quality of "wet blue," skin/hide is re-tanned to improve the chromium content of the skin (necessary quality control parameter).
- *Neutralization*: After washing the excess (acidic) chromates from the skin/hide, this is then added to a new bath with alkaline media in order to restore the pH value to a less acidic range (e.g. 4.5–6.5).
- *Dyeing*: The skin/hide can be colored by pigments, such as anionic, metallic, acidic or basic dyes.
- *Fat-liquoring*: The skin/hide is lubricated with an oil coating (from animal, vegetable or synthetic origin), aiming to separate the fibers, thereby to reduce the friction between them, and to result in better softness [5].
- *Finishing*: The final grooming is performed, according to the specific type of final leather product by applying a series of surface processes, chemical and/or mechanical, aiming in the coverage of possible defects, the delivery of the desired shade and the required durability, while the leather becomes more attractive in appearance and touch [4].

3 Major Environmental Impacts of the Tanning Process

The negative environmental impact of tanneries can be attributed to the formation of several liquid, solid and gaseous waste streams, as a result of raw skin/hide treatment by the use/consumption of chemicals, energy and water. Consequently, the nature of treatment processes and the associated use of dangerous materials/components have the potential to contaminate severely soil and water. It has to be noted, however, that the application of specific alternative techniques ("Green technologies") to the production process and the appropriate treatment of produced wastes can lead to the significant reduction of these impacts. However, this is a quite difficult task, since the quantity and quality of production wastes can vary not only from region to region, due to bigger or smaller differences in the applied tanning process technologies, but even within the same tannery industry from time to time, e.g., by the treatment of different raw materials.

Table 1 Range of water consumption during the tanning process [6]

Treatment step	Water consumption	
	Hide (m³/tn)	Skin (m³/tn)
Pre-treatment (stage 1)	7–25	65–150
Tanning (stage 2)	1–3	30–70
Post-tanning (stage 3)	4–9	15–45
Total	12–37	110–265

3.1 Water Consumption

Water consumption can be classified in two main categories, i.e., the requirements for the tanning procedure and for the overall technical part, such as house cleaning, energy production and sanitation. The latter cases are estimated to represent about one-fifth of the total water consumption by the tannery plants. According to the production monitoring of batch operating plants, it is possible to calculate the average daily needs for water and to estimate the volume of wastewater produced per raw material type and weight for each stage/steps of treatment process. It should be noted, however, that the water requirements during the tanning procedure may vary considerably between tanneries, depending on the peculiarities/differences of used processes, the raw materials (e.g., hides or skins) and the specific final product specifications. As Table 1 shows, the water consumption is higher in the early stage of tanning process and then, it is significantly reduced [5].

In addition, the constant attempts for improvement globally, have led to reduce the water consumption in tanneries to the absolute necessary quantities. For example, by the application of low-volume bath technologies along with advanced reactors, even in processes that seemed particularly difficult to be modified (e.g. calcination, tanning etc.), and also through the better organization of product lines. Nevertheless, it is estimated that about 75% of total water volume, required daily for the various treatment stages/steps and needs, may end up in the wastewaters' stream. According to the previously mentioned facts, perhaps the term "consumption" is not the right one to describe the water used by a tannery, though it has been prevailed, because it should be then treated as high-strength wastewater. Regarding also the cooling water of the dryers, and also that used in specific machinery, this represents a rather good-quality stream and it can be easily recycled and reused into the same plant for similar purposes.

3.2 Energy Consumption

The energy consumed by a tannery plant is mainly electric and thermal, with the majority being spent on the second one (~80%). This consumption depends mainly on the applied tanning methods and the specific equipment used, e.g., chrome,

Table 2 Estimations of energy consumption for tanneries [7]

Tanning steps	Electric and thermal energy consumption
	GJ/t
Skin/hide to wet blue (stage 1)	<3
Skin to leather (stage 1–3)	<6
Hide to leather (stage 1–3)	<14

vegetable or combined tanning procedures follow different operating parameters, such as number of treatment stages and temperatures, whereas during drying, which is necessary for all these cases, an equally high proportion of the required thermal energy (up to 35%) is usually necessary. Regarding the electric energy requirements, as in all industries, this is highly depended on specific equipment/machinery specifications and maintenance. Plant capacity/size, age, complexity and potential heat losses (or heat recovery) are considered as the major factors linked with energy consumption. In this category, it is also included the raw materials' transportation inside the tannery and the proper ventilation, in order to ensure safe working conditions. It is clarified that the previous data does not take into account the waste management/treatment aspects, which, when will be carried out inside the tannery, expected to further increase the overall energy consumption of following Table 2 [5].

3.3 Waste

Only 20–25% of the total weight of raw skin/hide ends up in the final product. The exact rate depends on the type and specifications of the desired leather. During tanning, residues can be classified as by-products, non-hazardous or hazardous wastes, which may occur mainly in liquid, but also in solid and gaseous phases. The residual weight and the excessive quantities of not-used chemicals, arising at specific steps of the tanning process, are either discharged into waste streams for further treatment, or they can be reused (after appropriate pre-treatment). A list of the three major waste categories, resulting from the tanning procedures, is following:

3.3.1 Gaseous Emissions

Gaseous pollution from a leather industry presents sometimes an important environment impact, and it can also create a hazardous workplace, due to the presence of problematic odors, gases, vapors, fumes, smoke, powders and suspended particles. The main pollutants are considered to be the airborne particles and chemicals (mainly the volatile organic compounds, VOCs). Especially, the particulate emissions may contain dust and chemicals, produced by the mechanical operations (e.g., shaving, milling, staking), as well as by handling powdery reagents, such as chromium sulfate [8]. In addition, volatile chemicals, such as hydrogen sulfide, ammonia, sulfur dioxide, amines, aldehydes and other VOCs may be also emitted into the atmosphere from the tannery plants. These volatile components are released during the various tanning processes as by-products from the corresponding residuals/wastewater, e.g., when the pH value of hair removal bath falls below 8, the hydrogen sulfide can be emitted from the sulfides residues [9]. Odors can be caused by either the aforementioned chemicals (e.g., ammonia, hydrogen sulfide), or by the improper handling of raw materials (e.g., during storage), as well as from the produced wastes or wastewater. Regarding the flue gases (e.g., CO, CO_2, SO_x, NO_x), which can be derived from boiler and generators operation to meet the energy needs of the tannery plant, usually these emissions are not burden the tanneries, as the energy is produced outside of them.

3.3.2 Wastewater

Tanneries produce wastewaters, containing high concentrations of organic and inorganic pollutants; these wastewaters are generally classified as hazardous (Table 3). As the tanneries can apply a wide range of processes and different raw materials, the respective wastewaters present a quite complex nature, showing differences of physicochemical characteristics from time to time, process to process and even from tannery to tannery [10].

As aforementioned, approximately 75% of water consumption ends up in wastewaters, and therefore, large amounts are produced during the tanning process (around 30–35 m^3/t hide/skin) [11] Tannery wastewaters are slightly alkaline (i.e., pH value is in the range 7–9), dark brown colored, having rather high concentration values of COD, BOD, TDS, Cr(III) and phenolics with strong odor. The most common physicochemical parameters, which are monitored in such effluents, are presented in the following Table 3, along with their respective ranges. In addition, nearly 90% of

Table 3 Range of wastewater pollutants' load in tanneries, depending on the raw material treated [6]

Pollutant	Hide	Skin
	Kg/t	
COD	145–230	140–320
BOD$_5$	48–86	52–115
Suspended solids (SS)	85–155	70–135
Total dissolved solids (TDS)	300–520	180–500
Total Kjeldahl nitrogen (TKN)	10–17	12–20
Chromium (Cr)	3–7	3–6
Sulfide (S^{2-})	2–9	3–7
Chloride (Cl^-)	145–220	80–240
Sulfate (SO_4^{2-})	45–110	40–100
Grease	9–18	34–71

this wastewater is attributed to the first two stages of the overall tanning procedure (i.e., pre-treatment and tanning). Especially, the wastewater produced during the pre-treatment steps, such as soaking, liming, de-liming, is highly alkaline, presenting also high values of BOD and COD parameters. Instead, wastewater from the tanning stage is acidic and colored, noting, however, that only around 60% of the initial amount of added chromium salts reacts with the treated hide/skin, while the rest 40% remains in the spent tanning solution [12]. In contrast, the wastewater produced during the vegetable tanning contains mainly high organic matter concentrations [4].

3.3.3 Solid Waste

The quantity and quality/composition of produced solid wastes in tanneries may vary significantly even for the same tannery plant, just like wastewater. Therefore, there are not specific control parameters, which can depend on the chemical and mechanical production processes. The type of raw material also affects the quantity of wastes and the ratio content between inorganic and organic pollutants [7]. Solid wastes may include salts, hair or wool, powders/dust, air filtration solids, packaging materials, grease, oils, fleshing and trimming residues. In addition, this category can also include the tannery sludge, which results from the corresponding wastewater treatment plant, as it is described in the following section. Table 4 presents the average range of produced solid wastes, depending on the different origin of raw materials and the applied overall tanning procedure. It is clarified that the main step, i.e., the chrome tanning bath, does not produce directly solid wastes, since the main potential pollutants, such as $Cr(III)$, are mostly soluble, due to the applied acidic pH conditions. As a result, these pollutants pass into the produced wastewater stream, which, after the appropriate treatment, produces a significant amount of sludge [13].

However, some of the aforementioned (mainly those considered as non-hazardous) solid wastes may be better classified as by-products (e.g., fat, scrap), since they can be sold, or supplied to other industries as raw materials, or even reused by the same tanneries, possibly through the recovery of useful chemical reagents and energy (after incineration). On the other hand, the produced (after treatment) tannery sludge is generally considered as being hazardous, mainly due to its high (trivalent) chromium content, as a consequence of the relevant high concentrations in the wastewater to be treated. At this point, it is worth mentioning that the trivalent

Table 4 Typical range of solid waste quantities produced in tanning plants [7]	Source	Kg/t
	Pre-treatment	220–650
	Post-tanning	200–300
	Tannery sludge	400–500
	Dust	2–10
	Packing	15

chromium is the dominant stable Cr form in the sludge, without being easily oxidized to the much more toxic hexavalent species, due to the high organic carbon content of sludge, acting as reductant agent [14].

4 Waste Management

4.1 The Good Housekeeping Concept

In order to minimize the environmental impact of tanning process, it is recommended by European Union the good housekeeping principles, regarding waste management, by applying a combination of the following measurements [7]:

- careful selection and testing of chemical reagents and raw materials (e.g., quality of hides/skins and chemicals),
- input–output analysis (by an inventory) and control of used chemicals, including their quantities and toxicological properties,
- minimizing the use of chemicals to the lowest possible level, as required by the final product quality specifications,
- careful handling and storing of raw materials and finished products to reduce/prevent leaks, accidents and production of excess wastewaters,
- separation of waste streams, whether this is possible, in order to allow their easier proper recycling,
- close monitoring of the major critical operational parameters in order to stabilize/improve the production process,
- regular maintenance of wastewater treatment systems,
- review the possibilities for specific water reuse options,
- review the respective major waste disposal options.

More specifically, regarding the gaseous, wastewater and solid wastes emissions from the tanning industries, the following issues should be also carefully considered.

4.2 Gaseous Emissions

Odors can result from the decomposition of improperly treated or stored raw skins/hides, as well as from the accumulated, locally disposed wastes, from the pre-tanning processes and from the wastewater treatment plants that are not appropriately controlled and maintained. Some of the toxic substances involved in these problems may include hydrogen sulfide, thiols, ammonia, amines, aldehydes, ketones, alcohols and organic acids. The treatment of odors involves mainly the reduction of ammonia and hydrogen sulfide emissions, by replacing the relevant chemical reagents, as well as by the purification of exhaust (ventilation) air from the production area, commonly using scrubber and/or biofiltration, as well as by the appropriate salting of skins/hides to avoid decomposition and by the continuous pH control at the various production stages/steps.

Due to the limited ability to deal with specific pollutants in air pollution abatement techniques, the best option for reducing VOC emissions is also the application of wet-scrubbing systems and the optimization of proper technical processes implementation. In this case, however, the pollution problem is usually transferred from the air phase to the, respectively, created aqueous phase. Therefore, there is a need to recover and reuse the (mostly volatile) organic solvents, which is possible, however, only when certain solvents would be applied. In addition to wet scrubbing, other available technologies may include the combustion of gaseous emissions, the adsorption onto activated carbon, and the biodegradation. The respective limitations during the implementation of the aforementioned technologies are mainly the high concentrations of gaseous pollutants and the operational and capital costs.

Airborne particles can arise not only from the applied mechanical operations, such as grinding and polishing, but also during the handling of chemicals and of dust. The major parameters for the estimation of particulate emissions are their concentration, chemical content and particle size. For the most effective control of dust and the prevention of relevant emissions, it is recommended the use of liquid reagents, rather than their solid/powder form. Furthermore, the airborne particle-producing machines and operations should be grouped in the same area in order to facilitate the collection of dust, e.g., by using an appropriately equipped ventilation system with bag filters, or applying liquid-type scrubbers [7].

4.3 Wastewater Treatment

The effective purification of hazardous wastewaters from tanneries, before their discharge into a water body, or being reused, is usually a multi-stage process. The main aim is to reduce or remove the organic matter content, as well as the solids, nutrients, chromium and other pollutants, as every receiver can accept specific (small) amounts of pollutants without being degraded. Therefore, each wastewater treatment plant must comply with the respective effluent standard limits, usually issued by the regional environmental authorities, which are expressed mainly as maximum permitted concentration levels of BOD_5, COD, suspended solids (SS), Cr, total dissolved solids (TDS), etc. quality parameters.

Moreover, each tanning step produces a specific, more or less, wastewater stream with distinct characteristics, resulting to the following three corresponding categories:

- Pre-treatment (e.g., calcination) wastewater stream from cutting and tearing machines, which may contain sulfates and has rather higher (alkaline) pH value, but without containing chromium.

- Tanning wastewater stream that usually contains quite high chromium concentrations and presents an acidic pH value.
- Post-tanning wastewater stream (e.g., by softening, lubrication, painting), which may contain significant amounts of organic pollutants, but rather lower chromium content [15].

It is crucial for these wastewater streams to be separated and treated properly, according to their specific characteristics in order to avoid potential hazards (e.g., hydrogen sulfide formation) and, also, to reduce the respective treatment costs. An example of poor management is the mixing of pre-treatment lime with the tanning streams, resulting in the creation of poisonous hydrogen sulfide (H_2S), besides the annoying smell that characterizes tanning. This gaseous pollutant is still nowadays the most common cause of tannery health problems and accidents, especially when insufficient ventilation conditions are occurred. Therefore, the wastewater treatment, especially during the early stages, can be applied by different techniques, followed by the more or less convenient wastewater treatment plant.

In order to reduce the emissions of pollutants into the receiving water bodies, a combination of the following techniques is recommended to treat the tanning wastewaters, within and/or outside the respective facility (Fig. 2):

Pre-treatment

Usually, in the case of centralized tannery wastewater treatment plants, serving the corresponding cluster installations, commonly found in developing countries, it is necessary to operate appropriate pre-treatment systems in the individual enterprises. Their role is to remove the larger particles, sand/gravel and grease, but also to reduce significantly the chromium and sulfide content, due to the aforementioned hazardous risks, before this (partially) pre-treated effluent would be discharged into the central collection network for further treatment. Regarding the chromium-rich stream, the respective content of metal can be reduced by applying a simple alkali precipitation

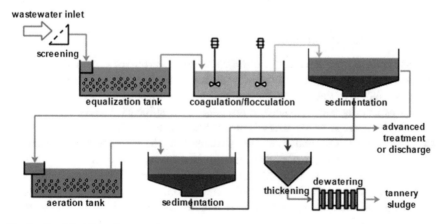

Fig. 2 Simplified flow diagram of a common tannery wastewater treatment plant

method, whereas in the pre-tanning stream a catalytic oxidation method of sulfide to thiosulfate can be applied, mainly by adding $MnSO_4$ under a mixing/aeration system [15]. This specific method may be also performed in the equalization tank of centralized wastewater treatment plant for a tanneries cluster.

Physicochemical treatment (primary)

The aim during the primary treatment is to remove most of organic and inorganic solids' content by the application of settling/sedimentation, as well as other materials that may be floating (e.g., fats, oils), using the same (combined) or different treatment tanks. About 25–50% of incoming biochemical oxygen demand (BOD_5), 50–70% of total suspended solids (SS) and 65% of oil/fats content can be usually removed during this treatment. Additionally, enhanced sedimentation can be also applied by the supplementary use of coagulation/flocculation process. The most commonly applied chemical reagents in this case are $Al_2(SO_4)_3$, $FeSO_4$, $FeCl_3$ and $Ca(OH)_2$, along with proper polyelectrolytes. Then, the sedimentation tank follows, where the appropriate bottom and top skimmers can separate the solid and supernatant fats/oils from the liquid phase, respectively [15].

Biological treatment (secondary)

In most cases, the primary treated effluent cannot meet the respective legisla-tion/regulation limits (i.e., the maximum allowable concentrations) for the direct discharge into a water body; therefore, a secondary treatment stage is usually required in order to supplementary reduce mainly the biodegradable dissolved and colloidal organic compounds, by using an aerobic biological treatment process. This process is carried out in the presence of oxygen and aerobic microorganisms (mainly bacteria) that can effectively metabolize the organic matter content of wastewater, thereby producing more microorganisms and several inorganic by-products (mainly CO_2, NH_3 and H_2O). On the other hand, the application of anaerobic treatment seems not to be quite suitable in this case, due to the specific characteristic of these wastewa-ters, such as relatively high toxic substances content, possibility of hazardous gaseous emissions [16].

Advanced treatment (tertiary)

If neither primary nor secondary wastewater treatment is proved enough to yield final (treated) pollutants' concentration below the respective regulation limits (as imposed by legislation), a tertiary treatment can be also supplementary applied, mainly when the corresponding tanneries need to discharge their effluents directly in a sensitive water body receiver. The aim of advanced treatment is to reduce further the carbon oxygen demand (COD) and/or the nitrogen content by the nitrification/denitrification process, i.e., the oxidation of organic nitrogen and ammonium to nitrate compounds, followed by the reduction of nitrate to harmless nitrogen gas. The relevant applicable tertiary treatment methods are usually the sand filtration, the activated carbon adsorp-tion, the use of ion-exchange resins, and the oxidation by ozonation with/without the presence of H_2O_2.

4.4 Solid Waste Treatment

The conventional method for the solid waste management of tanneries is landfilling, usually after the application of dewatering (volume reduction) and the proper stabilization [17–19]. However, as the solid wastes can be created in almost all stages/steps of tannery production, having various characteristics, they also require different treatment/stabilization and disposal options. Instead, several environmental organizations have recommended a further exploitation of these wastes, to avoid the possible risk of secondary pollution, through the leaching of toxics (e.g., Cr(III), or even HCN) and/or gaseous emissions (e.g., NH_3, NO_x) [9]. Viable alternatives may be their use for glue production and the collagen recovery from hides (raw and limed trimmings), which can be subsequently used in the manufacturing of biomaterials for medical, etc., applications. Also, inorganic salts used in softening and pickling processes may be recovered and reused, while hair/wool components can be recycled, e.g., as fertilizers by hydrolyzing their proteins, or further used for making low-cost textiles and fillings [20].

As aforementioned, rather large amounts of sludge can be produced by the tanneries wastewater treatment plants. The primary objective is to reduce the solids and the potentially hazardous substances content, since the biodegradable organic substances will be eventually transformed into bacterial cells and the latter can be rather easily removed/separated from the treated wastewater by simple settling. The main treatment that tannery sludge (from the primary and secondary sedimentation tanks) is usually undergoes is thickening and dewatering in order to reduce its volume and to achieve the proper moisture content, required for final disposal in specific landfills (Fig. 2).

However, a number of alternatives technologies have been proposed, tested and implemented for the further utilization and/or safe disposal of tannery sludge in pilot or industrial scale, such as the soil application after stabilization [21]), the composting, the anaerobic digestion, the combustion for energy recovery [22], the pyrolysis, the chromium recovery [23] and the use as additive in ceramics [24]. Nevertheless, there is not a universally accepted solution for the tannery sludge utilization/application, with the main obstacle being its rather high chromium content, while the specific legislation and practice can vary widely between different countries.

4.5 Legislation

The raw tanning effluents are considered as hazardous threat both for the environment and for the humans [25]. Therefore, several national and international organizations have tried numerous times to establish a global common legislation framework, regarding the respective regulation limits. However, since the tanning methods and the available discharge receivers may widely vary among different regions/countries,

Table 5 Range of regulation limits as imposed by different countries legislations, regarding the direct discharge of treated tanning effluents in a water body [5]

Pollutant	Regulation limit
	mg/L
COD	200–500
BOD$_5$	15–25
Suspended solids	<35
NH$_4$–N	<10
Total Cr	<0.3–1
Sulfates	<1

this was not achieved until now. In other occasions, the blame falls on the environmental organizations themselves, as the proposed standards were quite ambitious and unrealistic, due to the required level of applied advanced treatment technologies in order to achieve them and to the significant increase of production costs [26]. There were also certain cases, where the cost of waste/water treatment may exceed that of the corresponding fines threshold, imposed for the environmental pollution.

Instead, each country seems to adapt better a specific national legislation strategy, mainly according to its respective technological and economical status. Another approach, regarding the monitoring of pollution and its prevention, is the establishment of regulation limits in the chemical reagents used by tanneries (e.g., chromium) and their content in the final leather products [9]. The mentioned strategy involves mainly the European Union countries [27]. In any case, the legislative authorities promote, and the governments finance, the centralized organization of tanneries in specific clusters of organized industrial areas, in order to provide them with common facilities (e.g., wastewater treatment plants). Thus, it would be easier to establish the treated effluents limits for the centralized tanneries wastewater treatment plants and not for each individual enterprise (Table 5).

5   Chemistry of Chromium

A wide number of chemical reagents may be possibly applied during the tanning procedures. However, when practical criteria would be taken into account, such as effectiveness, availability, toxicity and cost, the relevant choices are rather limited and applied mainly in industrial scale chemicals' use, containing Cr(III), Al(III), Ti(IV) and Zr(IV) [28].

Nevertheless, in practice worldwide 80–90% of tanneries use trivalent chromium salts in the tanning process. Noting that Cr(III) is not included in the Annex X of the Water Framework Directive 2000/60/EC, as amended by the Directive 2008/105/EC, regarding the Priority Substances for pollution control. Also, certain tannery wastes, containing Cr(III), are not included in the European general list of hazardous waste, due to its specific characteristics (absence of Cr(VI), prevailing its classification to be considered as hazardous.

5.1 Reagent Production

Even though the chromium content of tanning solutions is usually expressed as Cr or Cr_2O_3 percentage, the majority of its chemical forms are very insoluble in the, respectively, applied conditions. Among the several chromium salts, the highest solubility is presented by the Cr(III) sulfate ($Cr_2(SO_4)_3$) [29]; when this compound is dissolved in water, it forms the corresponding hydroxide sulfate ($Cr(OH)SO_4$) [7].

This salt, which is suitable for tanning, is usually produced from chromite ore. Cr(III) content and its purity may be increased by applying a multi-step hydrometallurgical process. Initially, the chromite ore is subjected to combustion in a rotary furnace at 1200 °C under oxic conditions (i.e., with air supply) and in the presence of an alkaline media:

$$Cr_2O_3 + Na_2CO_3 + 1/2O_2 \leftrightarrow Na_2Cr_2O_7 + CO_2 \tag{1}$$

According to reaction (1), chromium(III) is oxidized to the respective hexavalent form, i.e., as bichromate salt (termed as chromite ash). Cr(VI) compounds, even though they are considered as very toxic, are also highly soluble in aquatic media. The chromite ash is mixed with an acidic media (sulfuric acid) so that the dichromate salt is selectively solubilized.

$$Na_2Cr_2O_7 + H_2SO_4 + H_2O \leftrightarrow 2H_2CrO_4 + Na_2SO_4 \tag{2}$$

Then, in the produced chromic acid solution (Reaction 2), a reducing reagent (mainly sulfur dioxide) is added under acidic conditions in order to form the Cr(III) species (Reaction 3).

$$2H_2CrO_4 + 3SO_2 \rightarrow Cr_2(SO_4)_3 + 2H_2O \tag{3}$$

The desired Cr(III) sulfate salts are separated from the aqueous phase, using the precipitation method by simply increasing the pH value. As aforementioned, Cr(III) generally forms insoluble compounds; therefore, the application of mild alkaline conditions is considered as adequate. Specific attention should be directed during the preparation process, since the co-existing impurities may decrease the purity of product/precipitate, regarding the Cr(III) content, such as the case of Na_2SO_4 (as the relevant reported content of this salt was up to 50%), co-produced during Reaction 2 [30].

5.2   Use of Cr(III) Salts in Tanning Baths

In tanning baths, several processes/reactions are designed to interact chromium species with the skin's collagen for the tanning procedure. After pickling, the readjustment of pH value takes place by the addition of an alkaline media, mainly sodium bicarbonate (at optimum pH range 3.5–4). This process is called basification and serves two main purposes, i.e., the increase of hydroxyl groups, combined with chromium(III), and also the activation of the carboxyl groups of proteins (the main components of collagen). Due to basification, Cr(III) species show a tendency to form polynuclear complexes (clusters) through the creation of Cr–O–Cr bridges, due to their lower stability (Fig. 3) [20]. Then, chains of these structures are created, mainly by the olation complexes, since oxalation is slow and irreversible. These chains have the ability to favor the coordination with collagen's carboxyl groups by covalent bonds. In addition, it was proved that the bidentate sulfate groups improve the coordination's stability, since they remain in the chromium clusters of final product (leather), even after the application of finishing process [31].

Regarding the presence of collagen in hides and skins, the dominant form is Type I (i.e., densely packed fibers); hence, large chromium clusters may deposit on the relevant fabrics, while smaller clusters can penetrate the microfiber structure, resulting in intra-single and inter-triple helix cross-links (Fig. 4). It is noted that the

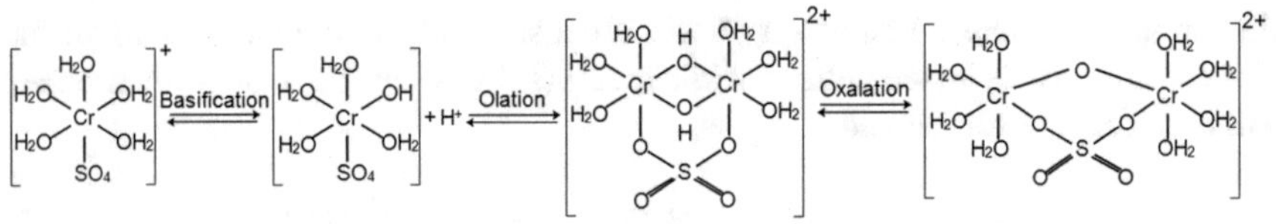

Fig. 3 Chromium(III) species transformation in a tanning bath [30]

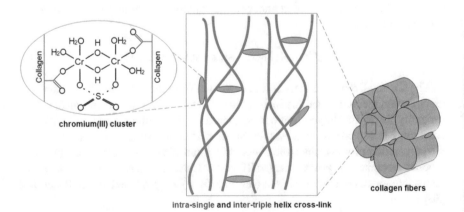

chromium(III) cluster

intra-single and inter-triple helix cross-link

collagen fibers

Fig. 4 Chromium(III) clusters complexation with collagen

addition of these clusters does not affect the structure of collagen [32]. Furthermore, a balance between the reaction rate and the penetration rate should be achieved, according to the desired requirements. Low reactivity conditions favor the clusters penetration, due to their small size. Instead, by increasing slightly the pH value, the reactivity will be also increased and therefore, the penetration rate will be reduced, due to the formation of larger clusters [30]. On the other hand, some of the formed Cr(III) species, i.e., the olation and the oxalation complexes, are the main forms (although not the only ones), which are presenting poor affinity with the collagen proteins and hence, lower uptake.

5.3 Tannery Wastewater

Chromium(III) in tannery baths exists in soluble form, but when it is discharged in a general wastewater stream, it may react with other co-existing substances, such as proteins, and subsequently can form precipitates or sludge [13].

5.4 Tannery Sludge

Tannery sludge, which is produced during the wastewater treatment, is a considered as a hazardous waste and contains very high chromium concentrations. In fact, the majority of residual (not reacting/consumed) tanning Cr(III) is connected within this sludge. As a result, in the literature has been reported that the average Cr(III) concentration rates may not exceed 10%, although there are also reports presenting even higher values [33]. As is reasonable, the Cr(III) concentration values in the tannery wastewaters are mainly depended on the performance of chromium's clusters deposition/reaction and uptake in the tanning bath. Cr(III) bounds mainly to oxides and hydroxides, due to its low solubility. Other notable bounds of Cr(III) may be with the organic matter content, as well as with the carbonates, while other cases such as water soluble or exchangeable species are not usually observed [29].

5.5 Environmental Fate After Waste Landfilling

The low availability and mobility of Cr(III) salts in the environment are well-studied. In natural waters, it may form insoluble complexes, or it can be adsorbed onto particulate matter and as a result, it will be deposited in sediments. Following this assumption, worldwide the landfilling is the main disposal method for the corresponding sludge, produced from the relevant tanning wastewaters treatment. However, according to recent reports, the presence of Cr(VI) was also detected in some soils of tannery sites [34]. The oxidation of chromium may be attributed mainly to the presence of Mn oxides and Fe^{2+} hydroxides [35]. In contrast with Cr(III), Cr(VI) is characterized by substantial higher bioavailability and mobility, since its compounds are very soluble, and eventually may lead to ground/water contamination [36].

6 Chromium Recovery

The sustainable waste management in tanneries is promoted by various international and national environmental organizations through specific guidelines, regarding realistic case studies [7]. According to these reports, the focal point in most suggested management strategies is considered to be the further exploitation of any waste stream, aiming to the recovery and reuse of valuable substances content, such as in this case chromium. The technologies for chromium recovery can be applied either directly in the tanning liquor, or indirectly in the respective wastewater and the resulting sludge after its treatment. It is noted that Cr(III) content in these streams is high enough to make its recovery economically feasible, depending also on the applied method.

6.1 From Wastewaters Through the Application of Different Technologies

The liquor obtained from the tanning bath is considered to be the main wastewater stream in which chromium recovery is worth to be examined. Noting, however, that it is necessary to distinguish the relevant proposed methodologies referred in the literature, between those that are able to be implemented in the tanning plant and those which are still at the research/laboratory/pilot level. Therefore, the following criteria may be used for grouping the relevant treatment methodologies described in this subsection and when any particular case responds positively to at least one of them, it may be considered currently as a research study:

- The treated sample was diluted.
- The reference sample was synthetic.
- A pre-treatment stage is required.
- Limited efficiency for the higher Cr(III) concentrations.
- Requirement of higher processing time.
- Resulting in a low purity Cr(III) residue.
- The resulting Cr(III) residue is contaminated with several impurities.
- The recovered chromium is unsuitable for reuse in the tanning plant, due to the reduced quality of final product, i.e., leather.

- High capital and operational costs have to be implemented, as the method for Cr(III) recycle/recovery from wastewaters is usually applied within the tanneries; hence, an additional cost has to be considered for the respective enterprises.

Nevertheless, the following main treatment technologies, aiming to Cr recovery from tanning wastewaters/sludge for potential reuse, have been proposed/applied.

6.1.1 Sedimentation

The chromium-rich wastewater stream from the tanning bath should not be discharged in the corresponding general sewage network, due to the expected chromium sulfide reaction health risk. In that case, a pre-treatment step is applied, usually in the tanneries, in order to recycle/recover most of the chromium content. This direct approach requires the preliminary removal of undissolved materials, as well as of the floating materials and of the other precipitates (e.g., leather fibers, fats). These materials are led to the general solid waste stream for further filtration and final appropriate disposal or recycle. The liquid phase, without any reagent addition, is allowed to rest in a sedimentation tank, where chromium can be precipitated after the appropriate pH regulation. However, this step may require high residence times, whereas over 90% of the initial Cr(III) content can be recovered and potentially reused with this rather simple treatment method [37].

6.1.2 Precipitation

A more rapid approach for Cr(III) recovery from the tanning liquor is the precipitation method by the addition of an alkaline media (Fig. 5). The precipitation of Cr(III) by increasing the solution pH is the most widespread method in order to separate Cr(III) from a liquid phase, due to its low solubility, and it can be also applied in the tannery plant. Similarly as before, in the first stage the particulate solids and oils should be removed by screening and skimming, respectively. It follows the addition of an alkaline media with the most widespread used being NaOH, MgO, CaO and Ca(OH)$_2$ [38]. In mild alkaline conditions (i.e., pH range 8.5–10), potassium and calcium oxides/hydroxides may present higher efficiency, regarding Cr(III) precipitation (up to 100%), while MgO requires a pH value above 10 [39]. Selectivity, settling rate and sludge volume are the main characteristics used in order to compare the examined reagents. Among them, the selectivity is considered to be the major disadvantage of this method, since the resulting precipitate shows rather low purity (< 20%), regarding Cr(III) content, and hence, its reuse may degrade the quality of the final product (leather) [40].

Regarding its implementation in the plant (Fig. 5), the pH value of wastewater from the tanning bath is adjusted in the range 8.5–9 in the precipitation tank, while the temperature is set in the range 35–40 °C. Cr(III) is precipitated and recovery efficiencies over 97% have been observed. An alternative to enhance further the

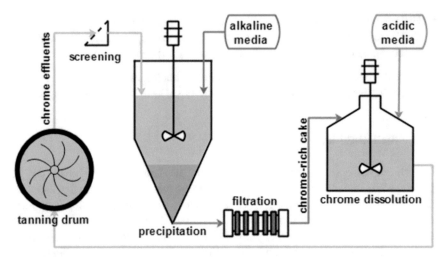

Fig. 5 Flow diagram of Cr(III) recovery and reuse from tanning bath by precipitation

efficiency (up to 99%) is to set the temperature in the range 60–80 °C, but this adjustment increases the application cost of the method. Afterward, the resulting sludge is dried and re-dissolved in sulfuric acid to form the desired chromium sulfate. In order to limit the negative effects of recovered chromium on tanning and leather's quality a 10:1 ratio of fresh to recovered Cr(III) is commonly applied [41].

The further improvement of precipitation, regarding efficiency and chemicals' consumption, has been studied at laboratory/pilot-scale level, using natural materials, rich in tannins. Especially, when the addition of wattle extract was examined along with NaOH addition, there was a slight increase in the percentage of chromium precipitated, due to adsorption. However, the precipitate may contain a mixture of chromium and natural tannins (i.e., wet blue and white wet processes), thus potentially affecting the quality of the final product (leather) [42].

6.1.3 Electrochemistry

Another treatment method, tested in laboratory that may also lead to Cr(III) precipitation, was electrochemistry, and more specifically electro-oxidation [43] and electro-flotation [44]. The main advantage considered to be in these cases the organic matter limitation in the produced precipitate, although on the other hand, the oxidation of Cr(III) to the more toxic Cr(VI) form is the major drawback. As a result, an additional reducing agent (e.g., gaseous sulfur dioxide, sodium sulfite, sodium metabisulfite or ferrous sulfate) should be applied in order to obtain the desired reduced form of chromium (i.e., Cr(III)); thus, the later can be reused in the tanning bath.

Instead, when electro-precipitation [45] was examined, the oxidation of Cr(III) was prevented, by sacrificing the anode electrode. As a result, the anode was corroded, provoking the possible secondary release of metal constituents from the anode into the aqueous phase, e.g., Cu^{2+}, Al^{3+}. On the cathode, the hydrolysis of water took place, producing hydroxyl ions and gaseous hydrogen. Due to hydroxyl groups, the pH of solution was increased and the Cr(III) was precipitated, along with the other metals from the anode. The consequence was the contamination of precipitate with metals that were not present in the initial wastewater, but the resulting aqueous phase was found to contain residual chromium concentrations below the corresponding threshold to discharge in an water body, namely less than 0.5 mg/L [46].

Despite the fact that by the application of electro-precipitation almost the entire Cr(III) may be recovered (>99%), the method has certain limitations/disadvantages. In particular, the extremely high concentrations of chromium in the tanning wastewater may cause the deposition of this metal onto the surface of electrodes, as the pH value increases. As the regular maintenance/cleaning and/or replacement of the electrodes are not recommended, it is necessary to dilute appropriately the raw wastewater in order to be operational this process [47]. Furthermore, the electrochemical treatment of wastewater with high organic matter content, such as the tanning liquor, may cause foaming and thus, an anti-foaming reagent needs to be added [48]. The resulting chrome-rich precipitate is marginally accepted for reuse in the tanning process, since a decrease in the quality of produced leather was observed; therefore, its mixture with fresh Cr(III) salt solution is preferred, as aforementioned [49].

6.1.4 Adsorption

The application of adsorption was also tested for the recovery of Cr(III) from tannery wastewaters by applying several sorbents, including biomaterials. However, it is commonly accepted that the adsorption process is suited better to water treatment in order to make it drinkable, rather than wastewater treatment. Due to the high concentrations of pollutants, the selectivity was reduced and in addition low removal capacities of Cr(III) were observed (e.g., 12.7 mg Cr(III)/g) [50]. Notwithstanding notable capacities of sorbents were obtained, when modified agricultural biomass, such as corn cob [51], or chitosan whiskers [52] were examined (277 mg Cr(III)/g and 180 mg Cr(III)/g, respectively); however, due to the aforementioned limitations, these biomaterials were studied only by using synthetic wastewater. Moreover, an additional leaching stage is a common pre-requirement in order to re-extract/leach the metal from the adsorbent, during which the application of a strongly acidic solution is necessary, due to the low solubility of chromium. As a result, the produced leachate may also contain various substances that have been retained by the adsorbent and even part of the adsorbent itself, due to its partial/gradual decomposition.

Another approach regarding the implementation of adsorption for Cr(III) recovery is its combination with precipitation. By applying biomaterials rich in alkaline substances (e.g., $CaCO_3$), such as shrimp shell [53], or thermally treated bark [54], the acidic pH value of tannery wastewaters (pH ~ 4) was increased in the mild alkaline range (pH 7–8). As a result, over 99% of the Cr(III) may be recovered as precipitate (noting that the maximum capacity data, due to a mixed mechanism is not provided);

mentioning also that the reuse of produced sludge was found to reduce the quality of produced leather. In comparison, when the pH was adjusted in the value 5 by applying activated carbon generated from sugar industry waste, then the maximum obtained capacity was significantly lower, than before (i.e., 41 mg Cr(III)/g; [55].

The removal of Cr(VI), following the possible oxidation of Cr(III), by adsorption using either inorganic or biosorbents, has been also extensively examined in the literature (e.g., [56, 57], respectively).

6.1.5 Ion Exchange

Similar to adsorbents, cation-exchange resins are also tested for Cr(III) recovery from tannery wastewaters. These resins may reach their maximum capacity at strongly acidic pH values, i.e., lower than the wastewaters (optimum pH range 2–4) [58]. Besides that, the ion-exchange method presents certain limitations for the potential implementation in the tannery plant. Especially, the low selectivity is a common disadvantage, since the tannery effluent may contain various other cations, and not only Cr(III), with the most important being Ca. Dissolved calcium shows almost equal content in wastewater with chromium and hence, an antagonistic action with it, considering the adsorption active sites [59]. Regarding the efficiency of this method, over 90% of Cr(III) may be removed, but not for the high initial concentrations usually found in tannery effluents [60]. Therefore, the appropriate dilution of wastewater is considered as necessary [61]. On the other hand, resins may be easily regenerated by the addition of a binary solution, containing NaOH and H_2O_2; Cr(III) can be oxidized to Cr(VI) by the presence of H_2O_2, whereas NaOH maintains an alkaline pH in order to increase extraction's selectivity, taking advantage of its very high solubility [62]. Noting also (again) that chromium in order to be reused in tanneries should be subsequently reduced to the original (trivalent) form.

6.1.6 Membrane Filtration

Filtration is widely used in wastewater treatment operations and as a technology presents high efficiency for the removal of various pollutants, although it is applied mainly for water recovery and reuse [63]. However, in the case of tannery wastewaters, due to high pollution load and in order to recover chromium, it is necessary to apply two or more filtration methods, or technologies. In the case of filtration methods application, the first step is ultra-filtration (UF) for the removal of organic matter, followed by nano-filtration (NF), where the chromium-rich wastewater is concentrated, leading to the formation of Cr(III) precipitates. The solid phase is re-dissolving, e.g., by adding an acidic media, but the resulting solution will contain rather lower content of Cr(III) (i.e., 9.2%, expressed as Cr_2O_3) [64]. Instead, the application of reverse osmosis may present higher efficiency/concentration of Cr(III) retention [65], and it may be applied in various (different) tannery effluents streams [66]. However, the observed low selectivity leads to a solution that contains the majority of initial pollutants and also results to severe fouling of membrane's surface, mainly from the presence of dissolved organic matter; these are considered as the major drawbacks, regarding the implementation of membranes for the treatment of these wastewaters [67]. The separation of Cr(VI), formed during Cr(III) oxidation, by the application of membranes has been also extensively examined in the literature (e.g., [68].

6.1.7 Bioaccumulation

The usage of plants and/or microorganisms in order to remove pollutants from wastewaters is a method that generally promotes the green chemistry principles. Especially, the Cr(III) removal from a tannery effluent was studied, by using both river [69], or sea weed [70], which presented a notable adsorption capacity, i.e., in both cases 88 mg Cr(III)/g. The mentioned efficiency is the maximum theoretical value, as calculated by the typical adsorption models (usually the Langmuir one), noting however that it may be achieved under specific experimental conditions. However, the yield is inversely proportional to the initial chromium concentration, and this combined with the usually low kinetic rate of the adsorption process is considered as the major disadvantages, regarding the application of this treatment technique. Furthermore, there is not provided yet a viable process for the subsequent recovery/separation of desired chromium from the plant/sorbent to be reused. In other case studies, Cr(III) was recovered from the tannery wastewater by precipitation and then, microorganisms can be used to remove the residual Cr(VI), e.g., by applying a binary chemical–biological method [71]. However, it should be clarified that the main form of chromium is the trivalent one, and therefore, this technology is referred only for specific cases, where the tanning process is problematic, resulting partially in the formation of hexavalent chromium.

6.2   From Tannery Sludge via the Application of Different Technologies

The dominant technology used to remove/recover (toxic) metals from sludge (as well as from soil) is the application of common hydrometallurgical methods, aiming to the successful leaching of them by the addition of an appropriate chemical reagent, while reducing simultaneously the original waste volume (Fig. 6). Therefore, the application of a suitable chemical reagent is considered as the major determining factor, influencing the effectiveness of this method. Solubility, cost, selectivity, regeneration (for reuse), etc., are considered usually as the main selection criteria for this reagent. Leaching/chemical extraction is followed by the application of a convenient solid–liquid separation, e.g., sedimentation, and filtration or centrifugation, whereas the applicable method depends mainly from the size of initial solid granules. The

Fig. 6 Flowchart of the
main hydrometallurgical
treatment stages for
Cr-containing sludge, aiming
to separate/recover Cr(III)

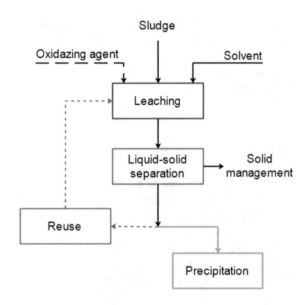

sludge residue, free from metals, may be subsequently safely applied even to agricultural land, or disposed of in landfills, or to proceed for further treatment, depending on the corresponding residuals concentration limits of control parameters.

In the leachate solution, containing the desired metal/chromium, the recovery stage of Cr(III) can be subsequently applied as a second extraction cycle, when possible. The separation at this stage can be usually achieved by the chemical precipitation of the metal (e.g., commonly by pH increase, adding an alkaline reagent). The selection criteria of the ideal alkaline reagent in this case, in addition to the cost, should be also selectivity, reusability and easily/rapid sedimentation of produced insoluble precipitate. As it is very difficult to meet all these criteria, mainly due to the presence of different impurities in the final product, this process is often preceded in order to increase purity with respect to the desired metal, e.g., by adsorption on activated carbon, or by using ion-exchange resins [72].

Other relevant methods suggested by the literature, regarding the extraction of Cr(III) from the tannery sludge, are the following: bioextraction (i.e., application of microorganisms, e.g., *thiobacillus ferrooxidans*), electro-kinetic extraction and supercritical fluid extraction. However, their yields may vary widely, as they are depended upon various parameters, such as sludge pre-treatment and composition, metal concentration and its major speciation forms, and also the specific extraction conditions.

6.2.1 Hydrometallurgical Treatment

Chemical extraction (or as termed in the case of solid/sludge-liquid contact: leaching) is the initial and most important stage in the hydrometallurgical treatment/recovery process. An appropriate solution, containing at least one chemical reagent, is applied, aiming to transfer the desired metal/substance from the solid into the aqueous phase. Therefore, the choice of the appropriate leaching media is the major determining factor, regarding the effectiveness of this method. Various chemical agents have been evaluated for this scope, such as inorganic acids (H_2SO_4, HCl, HNO_3), organic acids (citric and oxalic acid), chelating solvents (EDTA, NTA) and inorganic chemicals ($FeCl_2$) [73].

This method offers high leaching efficiencies for Cr(III), but also for the most other metals possibly contained in the raw tannery sludge. In addition to the low selectivity, which is a common feature of all the aforementioned methods, it requires the use of high acid concentration (i.e., pH value < 3), appropriate corrosion-resistant equipment and the management of hazardous solutions. In particular, H_2SO_4 yields the highest Cr(III) recovery rate (>90%), due to the formation of more soluble $CrSO_4^+$ species, as compared with other extraction reagents [29]. Regarding selectivity, only a portion of calcium content is subjected to leaching, since it is precipitated as $CaSO_4$, which is very insoluble, even in strongly acidic conditions [23].

The final stage in order to obtain a chromium-rich solid, free from other impurities and therefore, capable to be reused in the tanning process, is (selective) precipitation. The pH value of the leachate is increased to the range 8–9 by the addition of an alkaline media (e.g., NaOH, $Ca(OH)_2$, MgO) and Cr(III) precipitates by forming the respective hydroxylated species [74]. Among the examined alkaline reagents, NaOH showed the best performance, regarding the content of chromium in the precipitate (~ 60% w/w), since $Ca(OH)_2$ or MgO may generate larger amounts of solids, containing also other co-existing components/impurities, such as Ca and Mg [75].

The leaching media may be also combined with an oxidizing agent, mainly H_2O_2; hence, forming the highly soluble hexavalent chromium, if desirable [76]. In this case, the aim is to increase further the selectivity, because when increasing the pH value, the anionic hexavalent chromium remains soluble, unlike the other possibly co-existing metal speciation/forms. Then, a common reducing agent (e.g., Na_2SO_3 or SO_2) can be applied in order to obtain the trivalent chromium [3]. The additional chemical reagents (besides H_2SO_4) and the low recovery efficiency act as limitations in this case, regarding the application of this method in larger scale. Alternatively, the literature suggests that chromium may be oxidized directly by the preliminary thermal treatment (incineration) of tannery sludge, taking advantage of the organics content, reducing the initial volume and hence, increasing the Cr content in the produced ash. In this case, prior to the chemical extraction step, the sludge is usually combusted under controlled conditions (air supply, temperature, time), resulting in the oxidation of Cr(III) to the hexavalent form.

The respective reaction's mechanism (presented in the following Sect. 6.2.2) depends mainly on the presence of atmospheric oxygen, as well as of CaO, which may be already exists in the raw sludge, due to the liming process and to thermal treatment. It can be also obtained from other calcium sources, in case the calcium content is insufficient; then, CaO or other reagents (e.g., Na_2O_2), which may lead to Cr(III) oxidation, will be preliminary mixed with the tannery sludge [12]. Due to the soluble Cr(VI) form, high extraction efficiencies may be achieved even with the application of mild acidic conditions. Moreover, during the thermal treatment the organic matter of sludge is combusted and as a result its volume is substantially reduced (~50%). On the other hand, from the remaining ash the alkaline ions can be also easily leached; therefore, a portion of the used extraction (acidic) media will be consumed in order to neutralize these [29]. The Cr(VI)-rich leachate may be reused in the tannery procedures after the appropriate reduction of chromium, or it may be fed as raw material in other facilities, such as in the electroplating industry.

6.2.2 Thermal Treatment

The oxidation of Cr(III) to Cr(VI) through the application of thermal treatment is favored only under aerobic conditions, since in the case of pyrolysis (i.e., by applying almost anaerobic conditions), chromium remains in its initial (trivalent) form [77], i.e.,

$$2Cr_2O_3 + 3O_2 \rightarrow 4CrO_3 \tag{4}$$

The reaction of Cr(III) with atmospheric oxygen [reaction 4] is spontaneous, and supplementary heating is not a requirement. However, it is governed by kinetic limitations, and consequently, an increase in temperature can cause an increase in both speed and efficiency of this reaction. In addition, due to the reverse process, namely the reduction of Cr(VI) to Cr(III), because of the presence of other sludge components, such as the organic matter acting as reductant, divalent iron and sulfide ions, may also co-existing [78].

In the tannery sludge, the dominant Cr(III) form is present with the form of hydroxylated species (e.g., $Cr(OH)_3$), as a result of precipitation process during the wastewater treatment. On the other hand, in the temperature range 150–250 °C the dehydroxylation mechanism can take place, and therefore, the corresponding oxides are formed (i.e., Cr_2O_3), which also favor the previous reaction (4). As a result, even in the early stages of thermal treatment (i.e., at 300 °C), notable portions of Cr(III) may be oxidized to the hexavalent form [79].

However, other metal oxides contained in the tannery sludge, such as CaO, FeO, MgO, K_2O and Na_2O, may also contribute to the oxidation of Cr(III) [80, 81]. The oxide with the highest efficiency, being also the most commonly used, is CaO [reaction 6], as calcium is usually the component with the highest content in the tannery sludge, since it is used during the tanning process (e.g., pH regulation or liming procedure), as well as during wastewater treatment (e.g., liming and coagulation). There is also published information that the mixing of this oxide (or other calcium reagents), with sludge before the application of thermal treatment, e.g., in the case when the calcium content is not considered as sufficient, may be able to increase the efficiency of chromium oxidation [82]. In that case, the Cr(VI) formation is initiated at 300 °C, but with limited efficiency, as compared to higher temperature values [83].

$$CaCO_3 \rightarrow CaO + CO_2 \tag{5}$$

$$Cr_2O_3 + 2CaO + 3/2O_2 \rightarrow 2CaCrO_4 \tag{6}$$

However, the major calcium species observed in the tannery sludge are $Ca(OH)_2$, $CaSO_4$ and $CaCO_3$ [84]. CaO will be involved in the oxidation reaction of Cr(III) during the early stages of thermal treatment, when the hydroxylated calcium species are existing in the initial sludge, namely $Ca(OH)_2$. Similar to Cr(III), in the range of 150–250 °C temperature, the hydroxyl groups are removed and CaO is formed. Nevertheless, the main form of Ca is $CaCO_3$ and thus, an additional step of the oxidation mechanism is essential to precede, i.e., as that described by reaction (5). The decomposition of $CaCO_3$ requires higher temperature range 550–750 °C; hence, the oxidation reaction takes mainly place at temperatures above 600 °C and in the presence of oxygen [85].

Noting also that an increase of temperature above 700 °C can result in the decrease of Cr(VI) content, even if an increase in kinetic rate and efficiency of oxidation is expected. This is attributed to the chromate(VI) salts decomposition and also to the simultaneous (partial) reduction of Cr(VI) to its trivalent form (Vogel et al. 2004). The reduction is due to the presence of MgO, since this is used during the tanning process (e.g., for pickling), forming the most stable $MgCr_2O_4$, compound, according to the following reaction (7) [83].

$$4CaCrO_4 + 2MgO \rightarrow 2MgCr_2O_4 + 4CaO + 3O_2 \tag{7}$$

6.2.3 Other Extraction Methods

In addition to chemical extraction, other relevant methods have been also examined in laboratory scale, as their application in full-scale tannery plants may present several limitations. For instance, the bioextraction by the presence of iron- and sulfur-oxidizing bacteria that they can be naturally present in the tannery sludge, does not require the addition of an acidic media (as the pH value will be reduced due to bacterial activity), resulting also to quite high chromium removal efficiencies, i.e., ~90% [86]. However, the specific requirement for high moisture in the sludge (regarding samples taken from the wastewater treatment aeration tank), the low selectivity and the long processing periods are some of the major limitations in this technique [87]. Furthermore, certain experimental conditions should be carefully adjusted to make the bioextraction more efficient, regarding mainly the bacterial growth, such as nutrient content, aeration, temperature and liquid/solid ratio.

On the other hand, the electro-kinetic extraction is based on the electrically induced transfer of the metal from the solid to liquid phase. As liquid phase, several organic reagents were tested, e.g., oxalic, citric and lactic acids, as well as biosurfactants, e.g., saponin [88]. The efficiencies are much lower than in the previous study (i.e., <80%) and when combined with the relatively higher energy requirements, this technique seems to be unattractive for a full-scale application.

The supercritical fluid extraction (SFE) method appears to provide good chromium removal performance (>98%), although it is also at the very early stage of practical application. Laboratory studies are performed in batch reactor and have shown certain limitations, due to the complexity of the process and the rather high costs. Specific demands, such as oxygen supply, higher temperature and pressure, are some of the requirements, limiting the wider applicability of this interesting technique [89].

7 Conclusions

The wastes generated by chrome tanning of the skins/hides have proven to be hazardous to the environment and to humans also. On the other hand, alternative tanning methods are not capable of replacing it. This results in a constant research for sustainable technologies for the management of these wastes. Principles of circular economy recovery–reuse are considered to be, by researchers and international organizations, as the best strategy to minimize the environmental impact of tanneries. Since tannery wastes are Cr(III)-rich, its recovery is economically profitable and provides a sustainable management. The most efficient methods for chromium recovery are from tanning bath liquor by precipitation and from tannery sludge, resulting by the corresponding wastewater treatment plant, by hydrometallurgy. In both cases, the recovered chromium may be reused in the tanning process, as it does not affect the quality of the final product (leather), and their residues can be discharged of in a sewage treatment plant or directly to a recipient.

Acknowledgements This work was supported by the project "INVALOR: Research Infrastructure for Waste Valorization and Sustainable Management" (MIS 5002495) which is implemented under the Action "Reinforcement of the Research and Innovation Infrastructure," funded by the Operational Program "Competitiveness, Entrepreneurship and Innovation" (NSRF 2014–2020) and co-financed by Greece and the European Union (European Regional Development Fund).

References

1. Zouboulis AI, Samaras P, Krestou A, Tzoupanos ND (2012) Leather production modification methods towards minimization of tanning pollution: green tanning. Fresenius Environ Bull 21:2406–2412
2. Zouboulis AI, Peleka EN (2019) Cycle closure. In: Waste management: tools, procedures and examples. Glob Nest J 21:1–6. https://doi.org/10.30955/gnj.002679
3. Raguraman R, Sailo L (2017) Efficient chromium recovery from tannery sludge for sustainable management. Int J Environ Sci Technol 14:1473–1480. https://doi.org/10.1007/s13762-017-1244-z
4. Environmental Impact Assessment (EIA), 2010. Technical EIA guidance manual for leather/skin/hide processing industry. By IL&FS Ecosmart Limited Hyderabad for The Ministry of Environment and Forest, Government of India

5. European Union (2013/84/EU) Commission Implementing Decision of 11 February 2013 establishing the best available techniques (BAT) conclusions under Directive 2010/75/EU of the European Parliament and of the Council on industrial emissions for the tanning of hides and skins
6. IUE (International Union of Environment Commission) 6 (2018) Typical pollution values related to conventional tannery processes. In: Technical guidelines for environmental protection aspects for the world leather industry. http://www.iultcs.org/environment-iue.php
7. European Commission (2013) Best available techniques (BAT) reference document (BREF) on tanning of hides and skins
8. UNEP (2015) Agenda item 7: draft guide towards a more sustainable tannery sector in the Mediterranean, United Nations Environment Programme Mediterranean Action Plan, Athens, UNEP(DEPI)/MED WG.417/11
9. Dixit S, Yadav A, Dwivedi PD, Das M (2015) Toxic hazards of leather industry and technologies to combat threat: a review. J Clean Prod 87:39–49. https://doi.org/10.1016/j.jclepro.2014.10.017
10. Lofrano G, Meric S, Zengin GE, Orhon D (2013) Chemical and biological treatment technologies for leather tannery chemicals and wastewaters: a review. Sci Total Environ 461–462:265–281. https://doi.org/10.1016/j.scitotenv.2013.05.004
11. Islam BI, Musa AE, Ibrahim EH, Sharafa SAA, Elfaki BM (2014) Evaluation and characterization of tannery wastewater. J For Prod Ind 3:141–150
12. Erdem M (2006) Chromium recovery from chrome shaving generated in tanning process. J Hazard Mater B129:143–146. https://doi.org/10.1016/j.jhazmat.2005.08.021
13. UNIDO (2016) Pollutants in tannery effluent: sources, description, environmental impact. United Nations Industrial Development Organization, Vienna
14. Kocurek P, Kolomazník K, Bařinová M, Hendrych J (2017) Total control of chromium in tanneries – thermal decomposition of filtration cake from enzymatic hydrolysis of chrome shavings. Waste Manage Res 35(4):444–449. https://doi.org/10.1177/0734242X16680728
15. UNIDO (2011) Introduction to treatment of tannery effluents. What every tanner should know about effluent treatment. United Nations Industrial Development Organization, Vienna
16. Doble M, Kumar A (2005) Tannery effluent (Chap. 12). In: Doble M, Kumar A (eds) Biotreatment of industrial effluents. Butterworth-Heinemann, pp 133–143. ISBN 9780750678384. https://doi.org/10.1016/b978-075067838-4/50013-0
17. Daniil A, Dimitrakopulos GP, Varitis S, Vourlias G, Kaimakamis G, Pantazopoulou E, Pavlidou E, Zouboulis AI, Karakostas T, Komninou P (2018) Stabilization of Cr-rich tannery waste in fly ash matrices. Waste Manag Res 36:818–826. https://doi.org/10.1177/0734242x18775488
18. Pantazopoulou E, Zouboulis A (2018) Chemical toxicity and ecotoxicity evaluation of tannery sludge stabilized with ladle furnace slag. J Environ Manage 216:257–262. https://doi.org/10.1016/j.jenvman.2017.03.077
19. Varitis S, Kavouras P, Pavlidou E, Pantazopoulou E, Vourlias G, Chrissafis K, Zouboulis AI, Karakostas T, Komninou P (2017) Vitrification of incinerated tannery sludge in silicate matrices for chromium stabilization. Waste Manag 59:237–246. https://doi.org/10.1016/j.wasman.2016.10.011
20. Thanikaivelan P, Rao JR, Nair BU, Ramasami T (2005) Recent trends in leather making: processes, problems, and pathways. Crit Rev Environ Sci Technol 35:37–79. https://doi.org/10.1080/10643380590521436
21. Alibardi L, Cossu R (2016) Pre-treatment of tannery sludge for sustainable landfilling. Waste Manag 52:202–211. https://doi.org/10.1016/j.wasman.2016.04.008
22. Abbas N, Jamil N, Hussain N (2016) Assessment of key parameters in tannery sludge management: a prerequisite for energy recovery. Energy Sources Part A Recover Util Environ Eff 38:2656–2663. https://doi.org/10.1080/15567036.2015.1117544
23. Shen SB, Tyagi RD, Blais JF (2001) Extraction of Cr(III) and other metals from tannery sludge by mineral acids. Environ Technol (United Kingdom) 22:1007–1014. https://doi.org/10.1080/09593332208618216

24. Abreu MA, Toffoli SM (2009) Characterization of a chromium-rich tannery waste and its potential use in ceramics. Ceram Int 35:2225–2234. https://doi.org/10.1016/j.ceramint.2008. 12.011
25. Reimann J, McWhirter JE, Cimino A, Papadopoulos A, Dewey C (2019) Impact of legislation on youth indoor tanning behaviour: a systematic review. Prev Med (Baltim) 123:299–307. https://doi.org/10.1016/j.ypmed.2019.03.041
26. Saxena G, Chandra R, Bharagava RN (2017) Environmental pollution, toxicity profile and treatment approaches for tannery wastewater and its chemical pollutants. In: Reviews of environmental contamination and toxicology. Springer, Uttar Pradesh. https://doi.org/10.1007/398_2015_5009
27. ECHA (2010) Candidate list of substances of very high concern for authorization. European Chemical Agency, Helsinki
28. Covington AD (1997) Modern tanning chemistry. Chem Soc Rev 26:111–126. https://doi.org/10.1039/cs9972600111
29. Kokkinos E, Proskynitopoulou V, Zouboulis A (2019) Chromium and energy recovery from tannery wastewater treatment waste: Investigation of major mechanisms in the framework of circular economy. J Environ Chem Eng 7:103307. https://doi.org/10.1016/j.jece.2019.103307
30. Covington AD (2009) Tanning chemistry: the science of leather. Published by The Royal Society of Chemistry, Cambridge. ISBN 978-0-85404-170-1
31. Imer S, Varnali T (2000) Modeling chromium sulfate complexes in relation to chromium tannage in leather technology: a computational study. Appl Organomet Chem 14:660–669. https://doi.org/10.1002/1099-0739(200010)14:10%3c660:aid-aoc55%3e3.3.co;2-g
32. Wu B, Mu C, And GZ, Lin W (2009) Effects of Cr^{3+} on the structure of collagen fiber. Langmuir 25:11905–11910. https://doi.org/10.1021/la901577j
33. Wystalska K, Sobik-Szołtysek J (2019) Sludge from tannery industries (Chap. 2). In: Vara Prasad MN, de Campos Favas PJ, Vithanage M, Venkata Mohan S (eds) Industrial and municipal sludge. Butterworth-Heinemann, pp 31–46. https://doi.org/10.1016/b978-0-12-815907-1.00002-7
34. Haque MA, Chowdhury RA, Chowdhury WA, Baralaskar AH, Bhowmik S, Islam S (2019) Immobilization possibility of tannery wastewater contaminants in the tiles fixing mortars for eco-friendly land disposal. J Environ Manage 242:298–308. https://doi.org/10.1016/j.jenvman.2019.04.069
35. Langlois CL, James BR (2015) Chromium oxidation-reduction chemistry at soil horizon interfaces defined by iron and manganese oxides. Soil Sci Soc Am J 79:1329–1339. https://doi.org/10.2136/sssaj2014.12.0476
36. Oruko RO, Selvarajan R, Ogola HJO, Edokpayi JN, Odiyo JO (2020) Contemporary and future direction of chromium tanning and management in sub Saharan Africa tanneries. Process Saf Environ Prot 133:369–386. https://doi.org/10.1016/j.psep.2019.11.013
37. Buljan J, Kral I (2019) The framework for sustainable leather manufacture, 2nd edn. United Nations Industrial Development Organization (UNIDO)
38. Zhang C, Xia F, Long J, Peng B (2017) An integrated technology to minimize the pollution of chromium in wet-end process of leather manufacture. J Clean Prod 154:276–283. https://doi.org/10.1016/j.jclepro.2017.03.216
39. Minas F, Chandravanshi BS, Leta S (2017) Chemical precipitation method for chromium removal and its recovery from tannery wastewater in Ethiopia. Chem Int 3:291–305
40. Guo ZR, Zhang G, Fang J, Dou X (2006) Enhanced chromium recovery from tanning wastewater. J Clean Prod 14:75–79. https://doi.org/10.1016/j.jclepro.2005.01.005
41. Khan K, Khan MdIH, Khan II, Al Mahmud A, Hossain MdD (2018) Recovery and reuse of chromium from spent chrome tanning liquor by precipitation process. Am J Eng Res (AJER) 7:346–352
42. Kanagaraj J, Chandra Babu NK, Mandal AB (2008) Recovery and reuse of chromium from chrome tanning waste water aiming towards zero discharge of pollution. J Clean Prod 16:1807–1813. https://doi.org/10.1016/j.jclepro.2007.12.005

43. Selvakumar AM, Vimudha M, Lawrance I, Sundaramoorthy S, Ramanaiah B, Saravanan P (2019) Recovery and reuse of spent chrome tanning effluent from tannery using electro-oxidation technique. Desalin Water Treat 156:323–330. https://doi.org/10.5004/dwt.2019. 23825
44. Selvaraj R, Santhanam M, Selvamani V, Sundaramoorthy S, Sundaram M (2018) A membrane electroflotation process for recovery of recyclable chromium(III) from tannery spent liquor effluent. J Hazard Mater 346:133–139. https://doi.org/10.1016/j.jhazmat.2017.11.052
45. Ramírez-Estrada A, Mena-Cervantes VY, Fuentes-García J, Vazquez-Arenas J, Palma-Goyes R, Flores-Vela AI, Vazquez-Medina R, Altamirano RH (2018) Cr(III) removal from synthetic and real tanning effluents using an electro-precipitation method. J Environ Chem Eng 6:1219–1225. https://doi.org/10.1016/j.jece.2018.01.038
46. European Commission (2018) Implementing Decision (EU) 2018/1147 of 10 August 2018 establishing best available techniques (BAT) conclusions for waste treatment, under Directive 2010/75/EU of the European Parliament and of the Council of 24 November 2010 on industrial emissions was published in the Official Journal on 17 Aug 2018
47. Sirajuddin KL, Lutfullah G, Bhanger MI, Shah A, Niaz A (2007) Electrolytic recovery of chromium salts from tannery wastewater. J Hazard Mater 148:560–565. https://doi.org/10. 1016/j.jhazmat.2007.03.011
48. Da Silva GS, Dos Santos FA, Roth G, Frankenberg CLC (2020) Electroplating for chromium removal from tannery wastewater. Int J Environ Sci Technol 17:607–614. https://doi.org/10. 1007/s13762-019-02494-1
49. Mella B, Glanert AC, Gutterres M (2015) Removal of chromium from tanning wastewater and its reuse. Process Saf Environ Prot 95:195–201. https://doi.org/10.1016/j.psep.2015.03.007
50. Vilardi G, Ochando-Pulido JM, Stoller M, Verdone N, Di Palma L (2018) Fenton oxidation and chromium recovery from tannery wastewater by means of iron-based coated biomass as heterogeneous catalyst in fixed-bed columns. Chem Eng J 351:1–11. https://doi.org/10.1016/ j.cej.2018.06.095
51. Manzoor Q, Sajid A, Hussain T, Iqbal M, Abbas M, Nisar J (2019) Efficiency of immobilized Zea mays biomass for the adsorption of chromium from simulated media and tannery wastewater. J Mater Res Technol 8:75–86. https://doi.org/10.1016/j.jmrt.2017.05.016
52. Eladlani N, Dahmane EM, Ouahrouch A, Rhazi M, Taourirte M (2018) Recovery of chromium(III) from tannery wastewater by nanoparticles and whiskers of chitosan. J Polym Environ 26:152–157. https://doi.org/10.1007/s10924-016-0926-9
53. Fabbricino M, Naviglio B, Tortora G, d'Antonio L (2013) An environmental friendly cycle for Cr(III) removal and recovery from tannery wastewater. J Environ Manage 117:1–6. https://doi. org/10.1016/j.jenvman.2012.12.012
54. Hashem MA, Momen MA, Hasan M, Nur-A-Tomal MS, Sheikh MHR (2019) Chromium removal from tannery wastewater using Syzygium cumini bark adsorbent. Int J Environ Sci Technol 16:1395–1404. https://doi.org/10.1007/s13762-018-1714-y
55. Fahim NF, Barsoum BN, Eid AE, Khalil MS (2006) Removal of chromium(III) from tannery wastewater using activated carbon from sugar industrial waste. J Hazard Mater 136:303–309. https://doi.org/10.1016/j.jhazmat.2005.12.014
56. Loukidou MX, Zouboulis AI, Karapantsios ThD, Matis KA (2004) Equilibrium and kinetic modeling of chromium(VI) biosorption by Aeromonas caviae. Colloids Surf Part A: Phys-chem Eng Asp 242:93–104. https://doi.org/10.1016/j.colsurfa.2004.03.030
57. Zouboulis AI, Kydros KA, Matis KA (1995) Removal of hexavalent chromium ions using pyrite fines. Water Res 29(7):1755–1760. https://doi.org/10.1016/0043-1354(94)00319-3
58. Ahmad T, Mustafa S, Naeem A, Anwar F, Mehmood T, Shah KH (2014) Ion exchange removal of chromium (III) from tannery wastes by using a strong acid cation exchange resin amberlite IR-120 H+ and its hybrids. J Chem Soc Pakistan 36:818–828
59. Alanne AL, Tuikka M, Tõnsuaadu K, Ylisirniö M, Hämäläinen L, Turhanen P, Vepsäläinen J, Peräniemi S (2013) A novel bisphosphonate-based solid phase method for effective removal of chromium(III) from aqueous solutions and tannery effluents. RSC Adv 3:14132–14138. https://doi.org/10.1039/c3ra41501e

60. Sahu SK, Meshram P, Pandey BD, Kumar V, Mankhand TR (2009) Removal of chromium(III) by cation exchange resin, Indion 790 for tannery waste treatment. Hydrometallurgy 99:170–174. https://doi.org/10.1016/j.hydromet.2009.08.002

61. Meshram P, Sahu SK, Pandey BD, Kumar V, Mankhand TR (2012) Removal of Chromium(III) from the waste solution of an Indian Tannery by Amberlite IR 120 Resin. Int J Nonferr Metall 1:32–41. https://doi.org/10.4236/ijnm.2012.13005

62. Cetin G, Kocaoba S, Akcin G (2013) Removal and recovery of chromium from solutions simulating tannery wastewater by strong acid cation exchanger. J Chem. https://doi.org/10.1155/2013/158167

63. Zouboulis AI, Peleka EN, Ntolia A (2019) Treatment of tannery wastewater with vibratory shear-enhanced processing membrane filtration. Separations 6:20. https://doi.org/10.3390/separations6020020

64. Cassano A, Molinari R, Romano M, Drioli E (2001) Treatment of aqueous effluents of the leather industry by membrane processes: a review. J Membr Sci 181:111–126. https://doi.org/10.1016/s0376-7388(00)00399-9

65. Hintermeyer BH, Lacour NA, Perez Padilla A, Tavani EL (2008) Separation of the chromium(III) present in a tanning wastewater by means of precipitation, reverse osmosis and adsorption. Lat Am Appl Res 38:63–71

66. Ranganathan K, Kabadgi SD (2011) Studies on feasibility of reverse osmosis (membrane) technology for treatment of tannery wastewater. J Environ Prot 2:37–46. https://doi.org/10.4236/jep.2011.21004

67. Mohammed K, Sahu O (2015) Bioadsorption and membrane technology for reduction and recovery of chromium from tannery industry wastewater. Environ Technol Innov 4:150–158. https://doi.org/10.1016/j.eti.2015.06.003

68. Lazaridis NK, Jekel M, Zouboulis AI (2003) Removal of Cr(VI), Mo(VI) and V(V) ions from single metal solutions by (i) sorption, or (ii) nano-filtration. Sep Sci Technol 38(10):2201–2219. https://doi.org/10.1081/ss-120021620

69. Rahman MA, Rajeeb S, Alam AS (2015) Removal of Chromium(III) from tannery wastewater by bioaccumulation method using *Vallisneria* sp. River-weed. Dhaka Univ J Sci 63:91–96. https://doi.org/10.3329/dujs.v63i2.24442

70. Aravindhan R, Madhan B, Rao JR, Nair BU, Ramasami T (2004) Bioaccumulation of chromium from tannery wastewater: an approach for chrome recovery and reuse. Environ Sci Technol 38:300–306. https://doi.org/10.1021/es034427s

71. Ahmed E, Abdulla HM, Mohamed AH, El-Bassuony AD (2016) Remediation and recycling of chromium from tannery wastewater using combined chemical–biological treatment system. Process Saf Environ Prot 104:1–10. https://doi.org/10.1016/j.psep.2016.08.004

72. Habashi F (1993) A textbook of hydrometallurgy. Metallurgie Extractive Quebec, Enr., Quebec City

73. Babel S, del Mundo Dacera D (2006) Heavy metal removal from contaminated sludge for land application: a review. Waste Manag 26:988–1004. https://doi.org/10.1016/j.wasman.2005.09.017

74. Kokkinos E, Zouboulis A (2020) Hydrometallurgical recovery of Cr(III) from tannery waste: optimization and selectivity investigation. Water 12:719. https://doi.org/10.3390/w12030719

75. Pantazopoulou E, Zouboulis A (2020) Chromium recovery from tannery sludge and its ash, based on hydrometallurgical methods. Waste Manag Res 38:19–26. https://doi.org/10.1177/0734242x19866903

76. Kilic E, Font J, Puig R, Colak S, Celik D (2011) Chromium recovery from tannery sludge with saponin and oxidative remediation. J Hazard Mater 185:456–462. https://doi.org/10.1016/j.jhazmat.2010.09.054

77. Kavouras P, Pantazopoulou E, Varitis S, Vourlias G, Chrissafis K, Dimitrakopulos GP, Mitrakas M, Zouboulis AI, Karakostas T, Xenidis A (2015) Incineration of tannery sludge under oxic and anoxic conditions: study of chromium speciation. J Hazard Mater 283:672–679. https://doi.org/10.1016/j.jhazmat.2014.09.066

78. Apte AD, Tare V, Bose P (2006) Extent of oxidation of Cr(III) to Cr(VI) under various conditions pertaining to natural environment. J Hazard Mater 128:164–174. https://doi.org/10.1016/j.jha zmat.2005.07.057
79. Yang Y, Ma H, Chen X, Zhu C, Li X (2020) Effect of incineration temperature on chromium speciation in real chromium-rich tannery sludge under air atmosphere. Environ Res 183:109159. https://doi.org/10.1016/j.envres.2020.109159
80. Stam AF, Meij R, Winkel HT, Van Eijk RJ, Huggins FE, Brem G (2011) Chromium speciation in coal and biomass co-combustion products. Environ Sci Technol 45:2450–2456. https://doi.org/10.1021/es103361g
81. Verbinnen B, Billen P, Van Coninckxloo M, Vandecasteele C (2013) Heating temperature dependence of Cr(III) oxidation in the presence of alkali and alkaline earth salts and subsequent Cr(VI) leaching behavior. Environ Sci Technol 47:5858–5863. https://doi.org/10.1021/es4001455
82. Kirk DW, Chan CCY, Marsh H (2002) Chromium behavior during thermal treatment of MSW fly ash. J Hazard Mater 90:39–49. https://doi.org/10.1016/s0304-3894(01)00328-4
83. Mao L, Gao B, Deng N, Zhai J, Zhao Y, Li Q, Cui H (2015) The role of temperature on Cr(VI) formation and reduction during heating of chromium-containing sludge in the presence of CaO. Chemosphere 138:197–204. https://doi.org/10.1016/j.chemosphere.2015.05.097
84. Tahiri S, Albizane A, Messaoudi A, Azzi M, Bennazha J, Younssi SA, Bouhria M (2007) Thermal behaviour of chrome shavings and of sludges recovered after digestion of tanned solid wastes with calcium hydroxide. Waste Manag 27:89–95. https://doi.org/10.1016/j.wasman.2005.12.012
85. Basegio T, Berutti F, Bernardes A, Bergmann CP (2002) Environmental and technical aspects of the utilisation of tannery sludge as a raw material for clay products. J Eur Ceram Soc 22:2251–2259. https://doi.org/10.1016/s0955-2219(02)00024-9
86. Zhou SP, Zhou LX, Wang SM, Fang D (2006) Removal of Cr from tannery sludge by bioleaching method. J. Environ. Sci. (China) 18:885–890. https://doi.org/10.1016/s1001-0742(06)60009-0
87. Zeng J, Gou M, Tang YQ, Li GY, Sun ZY, Kida K (2016) Effective bioleaching of chromium in tannery sludge with an enriched sulfur-oxidizing bacterial community. Bioresour Technol 218:859–866. https://doi.org/10.1016/j.biortech.2016.07.051
88. Prakash P, Chakraborty PK, Priya T, Mishra BK (2019) Performance evaluation of saponin over other organic acid and tap water for removal of chromium in tannery sludge by electrokinetic enhancement. Sep Sci Technol 54:173–182. https://doi.org/10.1080/01496395.2018.1467449
89. Zou D, Chi Y, Dong J, Fu C, Wang F, Ni M (2013) Supercritical water oxidation of tannery sludge: stabilization of chromium and destruction of organics. Chemosphere 93:1413–1418. https://doi.org/10.1016/j.chemosphere.2013.07.009

Conviviality in Leather and Fashion Entrepreneurial Communities: Emerging Results from an Exploratory Research

Simone Guercini and Silvia Ranfagni

Abstract Conviviality is an interdisciplinary concept and a key phenomenon in entrepreneurial communities. Entrepreneurial communities are social units that share values, experiences, emotions, rituals and traditions. They give rise to personal contact networks that are sets of formal or informal individual relationships. Conviviality means sharing, openness and participation; in this sense, it can be a tool to foster, animate and amalgamate a community. Thus, it can increase social relations that stably bind individuals and, thus, become a source of business relations. Drawing on an analysis of the literature and a case fashion/leather entrepreneurial communities, we propose to investigate how conviviality relates with social sustainability by affecting social networks and the relations between them and business networks.

Keywords Conviviality · Leather · Fashion · Entrepreneurial communities · Social networks · Business networks

1 Introduction

Sustainable development can be explored in both ecological and economic dimensions. More specifically, sustainability implies (a) living in a way that is environmentally sustainable over the long term and (b) living in a way that is economically sustainable while maintaining living standards over the long term [1]. These dimensions can be added the social one [2]. Sustainability is also (c) living in a way that is socially sustainable, now and in the future [1]. Viewing sustainable development as "development which meets the needs of the present without compromising the ability for future generations to meet their own needs" [3], p. 43, the related needs of this development can also be social, thereby including education, recreation/leisure,

S. Guercini · S. Ranfagni (✉)
Department of Economics and Management, University of Florence, Via della Pandette 9, 50127 Florence, Italy
e-mail: silvia.ranfagni@unifi.it

S. Guercini
e-mail: simone.guercini@unifi.it

© The Editor(s) (if applicable) and The Author(s), under exclusive license to Springer Nature Singapore Pte Ltd. 2020
S. S. Muthu (ed.), *Leather and Footwear Sustainability*, Textile Science and Clothing Technology, https://doi.org/10.1007/978-981-15-6296-9_7

social relationships and self-fulfilment [4]. The satisfaction of these needs improves the quality of an individual's life and generates conditions of well-being in society. An intriguing opinion is put forward by Putnam [5], who points out that a social sustainability implies social capital, that is, social networks that enable collective action through the involvement of social institutions. At least in part, this role can be played by conviviality in business contexts. Conviviality can fuel social relationships that can orient collective entrepreneurial actions. Conceived as an interdisciplinary concept, conviviality is an integrating component of entrepreneurial communities. These communities are social networks made up of formal and informal relationships among individuals. They are social spaces where it is possible to share rules, traditions, experiences and values. Within them, conviviality generates sharing, openness and participation and animates business communities. It can act as a basis for social sustainability. By means of social relationships, conviviality produces practices to follow in business and improves the conditions of a community social being. In fact, it gives access to resources that can positively impact on personal life and entrepreneurial life. Some of them are revealed to be strategic assets that facilitate entrepreneurial processes.

In this chapter, we explore how generated conviviality can be a driver of business networks through social networks. We adopt the concept of conviviality proposed by the philosopher Ivan Illich, whose perspective envelopes that of other convivial community scholars. According to Illich, conviviality leaves ample space and capacity for individual intentions. Thus, it creates moments of self-revelation and of self-identification. For him "productivity [of the capitalist sort] is based on having, while conviviality on being" [6], p. 42. In line with this view, we aim to investigate how convivial tools by influencing social relations can activate and facilitate business relations and, thus, the relational capital circulating in business networks. In the first part of the work, we introduce the concept of conviviality as it emerges from interdisciplinary studies and from the consumer and business community literature. Then, we describe the paradigms of social networks and business networks as they relate to conviviality. Finally, we show the results of our exploratory study of three fashion entrepreneurial communities; two located in Italy (Tuscany) and one in China (Hangzhou). The methodology is based on case analysis employing ethnographic interviews with community directors and a substantial nucleus of entrepreneur-members.

2   Literature Overview

Conviviality: an interdisciplinary concept. Conviviality is quite a new topic for managerial disciplines [7]. Indeed, up to now, the main contributions to the topic have come from the fields of philosophy, sociology and anthropology, in which some authors have investigated conviviality as a mediator of cultural and tourist offerings [8, 9]. For example, Lloyd [8] explores the positive impact of conviviality on public fund raising policies in eighteenth-century London. Maitland [9] convincingly argues that, for tourists, "getting to know the city [London] was a convivial experience—local people and local places to drink coffee or shop were important" (p. 21). Other authors consider the politics of conviviality as one form of the "politics of the popular" that arises in contexts of rapid change, diversity and mobility. In this regard, Williams and Stroud [10], exploring linguistic practices as powerful mediators of political voice and agency, view "linguistic citizenship" as the foundation of a politics of conviviality. In other studies, conviviality emerges as a feature of a new cultural food movement. Germov et al. [11] analyze the portrayal of the slow food movement in the Australian print media. Some major aspects of conviviality that have emerged stem from analyzing the social pleasure associated both with sharing good food, which can in turn be linked to localism (the social, health and environmental benefits of local producers), and with romanticism (an idyllic rural lifestyle as an antidote to the time poverty of urban life). Some authors have even proposed measuring conviviality. In particular, Caire et al. [12] set forth formal measures of conviviality for networks using a coalition game theoretic framework. Among the contributions to the discussion, which goes furthest in defining, the concept of conviviality is *Tools for Conviviality* by philosopher Illich [6]. Illich considers conviviality as a "free space" of collective interaction, one in which people can exercise their right to autonomous action and, more generally, individual freedom, without being controlled. In this way, conviviality can also foster individual creativity by contributing to the reduction in regulation, standardization, dependence, and the abuses of capitalist societies. Illich argues that conviviality is based on being and not on having. In other words, its tools leave ample space for revealing personal and real intentions. Everyone can use them, effortlessly, whenever and as much as they wish to share individually defined goals. Thus, conviviality becomes a conductor of meaning and a translator of intentionality between people and society. In this chapter, we propose to investigate the role of conviviality, as Illich [6] has defined the notion, in business communities. In Illich's perspective (consistently with that of Marx and Hegel), people seek self-realization through communities, that is, through horizontal spaces of relations where individuals are equally free and supportive. Conviviality creates these spaces as social environments where people reveal themselves, talk about themselves, share and identify each other. In the next section, we specify the context where conviviality materializes assuming different forms. This context will be that of communities.

Community and conviviality. The relation between conviviality and community is a close one. Conviviality is seen as participation, and sharing is a founding element of consumer communities and thus of an aggregation of individuals who are together as they have something in common [13]. Brand communities are an example. Muniz and Guinn [14] describe them as "a specialized, non-geographically bound community, based on a structured set of social relationships among users of a brand" (p. 412). These communities form a fabric of social relationships where participation is based on an awareness of kind, rituals and traditions and a moral responsibility among members [15]. Thus, conviviality tacitly animates consumer communities, which, to the extent that they are rooted in a brand, find in its values the engine of member affiliation and socialization [16]. Recently, technologies have generated online communities [17] and thus contexts of brand observation [13] involving actors who assume different roles (current or potential customers, enthusiasts, experts). They foster virtual and convivial spaces by spontaneously shaping opinions, knowledge and feelings about their brand experience [18]. Schau et al. [15] have identified tools to consolidate and reinforce social ties in online brand communities. They include social networking, impression management, community engagement and brand use. All these tools are managed directly by community members. They generate *convivium* moments that increase social interactions, community involvement and brand perceptions of tangible attributes. Brand communities can take on the form of tribes. In tribes "the use value [of a brand] (functions and symbols at the service of the individual as a means of distinction) is being sought as much as the linking value (link with the other or with others and means of tribal symbiosis)" [13], p. 311. Tribes are groups of people emotionally connected as they share the same social linking value of a brand and reflect in it their identity. Thus, the social linking value acts as a convivial tool within a community. Its strength is manifested in the resulting aggregation and internal cohesion. From scientific contributions, it emerges how in consumer communities, conviviality is essentially self-sustaining and self-generated.

Although conviviality seems to be inherent to consumer communities, it would be intriguing to broaden the perspective and to investigate conviviality in business communities, which, after all, constitute one of the archetypes, along with segmentation, fragmentation and self-selection that entrepreneurs employ in their market representation [19, 20]. The community, from an entrepreneurial perspective, denotes both consumer and business groups. It has an external, but also an internal nature by indicating spaces where entrepreneurs can develop formal or informal relationships [21] and initiate processes of information sharing and knowledge production. While relations in consumer communities recall what Tönnies [22] defines *Gemeinschaft*, relations in a business community are often close to what he identifies as *Gesellschaft*. The first is a community characterized by informal social relationships and shared values that connect people and hold them together. The second takes the form of collective organizations that weave amongst themselves a web of social relations that are less intimate, more impersonal and based on formal rules and regulations governing appropriate behaviours. In a *Gesellschaft*, conviviality and thus community participation find its assumptions in "rational will", "regularity" and "shared social values" [22]. A community and its social context are the result of human will and exist, therefore, only through the will of individuals to associate and to develop a sense of belonging. A group life also requires regularity seen as order, law and morality. Order is based upon convention; law emerges from legislation and custom; morality is an expression of socially rooted and shared human conscience, reason and ideals. The social values aggregating organizations can be economic, political, intellectual or spiritual. The more shared they are, the greater the understanding, harmony and friendship among individuals within a community. All these elements can be found in the typical business community, that is, the industrial district. This is an ideal-typical model of a local productive system where a nucleus of people coexists with a localized industry. More specifically, it is a socio-territorial entity marked by the active presence of both a community of individuals and a population of firms situated in one naturally and historically bounded area [23]. The will to create aggregation is the result of an historical and social stratification of the community, where people share a homogenous system of views and values. The preservation of this system is an indispensable requirement for the development and the temporal reproduction of the district. Moreover, a set of internal rules and of institutions (family, school, etc.) have a substantial role: they spread common values throughout the district and transmit them across generations by fostering a social and economic community life. The resulting district community is shaped by the industrial atmosphere. This latter acts as the tacit engine of a local integration and coincides with "a set of shared cognitive, moral and behavioural attitudes drawing on locally-dense cultural interactions and which orientate technical, human and relational investments towards forms consistent with local accumulation" [24], p. 8.

Conviviality and, thus, a sense of collective participation underlay the consumer and business community, albeit in different forms: in the former, social life dominates economic life, while in the latter, both lives are more closely related. As we have said, in brand communities, conviviality is created around brand values and, recalling Illich's words, is based more on being than on having, whereas in the business community, conviviality can conceal an economic soul. In this case, conviviality is more business driven and, consequently, more closely tied to having than to being. In such communities, conviviality seems to lose one of its most intrinsic purposes that of creating free spaces of open dialogues. We believe that in a business community, a generated conviviality based not on formalisms but on social and open tools (e.g., dinners, informal meetings and seminars) is able to create such spaces and to lead to the growth of the social capital of a community by impacting economically its members [7]. In other words, a business community participation, if mediated by created convivial moments based on genuine tools, which leave greater room for the participants' perspective of being, may have a positive influence on the perspective of having. In order to better investigate the relation between an existing community and convivial tools in an ad hoc conviviality, we will examine the business community as business network, which can be broken down in turn into the social and the business components.

Bridging social and business networks. In our study, we explore conviviality within business communities as moments of self-revelation, and real occasions to get to know each other better, to share experiences and, thus, to generate individual relations. The idea is not to investigate in depth the tacit and spontaneous conviviality already existing in a business community (like that which distinguishes an industrial district), but to examine a generated conviviality—externally or internally organized by community members—whose tools can act as driver of social networks. Social networks are seen as the sum of social relations that form stable bonds between individuals and that become a source of social capital [25–27]. This is an "aggregate of actual or potential resources", resulting from the "more or less institutionalized relationships of mutual acquaintance or recognition" [25], p. 2. In a community, it stems from the interactions between individuals [28] and consists of all the available resources that arise both from the networks to which the individuals belong and from their position in each. Social networks can gain their members a privileged access to information and opportunities [29] or develop reciprocal obligations due to emerging feelings of gratitude, respect and friendships [30]. In cases of restricted memberships, they can produce social capital in the form of social status and reputation [31]. Nahapiet and Ghoshal [32] describe social capital as resources internally articulated in structural, relational and cognitive dimensions. The structural dimension concerns the overall patterns of connections between actors—that is, who you reach and how you reach them [31]. Facets of this dimension include the presence or absence of network ties between actors [33], network configuration [34] and appropriable organization [27]. This latter consists in reciprocal influences among networks thanks to social ties developed in one of these. The relational dimension embraces the kind of personal relationships (respect, friendship, etc.) people have developed with each other through their history of interactions [35]. This leads to trust, shared norms and reciprocal obligations. The cognitive dimension concerns collective representations, interpretations and systems of meaning among parties [36]. Its facets include common language and code and shared narratives. All these dimensions affect the production of social capital, and also resources exchange and combination generating intellectual capital. This capital can be defined as "the knowledge or knowing capability of a social collectivity such as organization, intellectual community or professional practices" [32], p. 245. Thus, intellectual capital identifies the collective knowledge of a social network. The social capital it embeds can foster the relational capital within business networks [37]. The relational capital has an economic value and takes shape in relations among organizations that make exchanges and share activities to generate joint value [38–40]. It populates the business networks, which consist of a set of "tangible and intangible investments that comprise the connected relationships between more than two businesses" [39], p. 236. Their structure includes a set of actors, activities, links, resources, ties and bonds. Business networks emerge from social networks whenever business relations result from personal contacts. In some emerging markets, such as the Chinese, the dependence of business relations on social relations is essential for success. In this case, some studies [41, 42] demonstrate that the development of social relationships is a prerequisite to penetrate Chinese business networks. Personal contacts act as mechanisms for reducing the cultural distance between individuals [43], for entering local social networks (based on social obligations' interchanges and on moral attitude to respect them) and for generating business exchanges. Social networks compensate, in fact, for the lack of codified and widely available public information and animate reciprocity conditions to develop local business relations. At the same time, social networks are an essential component of developed business networks. In this regard, Håkansson and Snehota [44] highlight that "the individuals involved in a business relationship tend to weave a web of personal relationships, and this appears to be a condition for the development of inter-organizational ties between any two companies" (p. 10). Thus, business relationships engage organizations, but they are managed by individuals. It follows that the social capital embedded in social networks can reinforce the relational capital of the related business networks. The interpenetration between social and business networks rests on the relational mechanisms produced by mutual trust and commitment in the social bonds. Mutual trust can be seen as "one party's belief

that its needs will be fulfilled in the future by actions undertaken by the other party"
[45], p. 312. In business relations, it is expressed as a willingness to be vulnerable
to each other; vulnerability is the consequence of beliefs that the actor involved in
the relation can develop. However, both of them believe: (1) in the good intent and
concern of exchange partners [46–48], (2) in their competence and capability [49],
(3) in their reliability [46, 50] and (4) in their perceived openness [46]. Trust impacts
on mutual commitment and, thus, on the "belief of an exchange partner that the
ongoing relationship with another is so important as to deserve maximum efforts
at maintaining it indefinitely" [51], p. 23. Now, we wonder whether conviviality
can produce similar relational mechanisms acting as converter of social relations in
business relations. Although it characterizes interactive environments, it is not to be
confused with the concept of atmosphere. According to the IMP Group approach, it is
a component embedded in business relations. Conceptually viewed as both a product
of the relationship and a factor contributing to future relationship development, the
atmosphere can be articulated in terms of "the power–dependence relationship which
exists between the companies, the state of conflict or cooperation and overall closeness or distance of the relationship as well as by the companies' mutual expectations"
[52], p. 29. Power is defined as the ability of one party to influence the actions of
the other [53]; thus, the relative dependence between the parties in the relationship
determines their relative power [54]. Cooperation is the willingness actors express to
collaborate and attain common benefits, while conflict implies a company attitude to
pursue individual goals to the detriment of its competitors [55]. Closeness entails a
restricted nucleus of relations and can depend upon psychological, social and cultural
distances. Instead, openness requires a willingness to understand the other party and
presumes a willingness to seek and develop collaborative relations. The expectation
depends on the evolutions characterizing all the other aspects of the atmosphere and
they emerge from past actions involving the actor parties [56]. Another trade-off,
more recently explored [57, 58], is that of trust versus opportunism. Trust presup-
poses a longer-term relational attitude together with the belief of positive relations
development, while opportunism implies a short-term interactive perspective stemming from opportunities that are grasped to reach individual interests at the expense
of hitherto accepted modes of behaviour. Each actor relates to the atmosphere in a
personal way in the sense that the resulting relational environment it produces can
also depend on individual perceptions. Now, we illustrate the main results emerging
from our empirical analysis, and in particular, we will focalize our attention on the
relations between convivial tools and social capital produced by social networks, and
on the mechanisms these tools activate to transform social into relational capital and,
thus, to build a bridge between social networks and business networks.

3 Research Methodologies and Objectives

Our study is exploratory in nature and based on ethnographic case research [59, 60]
of business communities and in particular of three important Italian entrepreneurial

communities belonging to the fashion system. Two are rooted in Italy (Tuscany) and one in China (Hangzhou). The communities situated in Italy include two associations that for reasons of confidentiality have been, respectively, denominated textile association and leather consortium. The first was founded in 1983 and includes about forty entrepreneurs operating in textiles and clothing. The second, on the other hand, emerged in the sixties and seventies and gave rise in 1997 to a consortium, which groups more than fifty leather entrepreneurs. Both consist of companies that have mainly business-to-business dealings. Their customers are multinational fashion and luxury goods companies. The leather producers have been monitored by one of the two authors since as early as 2005, and the implications of their functioning in terms of business networks have led to previous publications [61]. The third community we investigate is composed of Italian entrepreneurs and is localized in China, exactly in Hangzhou in the Province of Zhejiang, where a large part of the Chinese textile-clothing is located. Named Italian fashion association, its entrepreneur-members operate mainly in the textile and clothing industry and have as clients both local and international fashion companies. The purpose of this association is to develop and strengthen relations among Italian entrepreneurs who have decided to intensify their business activity in China trying to facilitate their social and business integration. The next section presents the main results of a series of ethnographic interviews [62] conducted with the President of the consortium and the Director of the textile association. These are combined with others realized in Hangzhou involving the Director of the Italian fashion association and six entrepreneur-members (Table 1). In all the three cases, we integrate personal interviews with occasions of immersion in the community life. The unit of our analysis are the convivial moments. Thus, the topics of each interview include: (a) the associations/consortium history, internal organization and activities; (b) convivial activities (which kinds, tools and management.); (c) conviviality and technologies; (d) experiences of conviviality (their impact on social relations); and (e) conviviality as mediator between social and business relations.

Table 1 Interviews in our exploratory analysis

Actors	Location	Role of interviewee(s)	Interview tools	Time (hours) for field research
Textile association	Florence (Italy)	Director	E-mail, qualitative analysis protocol, (audio and video) recorder	5
Leather consortium	Florence (Italy)	Director	E-mail, qualitative analysis protocol, (audio and video) recorder	7
Italian fashion association	Hangzhou (China)	Director and 6 entrepreneur-members	Skype, e-mail, qualitative analysis protocol, (audio) recorder	10

Each interview has been transcribed, discussed and interpreted by the authors. The aim of the research is to study conviviality in entrepreneurial communities. Specifically, with reference to conviviality, we propose to shed light on (a) the forms it takes and how it is managed; (b) its possible effects on social relations especially in terms of resulting social capital; (c) the connective links it might produce between social and business relations.

4   Main Results

Leather Consortium. In the late 1990s, a group of leather goods producers who had been associated casually for years took the opportunity to create a consortium in response to "the need to counteract an economic policy that the country was pursuing, namely outsourcing production". In this consortium, conviviality was not only an instrument unifying the community, but also the reason underlying its very creation. In this regard, the director states: "... fifteen years of convivial meetings ensured that when the need to set up a consortium of some importance emerged, it was easier to agree because we could understand each other and therefore trust each other. Trust within the community developed over time by involving convivial activities. Over the years, conviviality... resulted in organizing family dinners ... at least twice a year (Christmas and pre-summer holidays) ... [which included overall] fifty entrepreneurs with their families". This conviviality was characterized and fostered by unflagging attendance and the variety of activities carried out together. "[It gave rise to a] real convivial celebration, in which the topics were not bags or leather, but rather the desire to be together and to know each other in depth". The director adds that "other convivial moments are the football tournaments which have been quite successful. We deepen our mutual knowledge professionally, and each of us knows the business experience of the other with all the problems and successes". What facilitated the development of the consortium was also the existence of competitive relations among businesses, which is a reciprocal understanding not only of a human nature but also in terms of business relations. "What leads to the creation of the consortium is a work relationship in which we are at times united, at other times direct competitors, genuine reciprocal understanding and profound respect at an interpersonal level". The results are the consortium, the Italian Superior School of leather goods, and a number of other projects never before attempted in Italy, such as those regarding the introduction of high technology to leather goods production ...". This case shows how human contact based on individual relationships cannot be replaced by new online and social media technologies. Specifically, "social networks are ... instruments. Just as there used to be envelopes with stamps, nowadays we can use new online social media ... they are faster, quicker, more penetrating, but they cannot replace convivial occasions. Human relationships are the core of real convivial situations and must be preserved through frequent meetings and the exchanges of ideas.

"Conviviality is a way of creating involvement and conveying passion and entrepreneurial values. In other words "conviviality [can be an] associative marketing tool … [to identify with and to transmit to] young people the idea that a business opportunity may come from the opening of a new business in addition to their passion for a job … [and this can be facilitated by] sharing some paths and convivial acquaintances". What is important is to avoid the (once feared) risk of making convivial activities self-referential. The director asserts: "Convivial life is still as active today as it was in the past, but it has changed a lot, and nowadays it is completely different. I think that conviviality today is partitioned off, fragmented and has become much more self-referential, hence the need for transverse conviviality that is not shared (solely) among individuals in the same organization, but among individuals of different organizations. We have convivial acquaintances in the same business group for supply to and convivial relations with a large customer, but we do not enjoy transverse conviviality, which involves components of different groups (trans-conviviality). And I think this is a problem, because conviviality helps in (opening up) important business pathways through the transverse flow of thoughts and knowledge".

Textile Association. This association bases many of its activities on the organization of discussions concerning the future of the town where it is located. Issues are investigated through seminars, training courses and meetings. Conviviality is seen as a free, participatory and interdisciplinary dialogue that "makes it possible to dissect a problem, to go in detail, […] to communicate with people at the table about basic problems like the reorganization of the road network. There is no one who listens to you, there are no journalists". Conviviality is a direct and spontaneous comparison, usually organized around a "long table so that people look each other in the face, revealing themselves: [this is because] it is necessary to express our ideas in front of others". Convivial occasions require time, concentration and are not restricted to the community members. In this regard, the director says: "We organize residential meetings devoted to a specific subject, dedicating two days of full immersion to the problem … [these meetings] are open to a wide audience …". To power the intensity of participation, we try to share individual experiences that are filtered through cultural events. "In the summer months, the textile association was used to organize an important event … [it] consisted in the screening of a film on business issues and in a [follow-up] debate. All this is useful to rebuild the business experiences of the participants and make them shareable".

In the case investigated, it emerges how conviviality, if it is well managed, can affect the personal contact network of participants: it widens social networks and encourages individual growth processes within the community. "People that have been members of the textile association, and have then found themselves holding public offices, have become institutional figures, have gained awareness of the problems and learned to gain confidence in themselves and in their beliefs … All this animates our members and makes our association attractive". It follows that the identity and reputation of the business community can help to extend its boundaries and to involve new members.

Management of conviviality can involve an animator. "He makes sure that those who are at the table express their ideas, because people are not all equal. There are shy people who have difficulty stating their opinions. It is necessary to help them … so opinions come out, otherwise the dialectic, the debate is absent …". The human component is then embedded in the conviviality, which when compared to the new technologies can only be integrated.

Although "... the new technologies lead to an acceleration of the times ... there is still a need to strengthen the direct comparison, which cannot be replaced by technology". Moreover, social relations fostered by social occasions seem also open to business relationships. The conditions for this to happen are, on the one hand, mutual sympathy and, on the other, trust and an individual shared style". The members are chosen because there are sympathies, mutual understanding, a style they like and adapt to dialogue even in a more specific way". What, beyond assumptions, seems to trigger the development of business relations is the social position that the convivial has within the community. "[Business relations] emerge externally [outside of our association] and not all are well classified ... [we feel that] since professional associates (accountants and lawyers, for example) have entered, business relationships among companies of different sectors are much better developed, ... [but] what all this has actually generated cannot be recognized, it would be necessary to ask our participants."

Italian Fashion Association. The entrepreneurial community in Hangzhou was formed recently and organizes convivial occasions in the form of dinners and meetings whose purpose is that of exchanging information and generating support and mutual understanding in a high cultural distance context. The Director of the Italian fashion association argues that this group "is an important point of reference for the Italian business community ... many entrepreneurs attend it with a certain continuity. After registering, they take part in events, dinners and meetings and find them educational and informative ...". One entrepreneur points out that "before the dinners, we have the opportunity to meet, have exchanges, support each other, even revealing and sharing real problems ... we also talk with managers who are expressly invited ... and because of their skills they become interesting cases to listen to. These people can drive us in our business". Ultimately, the Italian fashion association creates a sort of island of social relations in which Italian immigrant entrepreneurs exchange views, recount their experiences by sharing successes and problems. The Director adds, "overall, we are a good number of people, there are about 70 of us, although the group that meets more often, at least five times a year, is smaller. Participation in the Italian Fashion Association is something takes place without commitments, obligations and supervision. The associations are supported by the Italian Chamber of Commerce in Shanghai ... and here, we organize everything that can foster aggregation".

The social relations that are developed involve the sharing of experiences and make it possible to find solutions both at work and in personal situations. In this regard, one entrepreneur relates that "during the dinner, you can meet other Italians working in the same area and you may often exchange life and professional experiences. We exchange opinions, information ... and very often, the resulting situations can become a problem solver. It is a personal pleasure because we speak Italian, which is not to be taken for granted ... then, if someone needs an attendant or a local supplier, they send a collective e-mail and ask if we have somebody to propose". New technologies (mailing lists) become a tool to give continuity to social occasions and to foster a constant exchange of information. The director states that "I created a mailing list and periodically send news related to Zhejiang … these messages can be useful from a professional perspective. When you are away from your native country, there are also human aspects that become important … maybe the kindergarten for children, the supermarket where you can find Italian products and so on. So, the technology allows you to spread information and to fuel the collective participation of the group".

Mutual knowledge, the sharing of personal and work situations, fosters mutual participation and trust. One entrepreneur, in fact, states that "conviviality helps you to live better and to face everyday life; given the distance from home, it is quite spontaneous to assume a participatory attitude to shared problems within the group … then, ultimately, if we live better, we work better and, what's more, we can find appropriate solutions to work problems". This state of emerging mutual empathy does not necessarily transform social relations into business relations. The director points out that "the community is scant, and members are unable to develop business together. We have often spoken of creating buying groups as an evolution of our association but for the moment, this is not happening. More specifically, we have considered the joint purchasing of packaging materials and, thus, of materials that are not competitive. It is something we have thought about, but that we have not yet put into action. It takes time, and we need someone who could organize business collaboration among community members. Some of us cannot do these things as it means taking time away from our work as entrepreneurs". What seems to be lacking to make the social relationships sources of business relations is the ability to organize business together. The fabrics of social relations become, however, a way to develop their business. They increase the contractual power of the community in cases of negotiation with the Chinese governmental institutions. Conviviality contributes to enhancing the relationships between the Italian community and the local market. "Until now, we have limited ourselves to forming a group in order to interface with the local government. This is one of the aspects that we would like to develop. Joining forces you have a different relationship with the government. If I, alone with 50 employees, show up at a government office a specific request, I find the door locked. If we present ourselves as a group of 20 companies, the government pays more attention".

The social relations among the association's members do not become socially transverse and, thus, they do not involve local entrepreneurs or, otherwise, workers. The Italian community, in fact, is limited to interacting with the Chinese community on specific occasions organized by Chinese employees working at Italian firms. "The only social relations that we have developed are those with some of our employees. Outside this circle, we have not created relationships with other people. Our family situation is very particular, and feeling a bit like foreigners, we tend to create our family and we have not really opened up towards the outside. So, we have held dinners, we have got to know the families of our employees with whom we try to create bonds. Beyond this … nothing. Moreover, the Chinese are closed, they always work, and their social life plays itself out within the family group".

Social relations with local people take place outside the community organizations and, thus, assume an individual rather than a collective form. "In Guangdong, textile entrepreneurs meet regularly, have their own textile associations and meet regularly for informal dinners ... these Chinese communities are very close, they are Chinese, and they want to stay with local people, and they want to speak in their native language". The limited social relations, although they foster processes of sharing and mutual trust, do not contribute to developing local business relationships. One way to create a sort of bridge between the Italian and Chinese entrepreneurial communities is to exploit the social relations that Italian entrepreneurs have developed with their employees (and a state of trust) and make them mediators between their personal social relations and local business relationships by producing a positive impact for the company where they work. One entrepreneur observes that "your employees should use their social relations to activate business relations and, thus, to create for us a local business network. The social relations we have with them are useful for our business. Personal relationships here in China are critical across the board, in any aspect of life, both personal and professional. Knowing the right people in the right place, you can do everything; it is still the network family, ex-classmates ... that allows you to open many doors in China. However, these relations are maintained by Chinese with Chinese. For us in the West, it is difficult to be able to integrate into their communities".

5 Discussion

The cases show that in the communities investigated ad hoc conviviality is based on common tools such as dinners, meetings and discussions. These tools can be internally self-managed or may involve animators whose role is to promote the intensity of community participation during the various social occasions. Conviviality tools contribute to creating or animating business communities to the extent that they are able to stimulate effective commitment on the part of members and to develop social relations, by creating a positive outlook in terms of collectively shared social capital production. The shared capital emerging from convivial occasions assumes a dimensional articulation that if compared with that proposed by Nahapiet and Ghoshal [32] is enriched and more focalized. Now, we try to explain better this articulation synthesized in Tables 2 and 3. The *structural dimension* emerging from the empirical cases includes an *introspective base*; convivial moments are, in fact, opportunities that allow their members to reveal the real self in their social relations. The members express themselves freely more than they could do in other relational situations. Instead, the *cognitive dimension* mainly consists of narrated and shared stories and thus focuses on *experiential contents*; there ensue social relations, which, being enriched by shared experiences, are able to foster exchanges of knowledge about the potential characterizing community members in terms of skills and competences. Then, the *relational dimension* takes form around a trust that is *empathy-based*. Conviviality tools, in fact, favour the development of trust through processes of mutual empathy and thus, of *social relations internalization* and of sharing of contexts, rules of judgment and choices among community members [63].

Table 2 Conviviality tools and management: their impact on social capital and on the relation between social and business networks

	Tools	Management	Technology	Social capital in convivial business network (some examples)	Bridging social and business networks	Bridging driver(s)
Leather consortium	• Dinners (Christmas and pre-summer holidays) • Football tournaments	Self-management of conviviality	Complementary and instrumental tool	"[It gave rise to a] real convivial celebration, in which the topics were not bags or leather, but rather the desire to be together and to know each other in depth" (*Introspective content*) "We deepen our mutual knowledge professionally and each of us knows the business experience of the other with all the problems and successes" (*Experiential content*) "Fifteen years of convivial meetings ensured that when the need to set up a consortium of some importance emerged, it was easier to agree because we could understand each other and thus trust each other" (*Empathetic content*) "Conviviality [can be an] associative marketing tool … [to identify with and to transmit to] young people the idea that a business opportunity may come from the opening of a new business in addition to their passion for a job" (*Empathetic content*)	• Mutual knowledge of reciprocal potential makes competition a source of collaboration (*competition as source of collaboration*) • Empathy-based trust is instrumental in making business decisions, collectively opportunistic (*trust as source of opportunism*) • The cross-community conviviality makes competences individual communities possess (internally protected) a way to trigger the sharing of transversal business experiences among communities (*closeness as source of openness*) → There may emerge conditions (potential knowledge, empathy-based trust) as drivers of business relations	Knowledge of reciprocal potential skills Empathy-based trust Cross-community conviviality

(continued)

Table 2 (continued)

	Tools	Management	Technology	Social capital in convivial business network (some examples)	Bridging social and business networks	Bridging driver(s)
Textile association	Seminars, training courses, periodic meetings	Master of ceremony	Complementary and instrumental tool	"Conviviality is a direct and spontaneous comparison, usually organized around a "long table so that people look each other in the face, revealing themselves: [this is because] it is necessary to express our ideas in front of others" (*Introspective content*) "To power the intensity of participation, we try to share individual experiences that are filtered through cultural events" (*Experiential content*) "The members are chosen among them because there are sympathies, mutual understanding, a style they like and adapt to dialogue even in a more specific way" (*Empathetic content*)	• Trust empathy-based intermediates specialization of some community members (lawyers, accountants, etc.) and makes it a source of openness in terms of interactive business perspectives (*closeness as source of openness*)	Empathy-based trust

Source Our elaboration from empirical data

Table 3 Conviviality tools and management: their impact on social capital and on the relation between social and business networks

	Tools	Management	Technology	Social capital in convivial business network (some examples)	Bridging social and business networks	Bridging driver
Italian fashion association	Seminars, training courses, periodic meetings	Self-management of conviviality	Technology as a means of preserving collective participation by giving continuity to convivial events	"Before the dinners, we have the opportunity to meeting, have exchanges, support each other, even revealing and share real problems" (*Introspective content*) "During the dinner you can meet other Italians working in the same area and usually you may exchange life and professional experiences" (*Experiential content*). "Given the distance from home it is quite spontaneous to assume a participatory attitude to shared problems within the group" (*Empathetic content*)	• *Intracommunity social relations mediator* → able to manage and channel the emerging empathy-based trust towards the development of business relations. It identifies, proposes and organizes collective business activities exploiting foreign companies' business motivations (*individualistic business attitude as source of collectivistic entrepreneurial attitude*) • *Intercommunity social relations mediator* uses his social relations as a source of business relations with local actors to the benefit of the foreign company where he works. He is a mediator of trust capable of regenerating the local mechanisms of confidence (exchange of social obligations, mutual respect) in the relations between the foreign company and the local business community. Thus, this reproduction creates openness in terms of business activities openness (*closeness as source of openness*)	Intra/Inter community social relations mediator

Source Our elaboration from empirical data

Certain decisions, such as the foundation of a consortium, take place thanks to the confidence generated by the mutual understanding of individual situations. The more conviviality is able to facilitate individual opening up in terms of *self-revelation*, the *sharing of experiences* and *state of identification*, the greater is the *social capital* embedded in the social network underlying the community.

Business communities animated by convivial tools may exploit the resulting social capital to develop business relations. This stage is not simple and cannot be taken for granted. Our exploratory research shows that it depends on the ability of convivial occasions to impact on the components of a relational atmosphere [52] making their constituting elements no longer as one opposed to the other, but rather as one the source of the other (Table 2). Of course, this is a preliminary interpretation, which will be interesting to investigate and deepen with subsequent empirical studies. In the leather consortium, for example, convivial moments generate *mutual knowledge* among competitive actors of their potential skills (Table 2). This has enabled to identify productive synergies among differentiate competences actors are bearers by transforming a competitive situation into a source of cooperation (competition as source of collaboration). Besides, among leather goods companies, *trust emerging from situations of empathy* (self-identification) has become instrumental in assuming collective business decisions, which are opportunistic-based (trust as source of opportunism). The consortium is, in fact, the result of a reciprocal knowledge but also of a community confidence. This latter has given rise to the choice of capturing the business opportunity not perceived by many competitors, that is, preserving manufacturing local roots instead of delocalizing abroad production processes. It follows that *mutual knowledge* and *empathy-based trust* can be seen as social capital resources, which act as basis to transform social relations into business relations. They are related to each other: mutual knowledge, if it leads to self-identification, allows making cooperation a source of community opportunistic behaviours, which arise from the need to further individual interests through the satisfaction of collective interests. The case of the consortium also shows that conviviality, if it is *cross-community*, makes competences (internally circumscribed); individual communities possess a source to trigger experiences sharing and, thus, transversal contamination among other communities (closeness as source of openness). What may follow is the development of conditions (mutual knowledge, empathy-based trust) to generate business relations. Therefore, also a *cross-community conviviality* can indirectly contribute to creating a fabric of business and socialization.

In the convivial moments of the Italian Fashion Association (Table 2), this plot emerges when the *trust based on mutual understanding* (empathy) can overcome cultural and professional barriers that can elapse between companies as bearers of different competences. This makes it possible to intensify the dialogue between actors, which would otherwise remain closed in their spheres of action. More specifically, expert service firms (lawyers, accountants, etc.) adapt their specialization to develop new collaborations with the community of fashion companies enlarging or intensifying their business relations. It follows that *empathy-based trust* not only coexists with forms of collective opportunism (see consortium), but can also intermediate specialization closed to some actors and make it a source of openness in terms of interactive business perspectives (closeness as source of openness). Cross-community convivial experiences together with empathy-based trust contribute to the coexistence of closeness and openness. The empathy-based trust can have a more direct impact on business relations to the extent to which self-identification already embeds the reciprocal knowledge of potential abilities.

In the case of the foreign (Italian) community in China (Table 3), trust, due to self-identification in experiences both professional and of real life, characterizes social relations, but it, by itself, is not able to convert them into business relations. The emerging *empathy-based trust*, in fact, has to be managed and channelled in the development of business relations. The conversion is therefore not spontaneous, but mediated. It is up to a collective organism (*intracommunity social relations mediator*), which identifies, proposes and then organizes shared business activities exploiting foreign companies' business motivations. Doing this, it can contribute to make an *individualistic* business attitude a source of a *collectivistic* entrepreneurial attitude. It derives that conviviality can enrich the relational atmosphere with another trade-off, that of individualistic vs collectivistic business culture. It further refines the trade-off between *closeness* and *openness*. Besides, in China, self-managed convivial moments can also develop social relations between foreign companies' managers and their local workers. However, these relations can produce local business relations if mediated through social relations workers who have locally built. Since they are members of a foreign community and of a local community, they can create bridges between communities. Their intercommunity mediation has not mainly an organizational nature like that of intracommunity. It depends on the trust they have developed in their community by participating to the local mechanisms of mutual respect and social obligations' exchange [43]. Thus, local workers can be mediators of trust; in doing this, they mediate the regeneration of local mechanisms acting as driver of confidence in the relation between the foreign company and the economic actor belonging to the local community (*intercommunity social relations mediator*). It is just this reproduction that generates a local openness in terms of business activities with foreign companies (closeness of the local community as source of openness). The trust it generates, considering the strong personal content, tends to take empathetic characters that is of mutual understanding.

The components of the atmosphere that do not emerge from the conversion of social into business relations within the communities investigated are those related to power/distance and expectations. The analysis of these factors will require specific insights with the community members. However, our exploratory results show that the social capital produced is transformed into the relational capital to the extent that the opposite elements constituting the atmosphere components coexist; their coexistence is as if it fuelled a *sort of zero-moment of relational conversion*. We wonder what will happen after this time-zero. In other words, it could be interesting to investigate whether the opposite elements break up and regain autonomy together with the development of the relation. If this were so, the poles trust/opportunism, openness/closeness and collaboration/competition could take shape again even if under new nuances. Thus, we ask ourselves if conviviality effects on the atmosphere components are short term or long term. However, the activation of the moment-zero

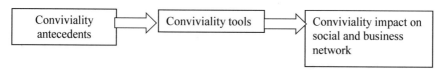

Fig. 1 Conviviality tools as a process unifying social and business networks

of conversion depends upon the social capital resulting from the convivial moments and upon the related emerging trade-offs. These may be formulated as revealing vs concealing oneself (the introspective dimension), narrating vs describing oneself (the experiential dimension) and recognizing vs alienating oneself (the empathetic dimension). These trade-offs lay the foundations for the generation of common, shareable resources that can lead to new business relations and, consequently, to important effects in terms of relation management. The community social capital generated by genuine friendliness could lead the participants, especially entrepreneurs, to rethink their business activities and hence reshape their business relations to modify the structure of the business networks they belong to. In the final analysis, aside from the results just described, it emerges that the relation between conviviality antecedents and tools, social network and relation network, is complex and may be approached from a range of analytical perspectives (Fig. 1).

6 Limitation and Conclusions

The current research is an attempt to delve into the mechanisms of sharing generated by conviviality and their impact on social and business networks. Although this brief discussion is the outcome of a purely exploratory analysis, it represents the motive force for continuing on to the next stages of research, which is to consist of ethnographic immersion [64] into the community under study as well as others.

Our chapter has some limitations. First, the limited number of interviews was carried out in depth. Then, it is also necessary to examine in greater detail the mechanisms that transform social relations into business relations. In particular, we aim to investigate the impact that different conviviality tools have on the relational atmosphere, the sources that make its opposite elements coexistent as well as the reciprocal relation among the sources that are capable of fostering the moment-zero of conversion.

Despite the path we are aware must be undertaken before reaching a scientific conceptualization of the phenomenon, we believe that our analysis of conviviality enables the identification of new approaches to the understanding of the mechanisms operating between social and business relations. Moreover, we think that the approach adopted can, in addition to assessing the actual impact of conviviality on business relations, also contribute to responding to the crisis, in terms of loss of identity and unifying values, now affecting our national business system. In this regard, we are quite convinced that conviviality, especially transverse and non-self-referential, can reduce current trends towards social individualism [65]. This has become an end in itself and likely represents a significant weakness in our industrial system, and opens up greater possibilities for a revival in both social and business practices based on a stronger sense of participation and collective sharing.

References

1. Dillard J, Dujon V, King MC (eds) (2008) Understanding the social dimension of sustainability. Routledge, UK
2. Ranfagni S, Guercini S (2016) Beyond appearances: the hidden meanings of sustainable luxury. Handbook of sustainable luxury textiles and fashion. Springer, Singapore, pp 51–72
3. World Commission on Environment and Development, WCED (1987) Our common future. Oxford University Press, Oxford
4. Griessler E, Littig B (2005) Social sustainability: a catchword between political pragmatism and social theory. Int J Sustain Dev 8(1/2):65–79
5. Putnam RD (2000) Bowling alone: the collapse and revival of American community. Simon and Schuster
6. Illich I (1973) Tools for conviviality. Harper and Row, New York
7. Guercini S, Ranfagni S (2016) Conviviality behavior in entrepreneurial communities and business networks. J Bus Res 69(2):770–776
8. Lloyd S (2002) Pleasing spectacles and elegant dinners: conviviality, benevolence, and charity anniversaries in eighteenth-century London. J Br Stud 41(January):23–57
9. Maitland R (2008) Conviviality and everyday life: the appeal of new areas of London for visitors. Int J Tourism Res 10(1):15–25
10. Williams QE, Stroud C (2013) Multilingualism in transformative spaces: contact and conviviality. Lang Policy 12:289–311
11. Germov J, William L, Freij M (2010) Portrayal of the slow food movement in the Australian print media. Conviviality, localism and romanticism. J Soc 47(1), 89–106
12. Caire P, Alcalde B, Sombattheera C (2011) Conviviality measures. In: Paper presented at the 10th international conference on autonomous agents and multiagent systems, vol 2, pp 895–902
13. Cova B (1997) Community and consumption: towards a definition of the "linking value" of product or services. Eur J Mark 31(3/4):297–316
14. Muniz AM, Guinn TC (2001) Brand community. J Consum Res 27(4):412–432
15. Schau HJ, Muñiz AM Jr, Arnould EJ (2009) How brand community practices create value. J Mark 73(September):30–51
16. Brown S, Kozinets RV, Sherry F (2003) Teaching old brands new tricks: retro branding and the revival of brand meaning. J Mark 67(July):19–33
17. Kozinets RV (2002) The field behind the screen: using netnography for marketing research in online communities. J Mark Res 39(1):61–72
18. De Valck K (2005) Virtual communities of consumption: networks of consumer knowledge and companionship. ERIM PhD, Series: Research in Management
19. Burresi A, Guercini S (January, 2002) La rappresentazione del mercato in funzione dell'innovazione di prodotto nelle imprese del tessile e abbigliamento. In: Paper presented at international marketing trends conference. Ecole Supérieure de Commerce de Paris, pp 25–26
20. Guercini S (August, 2003) Segmentation versus emerging archetypes in the market mapping process. In: Paper presented at the 63rd, annual meeting of the academy of management. Seattle, Washington, pp 1–6
21. Carson D, Cromie S, McGowan P, Hill J (1995) Marketing and entrepreneurship in SMEs. Prentice Hall, London

22. Tönnies F (1957) Community and society. Michigan State University Press, East Lansing
23. Becattini G (1987) Mercato e forze locali: il distretto industriale. Il Mulino, Bologna
24. Becattini G, Bellandi M, De Propris L (2009) Handbook of industrial districts. Edward Elgar, Cheltenham, England
25. Bourdieu P (1980) Le capital social. Actes de la recherche en sciences sociales 31(31):2–3
26. Bourdieu P (1985) Social space and the genesis of groups. Theory Soc 14(November):723–744
27. Coleman JS (1988) Social capital in the creation of human capital. Am J Sociol 94:95–120
28. Burt RS, Jannotta JE, Mahoney JT (1998) Personality correlates of structural holes. Soc Netw 20:63–87
29. Granovetter M (1973) The strength of weak ties. Am J Sociol 78(6):1360–1380
30. Bourdieu P (1996) The rules of art: genesis and structure of the literary field. University Press, Stanford
31. Burt RS (1992) Structural holes: the social structure of competition. Harvard University Press
32. Nahapiet J, Ghoshal S (1998) Social capital, intellectual capital, and the organizational advantage. Acad Manag Rev 23(2):242–266
33. Scott J (1991) Social network analysis: a handbook. Sage, London
34. Krackhardt D (1989) Graph theoretical dimensions of informal organization. In: Paper presented at the annual meeting of the academy of management. Washington
35. Granovetter M (1992) Economic institutions as social constructions: a framework for analysis. Acta Sociol 35(1):3–11
36. Cicourel AV (1973) Cognitive sociology: language and meaning in social interaction. Penguin, Harmondsworth
37. Bondeli JV, Havenvid MI, Solli-Sæther H (2018) Placing social capital in business networks: conceptualisation and research Agenda. J Bus Ind Mark 33(8):1100–1113
38. Easton G, Håkansson H (1996) Markets as networks: editorial introduction. Int J Res Mark 13(5):407–413
39. Håkansson H, Ford D, Gadde L-E, Snehota I, Waluszewski A (2009) Business in networks. Wiley, Chichester
40. Johanson J, Mattsson L-G (1988) Internationalization in industrial systems—a network approach. In: Hood N, Vahlne JE (eds) Strategies in global competition. Croom Helm, New York, pp 303–321
41. Björkman I, Kock S (1995) Social relationships and business networks: the case of Western companies in China. Int Bus Rev 4:519–535
42. Ranfagni S, Guercini S (2014) Guanxi and distribution in China: the case of Ferrero Group. Int Rev Retail Distrib Consum Res 24(3):294–310
43. Cunningham MT, Homse E (1986) Controlling the marketing-purchasing interface: resource development and organisational implications. Ind Mark Purchasing 1(2):3–27
44. Håkansson H, Snehota I (1995) Developing relationships in business networks. Routledge, London
45. Anderson E, Weitz B (1989) Determinants of continuity in conventional industrial channel dyads. Mark Sci 8(4):310–323
46. Ouchi WG (1981) Theory Z: how American business can meet the Japanese challenge. Addison-Wesley, Reading, MA
47. Pascale R (1990) Managing on the edge: how the smartest companies use conflict to stay ahead. Simon and Schuster, New York
48. Ring PS, Van de Ven AH (1994) Developmental processes of cooperative interorganizational relationships. Acad Manag Rev 19(1):90–118
49. Szulanski G (1996) Exploring internal stickiness: impediments to the transfer of best practice within the firm. Strateg Manag J 17(S2):27–44
50. Giddens A (1990) The consequences of modernity. Polity Press, Cambridge, England
51. Morgan RM, Hunt SD (1994) The commitment-trust theory of relationship marketing. J Mark 58(3):20–38
52. Håkansson H (1982) International marketing and purchasing of industrial goods: an interaction approach. Wiley, Chichester

53. Gaski JF (1984) The theory of power and conflict in channels of distribution. J Mark 48(3):9–29
54. Hallen L, Johannson J, Seyed-Mohammed N (1991) Interfirm adaptation in business relationships. J Mark 55(April):29–37
55. Hallen L, Sandstorm M (1991) Relationship atmosphere in international business. In: Paliwoda SJ (ed) New perspectives on international marketing. Routledge, London
56. Hedaa L, Törnroos JÅ (2007) Atmospheric disturbances in the IMP interaction model: introducing semiosphere into business interaction. In: Paper presented at 23nd IMP Conference. Manchester
57. Roehrich G, Spencer R (2001) Relationship atmosphere: behind the smoke-screen. In: Paper presented at the 17th IMP annual international conference. Oslo, Norway
58. Sutton-Brady C (2001) Relationship atmosphere—the final chapter. In: Paper presented at the 17th IMP annual international conference. Oslo, Norway
59. Woodside AG, Wilson EJ (2003) Case study research methods for theory building. J Bus Ind Mark 18(6/7):493–508
60. Yin R (2009) Case study research. Sage Inc, California
61. Guercini S, Woodside AG (2012) A strategic supply chain approach: consortium marketing in the Italian leatherwear industry. Mark Intell Planning 30(7):700–716
62. Spradley JP (1979) The ethnographic interview. Holt, Rinehart, New York
63. Guercini S (2012) New approaches to heuristic processes and entrepreneurial cognition of the market. J Res Mark Entrepreneurship 14(2):199–213
64. Le Compte MD, Schensul JJ (2010) Designing and conducting ethnographic research. Alta Mira Press, UK
65. Flint J, Robinson D (2008) Community cohesion in crisis? New dimensions of diversity and difference. The Policy Press, UK

Certifications for Sustainability in Footwear and Leather Sectors

P. Senthil Kumar and C. Femina Carolin

Abstract One of the most generating sectors in the world is footwear and leather sectors. These industries involve lots of chemical usage in order to transfer the raw material into finished product. Thus, footwear and leather sectors produce pollutants which are harmful and toxic to the human and environment. Awareness based on the toxicity of pollutants among the consumers is increasing rapidly, and also, they would like to select the environmentally products. In order to demonstrate, the performance of the industry certifications is assessed based on the social, economic safety issues adopted by the sectors. The sustainability is combined with the certification process to suggest that the enterprise fulfills the standards. This chapter explains some of the popular certification which is responsible for the footwear and leather sectors as an evidence for the sustainability.

Keywords Chemical usage · Toxicity · Economic safety · Certification · Sustainability

1 Introduction

Sustainability is a fundamental goal in the greater part of the manufacturing and sustainable practices in which commitment toward the customers and the environment is important beyond the company. Leather and footwear industries have been changed since the last few years. They play a particular position in the most recent patterns of style and are currently included into regular use. As the interest of these items has developed regularly, the necessity of its quality testing has additionally developed in the market. The quality fundamentally focussed around the element and is regularly associated with clients' comfort, whereas safety plays a major role in protecting purchasers from any sort of health hazard. Recently, sustainability is gaining importance in leather and footwear sectors to expand purchaser awareness and strict worldwide enactments [4, 6]. Sustainability is a key factor that helps to find

P. Senthil Kumar (✉) · C. Femina Carolin
Department of Chemical Engineering, SSN College of Engineering, 603110 Chennai, India
e-mail: senthilchem8582@gmail.com

© The Editor(s) (if applicable) and The Author(s), under exclusive license
to Springer Nature Singapore Pte Ltd. 2020
S. S. Muthu (ed.), *Leather and Footwear Sustainability*, Textile Science
and Clothing Technology, https://doi.org/10.1007/978-981-15-6296-9_8

new outcomes and chances to generate new products without creating impacts on the social, environmental and economic concern. Since the mid-1980s, purchasers looking for ecological friendly items and associations have progressively become conscious about environmental management systems, responsible behavior, practical development and advancement around the globe. In recent years, the interest for sustainable materials, products and administrations is expanding because of the developing health-based issues among the consumers [9]. Social sustainability has gained importance in management theory and practice [6, 12]. The idea of sustainability is increased in leather and footwear sectors because of pressure from international organizations and non-governmental organizations. Some of the entrepreneurs also decided that environmental sustainability has reduced the negative impacts arising from the manufacturing process. This is the key factor in gaining advantages to guarantee that any negative effect of the business on society is diminished. A few footwear and leather producers in the past have utilized unreasonable practices to meet the demand and gained the extra profits. As there is an expanded worldwide pattern toward sectors, huge numbers of the brands are currently focussing on sustainability. To fulfill the worldwide need of the universal brands, the leather and footwear sectors are focussing to accomplish certifications like Bluesign, Content Claim Standard (CCS), Cradle to Cradle, Fair-trade certification, Global Recycle Standard (GRS), Global Organic Textile Standard (GOTS) Certification, Oeko-Tex, Recycled Claim Standard (RCS), SA8000 Certification, Sustainable Fair Trade Management System (SFTMS), Higg Index, Source Map, Internationale Verband der Naturtextilwirtschaft (IVN) Natural Leather Standard, Worldwide Responsible Accredited Production (WRAP), National Science Foundation (NSF) and Leather Working Group (LWG). Oeko-Tex and GOTS come under the commercial category, and Bluesign comes under the economic category. Certification is the provision given by the independent body with written assurance that the product, system and manufacturing process meet the specific requirements. The certifications are utilized by a few organizations to direct purchasers toward environmentally capable products [2]. This chapter will focus on the list of certifications required for maintaining sustainability in leather and footwear sectors.

2 Certification for Sustainability

2.1 Bluesign

It is a certification concentrated on the parts of customer well-being, water, and air emissions and human health. Likewise, there is an emphasis on the decrease in harmful substance usage at the beginning stages of manufacturing. Bluesign in the product indicates the responsibility and sustainability of manufactured products [3]. It also represents that the input streams utilized for the production of products from the different resources are assessed based on the ecological impact. Bluesign traces each path from manufacturing till finished product to make improvements. It acts as a verifier to the finished product to get the environmental sustainability. Based on the strict aspects, services were developed to support the industries in sustainability. This Bluesign assures that the product manufacturers used responsible resources which created the lowest impact on the environment and human beings [18]. Each product has a comprehensive fabricating process which requires raw materials and many other components for this. Bluesign gives the best solution for the manufacturer; that is, they can create a product with Bluesign-approved process steps and composition of components. This can be achieved by the procedure called Intelligent Input Stream Management. Based on the environmental problems, Bluesign works on the five concepts, namely resource productivity, air emissions, occupational health and safety, water emissions and consumer safety. This technique is the best way to ensure the greatest efficiency and security in a cost-effective way. Even small companies with several competitors used the Bluesign standard to manage the chemicals for their products. The major goal of this standard is to eliminate hazardous chemicals in design and manufacture [17].

2.2 Content Claim Standard (CCS)

It is a voluntary certification that is used to follow a material through the supply chain. If other validation methods are not accessible, this method can be utilized to back up content claims. It tracks the progression of material from the source to the last item and is guaranteed by an authorized third party. It is a comprehensive and autonomous assessment and confirmation of material content claims on the final product. The Content Claim Standard (CCS) gives organizations a device to check the content of materials [13]. The target of this CCS is to certify the accuracy of content claims. This can be accomplished by analyzing the availability and quantity of the raw material in the finished product. It gives organizations a way to guarantee that they are selling quality items. From the source to the final product, it gives the strong custody system and provides the certification by third party. It is a business-to-business tool to provide evidence to products. This standard gives certification not only to the material but also to the production process. It does not discuss the inputs and environmental aspects like social issues, safety issues, legal compliance, etc. It tells that the product with this certification can be applied to the supply chain.

2.3 Cradle to Cradle

This accreditation is noteworthy among organizations as an approach to recognize eco-friendly products. Cradle to Cradle focussed on the manufacturing process of the final product with the goal that the reuse of material and additionally eliminating ineffective waste [10]. Principle of Cradle to Cradle leads to consider the waste to be another asset that works based on designing to loop materials. This infers items are designed such that make it more cost-effective to reuse them. The main focus of this certification is to limit degradation through a more sustainable manufacturing process [15]. The Cradle to Cradle certified product standard aids creators and makers through a constant improvement process to get this certification and that takes a final product through five classifications, namely material health, material reuse, sustainable power source and carbon management, water stewardship and social fairness with five levels: basic, bronze, silver, gold and platinum [11].

2.3.1   Material Health

It helps to guarantee that the products are created with the chemicals which are harmless for the surroundings and humans. It also ensures that the product is made by the experts through the process like inventorying, assessing and optimizing material chemistries. As a stage toward the full certification, makers may earn the individual Material Health Certificate that the manufactured product meets the Cradle to Cradle material health necessities. Platinum guaranteed items are completely Cradle to Cradle consistent, but the diverse certification levels are intended to reward the organization's effort in the continuous improvement of their products along the way toward eco-effectiveness. Product designers, manufacturers and several brands depend on this Cradle to Cradle certified product standard for manufacturing items without the negative impact on the environment and human health. The above-mentioned category is applied to the final product, and products with the lower category will also be certified with the level. Cradle to Cradle increases economic value with zero impacts to accomplish the objective like zero waste emissions, zero resource use and zero toxicity [1]. A significant difference between Cradle to Cradle and other numerous approaches is to accomplish a sustainable world and good certification framework. This is not just a theory on sustainability, but it also has its system for ensuring compliant items. This system is planned to help organizations making Cradle to Cradle items. The certification can be viewed as an award for the best outcomes. This certification is intended to be relevant to materials and finished products. It is not limited to specific companies and types of material. Because of their certification mark, organizations can get a certification mark for their product. Cradle to Cradle certification can be viewed as a significant eco-name and a valuable tool to evaluate the product quality.

2.3.2   Material Reuse

This class aims to dispose of the concept of waste generating from the cycle by allowing the waste to remain in the continuous cycle and reuse from one product and use it for the next cycle.

2.3.3 Sustainable Power Source and Carbon Management

It depicts that climate change and the release of greenhouse gases are reduced due to the usage of renewable energy in the fabrication of product.

2.3.4 Water Stewardship

It specifies that the water is perceived as a significant asset, watersheds are ensured, and clean water is accessible to individuals and every single other life form.

2.3.5 Social Fairness

It aims to design business tasks that respect all the humans and nature affected by the fabricated product.

2.4 Fairtrade Certification

Fairtrade Standards are intended to help the development of the sustainability of small producers. It contributes the practical improvement by offering better-exchanging conditions and safeguards the rights of makers and laborers. The principles contain advancement necessities focussed on enhancements that give benefits to the makers and their networks. It incorporates social, economic and environmental criteria. Fair Trade is the elective method of exchange, which gives support to the organizations in developing and underdeveloped countries. It complies with ten principles of fair trade that involves production and trade-related activities to generate income. Ten principles of Fairtrade are making opportunities for the producers, transparency and accountability fair trading practices, payment of fair trade, guaranteeing no child labor and forced labor, non-discrimination commitments, gender equality and women's economic empowerment, and freedom of association, guaranteeing good working conditions, giving capacity building, advancing fair trade and giving respect to the environment. To become Fairtrade certified, makers must hold to economic, social and natural guidelines of production. This certification is most popular for its financial measures, which expect purchasers to pay a minimum price and a social premium to makers, and suggest that purchasers give pre-financing and long-term contracts. It also includes specific natural protection standards to guarantee that makers have environmental practices that are safe and sustainable [5]. Fairtrade helps laborers and networks by spreading benefits to all the organizations more equally and enhancing the neighborhood economy. Many Fair Trade individuals work with makers to create items based on the sustainability of their normal assets. This gives the networks motivation to safeguard and maintain the environment.

2.5   Global Recycle Standard (GRS)

Initially, the Global Recycle Standard was evolved by Control Union Certifications in 2008. Its objective was to make a higher standard and to incorporate new chemical prerequisites. To modify the standard, an International Working Group (IWG) of certification bodies was created. This standard applies to all organizations that manufacture the leathers. It covers handling, fabricating, bundling, naming, exchanging and distribution of all items that are made with at least 20% reused material. Organizations should ensure that they have met a lot of necessities and reliable with the objectives of this standard and to be reviewed by a certification body on arbitrary premises. The target of GRS is to identify the requirements of content claims, increase the use of recycled materials and reduce the harmful ecological and chemical effects on humans. This standard is mainly focussed on

- Detection of the reused content in a material
- Environmental requirements
- Social responsibilities
- Characterize and maintain limitations in the utilization of synthetic substances.

All the phase of production is required to be certified starting at the reusing stage and final at the last seller in business-to-business transaction. Material assortment and material sites are dependent upon self-statement, report assortment and on-location visits. Its significance is based on denying certain synthetic compounds, requiring water treatment and maintaining laborer's rights. It follows some of the standards which were mentioned in the Global Organic Textile Standard like organizations must keep full records of the utilization of synthetic compounds, vitality, water utilization and wastewater treatment including the removal of sludge; all prohibited synthetic compounds recorded in GOTS should be excluded in the GRS; before removal, wastewater should be treated for pH, temperature, chemical oxygen demand (COD) and biological oxygen demand (BOD).

2.6   GOTS Certification

The Global Organic Textile Standard (GOTS) is the overall leading standard for natural fibers, including biological and social criteria. The point of the standard is to characterize overall perceived necessities that guarantee the natural status of materials, from harvesting of crude materials, through socially capable manufacturing up to labeling to confirm the end purchaser. This is the one acknowledged certification accepted in all the markets. The standard covers the preparing, fabricating, bundling, naming, exchanging and conveyance of all materials produced using at any rate of 70% guaranteed natural materials. But this GOTS does not set criteria for leather products. But the cotton utilized in the footwear is guaranteed by Global Organic Textile Standard (GOTS) program, additionally licensed by USDA's

National Organic Program. This standard also implies that the materials hold strict principles for cotton development, but also preparing (e.g., pre-treatment, washing, drying, coloring, waste treatment, finishing) to give a natural product. The GOTS-endorsed accreditation bodies are additionally effectively associated with the GOTS update process through the 'Certifiers Council.' It sets up ecological necessities along with a lot of obligatory social criteria.

2.7   Oeko-Tex

Leather standard by Oeko-Tex is an institutionalized overall testing and affirmation framework for leather and leather products from all phases of creation. This incorporates semi-completed items (wet-blue, wet-white, outside layer) up to completed items. It certifies leather fiber materials, accessories, leather finished products, leather gloves, bags and covers, etc. But it will not certify the leather products coming from animals like snakes, crocodiles and armadillos. This ensures that the product is made from the organic leather and checked all through the whole creation chain. The Oeko-Tex authentication is valid for one year to the companies. The Oeko-Tex Standard 100 is an all-around uniform testing and affirmation framework for leather raw materials, intermediate products and end products in all phases of creation. With this mark, makers guarantee to their users that their items have been upgraded for human ecology and their creation conditions are eco-friendly. Nowadays, all the manufacturers are trying to achieve the idea of green manufacturing, through the certifications, for example, Oeko-Tex [14]. This mark can be applied to the material or accessory giving the general and extraordinary conditions for giving authorization are fulfilled and if authorization to utilize this standard on an item has been conceded by an organization. The mark on the product denotes that the product has been checked for harmful that are prohibited or controlled by the standard. The certification covers various human–environmental characteristics, including unsafe substances that are restricted or managed by law, synthetic substances that are known to be a hazard to human beings. For effective certification, various parts of an article must meet the necessary criteria. The certification becomes a fixed segment in companies and quality frameworks throughout the manufacturing process which is a higher advantage to the producers, retailers and customers. The other components of leather materials like sewing thread, buttons, zip fasteners, linings, inserts, prints, labels are checked by the Oeko-Tex Standard 100. The criteria index, which depends on universal test principles and other perceived test forms, incorporates around 100 test parameters and is similarly authoritative for all approved Oeko-Tex organizations. The limit values are more stringent than the national and international specifications which include

- Significant statutory guidelines such as banned AZO coloring agents, chromium (VI), PFOS and lead (US-CPSIA)

Table 1 Limit value of specific items based on the Oeko-Tex standard [7]

Specific items	Product I	Product II	Product III	Product IV
Pentachlorophenol (PCP) (mg kg^{-1})	0.05	0.5	0.5	0.5
Tetrachlorophenol (mg kg^{-1})	0.05	0.5	0.5	0.5
Pb (lead) (mg kg^{-1})	90.0	90.0	90.0	90.0
Cd (cadmium) (mg kg^{-1})	50.0	100.0	100.0	100.0
As (arsenic) (mg kg^{-1})	0.2	1.0	1.0	1.0
Ni (nickel) (mg kg^{-1})	1.0	4.0	4.0	4.0
Hg (mercury) (mg kg^{-1})	0.02	0.02	0.02	0.02
Co (cobalt) (mg kg^{-1})	1.0	4.0	4.0	4.0

- Various chemicals are very hazardous to health regardless of whether they have not yet been legally managed
- Numerous naturally significant substance classes.

Products are investigated for the harmful compounds: cancer-causing colorants, legitimately controlled substances, for example, formaldehyde, plasticizers, metals or pentachlorophenol, etc. Oeko Standard identifies four types of product classes: (i) product for babies up to 3 years of age, (ii) product used to close the skin, (iii) product utilized away from the skin and (iv) decorative materials. Oeko-Tex also allows only certain skin to be used for leather product, namely sheepskin, lambskin, goatskin, cattle and cowhide, calfskin and horse leather. The standard sets very stringent conditions for child and baby material items; for instance, the most extreme estimation of formaldehyde content admissible is < 16 ppm. The real limit for formaldehyde content is 20 ppm; 75 ppm is the formaldehyde limit for items that interact with the skin, for example, bed cloth, clothing and shirts. Indeed, even items that do not come into constant contact with the skin, for example, outerwear and goods, are restricted to formaldehyde content lower than the value of 300 ppm. The limit value of specific items in product making according to the Oeko-Tex product classification is displayed in Table 1.

2.8 Recycled Claim Standard (RCS)

Recycled Claim Standard is used to follow recycled crude materials through the supply chain. The RCS confirms the presence and quantity of reused material in a final product. This occurs through info and verified through the third customer. It takes into account the predictable and extensive assessment and check of reused material substance on items. RCS can be utilized as a business-to-business tool to give organizations the way to guarantee that they are selling quality items. It is additionally utilized as an approach to guarantee exact and genuine communication with customers. Some of the objectives of RCS are: Track and follow recycled input materials and provide certification that materials are recycled and in a final product. The Recycled Claim Standard (RCS) is a worldwide standard that sets prerequisites for other parties. Associations engaged with the creation and exchange of RCS items are liable to RCS certification. Purchasers of the RCS item will be dependable to set any further necessities on the particular guidelines or prerequisites to which the info material will be ensured. These extra necessities are independent on the RCS and its certification procedure. The Recycled Claim Standard is expected for use with any item that contains at any rate of 5% recycled material. Each phase of creation is required to be affirmed, starting at the reusing stage and the last stage. The RCS does not give information based on social or ecological parts of preparing and manufacturing quality, etc. RCS clients are recyclers, makers, brands and retailers, certification bodies and associations that help reused material activities.

2.9 SA8000 Certification

The SA8000 Standard is the main social certification standard for leather and footwear sectors and associations over the world. It was built up by Social Accountability International in 1997. It is valued by brands and industry pioneers for its thorough way to deal with guaranteeing the highest quality of social consistency in their supply chains. It is a certification standard that is the organization that must be reviewed by an autonomous association, and it is one of the world's first and most broadly received measures for social issues [8, 16]. As the initial phase in the SA8000 Certification process, the association undertakes a self-assessment of the online management system. Self-assessment enables the association to comprehend the requirements of SA8000 and whether it is prepared to apply for certification. When the association considers its administration is enough to seek after certification, it chooses and works with one of the more than twenty SA8000 Certification bodies to begin the full assessment process. The certification body's assessment against the SA8000 Standard contains audits of documentation, working practices, worker responses and operational records. When the certification body has established that the association has executed the essential activities and upgrades to get consistent with the standard, it gives an SA8000 Certification, which might be utilized by the association to expose its accomplishment.

The normal period time for the certification procedure fluctuates based on the availability of the organization to meet the necessities of SA8000. The review procedure is a two-phase review process. The first phase is based on scoping and planning the certification audit to understand the organization. The second phase of the review is an assessment of compliance practices against the SA8000 against the necessities. Particularly, it measures the performance of organizations in eight areas like child labor, forced labor, health and safety, free association and collective bargaining, discrimination, disciplinary practices, working hours and compensation. SA8000 additionally takes existing global agreements and conventions from the International Labour Organization, the Universal Declaration on Human Rights and the United

Nations Convention on the Rights of the Child. SA8000 accreditation implies an association must consider the social effect of their activities under which their workers work.

2.10   Sustainable Fair Trade Management System (SFTMS)

The Sustainable Fair Trade Management System (SFTMS) is the primary accreditation approach that helps to overcome the disadvantages in the Fair Trade certification system. It is proposed to confirm that an association follows Fair Trade in all of its processes. When ensured, the association will have the option to utilize the name on the items over the entire items. SFTMS is the new overall standard for the free accreditation of associations that show Fair Trade strategic policies. The prerequisites of this standard are appropriate to any association applying Fair Trade standards in its business. This new overall standard structure supplements the current Fair Trade item naming methodologies. Sustainability in this certification means social, environmental and economic issues. The establishment and foundation of this standard are the World Fair Trade Organization (WFTO) Fair Trade Principles which give a strong support and value-based models.

2.11   Higg Index

Higg Index is a tool to ensure the sustainability and measure product sustainability performance of each company. It is a self-evaluation standard for surveying natural and social maintainability by the supply chain. It enables organizations to make significant enhancements in the protection of the environment and environment. This certification was propelled in 2012 and created by the Sustainable Apparel Coalition (SAC). SAC is implied for sustainable production in the footwear and leather industries. It gives a solitary way to deal with sustainability. It built up an estimation device called the Higg Index. It is the primary open rollout from the diverse group of makers, retailers, non-governmental associations. Higg Index acts as a tool for the footwear industries to maintain sustainability throughout the entire cycle of the product. The index asks practice-based questions to check social and ecological sustainability performance and helps to improve measurement and assess the social and environmental performance of footwear products at the brand and product. It measures the ecological and social impacts across the life cycle of the footwear product. Estimating this information can assist support with providing improvements. The Higg Index is a learning tool for organizations to distinguish difficulties and identify improvements. The Higg Index poses inquiries about the brand and facilities in the whole product making from packaging, manufacturing and transportation. This also helps to comprehend the effect of items and manufacturing process, while additionally emphasizes the opportunities for development and improvement.

2.12 Source Map

It gives support to the sustainable decision making in which makers share point-by-point data about their procedures with their purchasers and their purchasers and right to the end customer. Source Map ecolabel focusses on data on an item's segments and their starting points and natural and social footprints. A Source Map is an explorer in the production network.

2.13 Internationale Verband der Naturtextilwirtschaft (IVN) Natural Leather Standard

IVN was established in the year 1999 and has created two quality seals: Naturtextil IVN certified BEST and Naturleder IVN certified. These two seals are liable for the protection and evaluation of the whole material creation chain, as far as both environmental benchmarks and social responsibility. The point of IVN is to expand the awareness among the consumers and retail trade. It characterizes and actualizes explicit criteria to set natural and social responsibility in the creation process and high-quality standards in finished products. The IVN standard gives environmental quality and empowers eco-friendly leather items. Certified material may convey the IVN certified or IVN certified BEST label. IVN offers an affirmation procedure for all makers, merchants, producers and providers in the leather industry. The IVN Natural Leather Standard characterizes necessities for the creation of quality leather specialized, biological and human health parameters. The accreditation covers the entire preparing chain through naturally manufacturing process up to marking of the leather to give a trustworthy confirmation to the end consumer. By using this IVN, a business can reach sustainability easily and promptly in a conceivable way. IVN strategy necessitates that issuing labels to the product, testing organizations and certifications to the business is independent of each other. Institute testing is conducted regularly and nationally accredited. Each progression in the creation chain of material is checked on site. This ensures higher creditability to the organizations. This standard can be compared with the Global Organic Textile Standard (GOTS) for textiles as it contains comparable limitations on harmful substances and contains comparative social guidelines for all creation steps. It explains all the creation stages from raw material and the use of the finished item. IVN is known by the world due to its commitment to the GOTS in the materials. Certification offers leather sectors a few benefits initiating with believability and security in the sector and additionally effective sourcing.

2.14   Leather Working Group (LWG)

Leather working group rates (gold, silver or bronze) on leather sectors dependent on how their creation forms influence the environment. Reviews should be possible by a few outsiders utilizing a similar set of standards. They consider things like waste administration, vitality utilization, water use, detectability, hazardous substances. It promotes sustainable business form and ecological needs throughout the leather sectors. This group is comprised of brand and retailers; leather makers; dealers of finished product; product manufacturers, and technical experts have cooperated to build up an ecological stewardship protocol particularly for the leather fabricating industry. Brands are encouraged to review their providers through the Leather Working Group, to advocate suitable natural strategic policies within the leather sectors. This normally has a beneficial outcome for purchasers, indicating that leather utilized in the manufacture of a leather product has been made in a naturally proper manner. The LWG review to the LWG Environmental Protocol evaluates tanneries and leather makers. After the reviewing process, the business being examined will get an LWG Environmental Audit Protocol Responses Report containing their general accreditation. The major driving force of this group is to advance the improvement in the leather-producing industry, featuring and upgrading the best practices that will support improvements within the leather industry. This normally has the best outcome for buyers, demonstrating that items are fabricated in a supportable way.

2.15   National Science Foundation (NSF)

NSF standard was established in the year 1994. The primary assessment is Zero Discharge of Hazardous Chemicals Manufactured Restricted Substances List (ZDHC MRSL), screens for hazardous chemicals in the manufacture of leather. MRSL contains a list of chemical substances that are prohibited from global use in leather and footwear sectors. The second assessment is the ToxFMD Screened Chemistry Program incorporates the ZDHC MRSL screen yet, in addition, consolidates extra substance surveys. ToxFMD surveys the human and natural dangers related to every compound inside utilized in the leather sectors. The central goal of NSF is to secure and improve worldwide human well-being. Manufacturers, controllers and makers look into this NFC to encourage the improvement of general public health and certification that help to secure consumer products and environment. NSF accreditation gives all partners, industry, controllers and clients that a guaranteed item, material and segment follow the specialized prerequisites of the referenced standard. The NSF accreditation process is explicit to the item, procedure and kind of confirmation; however, for the most part, the following seven stages are displayed in Fig. 1.

NSF Certification

Fig. 1 Seven stages of NSF certification process

2.16 Worldwide Responsible Accredited Production (WRAP)

Overall Worldwide Responsible Accredited Production (WRAP) is the world's biggest sector-based accreditation program for manufacturers of leather and footwear items. WRAP was framed which helps to create an independent body to support footwear industrial facilities around the globe confirm that they are working in consistence with laws and universally accepted standards of ethical practices. WRAP has developed to turn into a worldwide pioneer in social consistence and a supply chain partner for organizations around the world. Its facility-based model has made it the world's biggest autonomous social consistence affirmation program for the footwear. The WRAP program ensures opportunities for consistence with the 12 WRAP Principles which guarantee protected, lawful and moral assembling forms. A WRAP accreditation review will assess the whole creation procedure to guarantee that the product is created with the WRAP Principles and gives the clients confirmation that the products are morally manufactured. The 12 principles of WRAP are

- Consistence with laws and work environment guidelines
- Restriction of forced labor
- Restriction of kid work
- Restriction of abuse
- Compensation and benefits as required by law
- Long-time of work as restricted by law
- Prohibition of discrimination
- Health and security
- Freedom of association and collective bargaining
- Environment
- Customs compliance
- Security.

WRAP is important to the footwear and leather sectors because it ensures that the laborers are treated humanely, helps to develop the brand among the customer in a reasonable way, increases the safety, improves worker commitment, develops efficiently and helps to avoid the damage of the product. WRAP comprises a three-level program, which grants confirmation to companies in understanding the level of consistence with the 12 WRAP Principles. Platinum (valid for a long time): Facilities exhibit full consistence for three sequential accreditation reviews. Gold (valid for 1 year): Facilities show full consistence with WRAP's 12 Principles. Silver (valid for a half year): Facilities exhibit significant consistence with minor issues. The certification process is mentioned in Fig. 2.

2.16.1   Silver Certification Level

A sector may demand a silver certification if a review sees it as insignificant consistence with WRAP's 12 Principles and, however, recognizes minor non-compliances in approaches, systems or manufacturing. Facilities requiring silver certifications must submit the action plan that incorporates any proof of remediation. The WRAP Review Board may prescribe a silver endorsement if a facility has exhibited other hazard factors that may be avoidable from supporting consistence for the full duration of a gold authentication.

2.16.2   Gold Certification Level

Gold accreditation is the standard WRAP affirmation level, granted to companies that exhibit full consistence with WRAP's 12 Principles.

2.16.3   Platinum Certification Level

Platinum accreditations are granted to companies that have shown full consistence with WRAP's 12 Principles for 3 continuous audition reviews. Platinum-awarded companies should effectively pass each review with no restorative activities and keep up persistent confirmation without any holes between certification periods.

3   Case Study

This case study is based on a tanning agent developed by wet-green GmbH for natural leather. Cradle to Cradle certification was given to the wet-green GmbH, and they produced a tanning agent for the production of natural leather. This is produced from plant extract like an olive leaf. As per EPEA, this is the first item which is gold affirmed concurring the new C2C standard. Tanning agents called OBE are

Fig. 2 Certification process
of WRAP

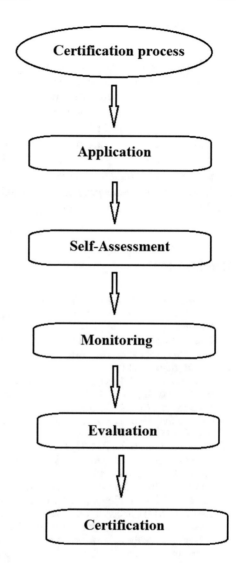

available in some of the cosmetic articles and olive oil. It is neither destructive nor a dangerous substance which thus makes for improved coordination. OBE tanning agent is additionally free from metals. In the case of natural and environmental aspects, the item is physiologically harmless to the whole worth chain. Contrasted with traditional tanning strategies, it not just ensures most extreme manageability and secures the earth, yet also outlays on acids, salts and suntans. For the first time, the vegetable tanning process is able to produce leathers with good shape retention and with softness and life span. The subsequent premium quality leathers are reasonable for applications in a wide range of regions like furniture for the private and business/open areas, shoes and attire.

4 Conclusion

This chapter has discussed the certification for sustainability in leather and footwear sectors. It also gives a brief outline of the concept of certification, and principles behind each standard were explained. The major challenge faced by these sectors is environmental pollution during the manufacture of the final product. The utilization of certification in the leather and footwear sectors influences the sustainable performance of the manufacturing process and final products. This implies assessing the ecological effects of a specific item through its whole life from cradle to grave. Any certified product should demonstrate to have brought down its environmental effect in every cycle stage. For a progressively sustainable industry, there is a need to extend the skill and data in the manufacturing stage concerning sustainability.

References

1. Braungart M, McDonough W, Bollinger A (2007) Cradle-to-cradle design: creating healthy emissions e a strategy for eco-effective product and system design. J Clean Prod 15:1337–1348. https://doi.org/10.1016/j.jclepro.2006.08.003
2. Caniato F, Caridi M, Crippa L, Moretto A (2012) Environmental sustainability in fashion supply chains: an exploratory case based research. Int J Prod Econ 135(2):659–670. https://doi.org/10.1016/j.ijpe.2011.06.001
3. Clancy G, Froling M, Peters, G (2015) Ecolabels as drivers of clothing design. J Clean Prod 99: 345–353. https://doi.org/10.1016/j.jclepro.2015.02.086
4. De Brito MP, Carbone V, Blanquart CM (2008) Towards a sustainable fashion retail supply chain in Europe: organisation and performance. Int J Prod Econ 114(2):534–553. https://doi.org/10.1016/j.ijpe.2007.06.012
5. Elder SD, Zerrif H, Le P (2013) Is Fairtrade certification greening agricultural practices? An analysis of Fairtrade environmental standards in Rwanda. J Rural Stud 32:264–274. https://doi.org/10.1016/j.jrurstud.2013.07.009
6. Fletcher K (2012) Durability, fashion, sustainability: the processes and practices of use. Fash Pract 4(2):221–238. https://doi.org/10.2752/175693812X13403765252389
7. Gersak J (2013) Quality requirements for clothing materials, design of clothing manufacturing processes, pp 250–294. http://doi.org/10.1533/9780857097835.250
8. Gilbert DU, Rasche A, Waddock S (2011) Accountability in a global economy: the emergence of international accountability standards to advance corporate social responsibility. Bus Ethics Q 21(1):23–44. https://doi.org/10.1017/S1052150X00013038
9. Islam S, Shahid M, Mohammad F (2013) Green chemistry approaches to develop antimicrobial textiles based on sustainable biopolymers-a review. Ind Eng Chem Res 52:5245–5260. https://doi.org/10.1021/ie303627x
10. Kopnina H (2019) Green-washing or best case practices ? using circular economy and Cradle to Cradle case studies in business education. J Clean Prod 219: 613–621. http://doi.org/10.1016/j.jclepro.2019.02.005
11. Llorach-Massana P, Farreny R, Oliver-Sola J (2015) Are cradle to cradle certified products environmentally preferable? analysis from an LCA approach. J Clean Prod 93:243–250. https://doi.org/10.1016/j.jclepro.2015.01.032
12. Longoni A, Golini R, Cagliano R (2014) The role of new forms of work organization in developing sustainability strategies in operations. Int J Prod Econ 147:147–160. https://doi.org/10.1016/j.ijpe.2013.09.009

13. McDonough W, Braungart M (2013) The upcycle, the upcycle: beyond sustainability-designing for abundance
14. Nayak R, Akbari M, Maleki S (2019) Recent sustainable trends in Vietnam' s fashion supply chain. J Clean Prod 225: 91–303. https://doi.org/10.1016/j.jclepro.2019.03.239
15. Peterson M (2004) Cradle to cradle: remaking the way we make things. J Macromark 24(1):78–79. https://doi.org/10.1177/0276146704264148
16. Rajabzadeh M (2013) Statistical analysis of certification process of international standard SA8000 on social accountability. J Int Soc Res 24(6):306–315. https://doi.org/10.4028/www.scientific.net/JERA.9.67
17. Scruggs CE (2013) Reducing hazardous chemicals in consumer products: proactive company strategies. J Clean Prod 44:105–114. https://doi.org/10.1016/j.jclepro.2012.12.005
18. Wakankar DM (2013) 5-regulations relating to the use of textile dyes and chemicals. In: Gulrajani M.L.B.T.-A. in the D. and F. of T.T. (ed) Woodhead Publishing Series in Textiles. Woodhead Publishing, pp 105–132. https://doi.org/10.1533/9780857097613.1.105

Brazilian Sustainability Outlook in Footwear Sector

Lais Kohan, Cristiane Reis Martins, Heloisa Nazare dos Santos, Palloma Renny Beserra Fernandes, Fernando Brandao, and Julia Baruque-Ramos

Abstract The footwear production in the world exceeded more than 21 billion/pairs in 2017. The largest producer (11.4 billion/year) and exporter (8 billion/year) is China, manufacturing a wide variety of models and being USA its main purchaser. Brazilian footwear production was 950 million/pairs in 2019, corresponding the fourth largest production in the world. While Brazilian footwear is mainly focused on national market (around 87% of the production), the first three largest manufacture countries (China, Vietnam and Indonesia) aim at exporting their products to European and North American brands. This study aimed to verify the panorama and the sustainable solutions being carried out by Brazilian footwear sector. The methodology was based in literature research and informations provided by Brazilian companies. Brazilian footwear industry is finding solutions to reduce production costs in order to diversify shoe models and compete in the international market. One

L. Kohan · H. N. dos Santos · P. R. B. Fernandes · J. Baruque-Ramos (✉)
School of Arts; Sciences and Humanities, University of Sao Paulo, Sao Paulo, SP, Brazil
e-mail: jbaruque@usp.br

L. Kohan
e-mail: laiskohan@hotmail.com

H. N. dos Santos
e-mail: heloisa.santos@uemg.br

P. R. B. Fernandes
e-mail: palloma_renny@hotmail.com

C. R. Martins
Institute of Environmental; Chemical and Pharmaceutical Sciences, Federal University of Sao Paulo, Diadema, SP, Brazil
e-mail: cr.martins@unifesp.br

H. N. dos Santos
Design School, University of Minas Gerais State, Belo Horizonte, MG, Brazil

F. Brandao
CIPATEX, Cerquilho, SP, Brazil
e-mail: fernando.brandao@cipatex.com.br

of the strategies is the gradual replacement of leather by fabrics, synthetic polymers and rubbers. In 2017, the correspondent amounts were: 49.0% plastic/rubber; 28.8% synthetic laminate; 17.7% leather and fabrics—only 4.5%. Another important issue to reduce cost and distance is the cluster agglomeration, whose there are 13 clusters in 6 Brazilian states. The employment of natural fibers in footwear corresponds to a small part in comparison of synthetic fibers. Cellulosic fibers are being researched as biodegradable materials, showing innovations for instance, the use of leather waste, coconut fiber and latex, multilayers fabrics and finishing process adding bactericide properties to cotton. Particularly, the employment of cotton, raffia and jute fabrics outstands in footwear summer collections. However, more consistent initiatives related to the research of alternative materials, waste reduction, reuse and recycling are being carried out. New standards in waste disposal have forced major changes in production processes, and national seals also have been supporting these initiatives to manage processes in shoe, components factories and tanning. There are alternatives to chromium in tanning employing vegetable tannin. In addition, the design conception assists the durability increasing and post-consumption recycling easiness of its components, for instance reducing the number of components through digital fabrication. Thus, the increasing of researches about recycling footwear components and the development of biodegradable materials and processes points a sustainability trend in this sector.

Keywords Brazilian footwear · Sustainability · Biodegradable · Cellulosic fiber · Design · Durability · Recycling · Waste reduction · Reuse · Vegetal tannin

1   Introduction

Brazilian footwear sector presents a huge potential to green sustainable products and processes, has been growing in terms of the development of technological materials, as well as their application in the market, and new guidelines are already focusing on strategies for improve the manufacture of conventional and sustainable footwear.

Brazil is the fourth largest shoe producer in the world, with a total production of 950 million pairs (2019) [1]. In per capita consumption, the country was only 31st place with an average of 4 pairs/year [2, 3]. The Brazilian import in 2019 corresponded to US$ 373.9 million, of which 49.2% of sports shoes and 47.6% made in Vietnam. While exports represented US$ 967.1 million in 2017, which mainly composed of shoes with synthetic uppers (45.4%), indicating that the country maintains a positive trade balance [3].

Together the conventional shoe's production, at the same time, small and micro-enterprises working in niche markets, including that sustainable ones, which incorporate ecodesign principles in their production, are arising and growing in Brazil.

Brazil is the second largest cattle creator in the world, behind India only [4, 5]. Despite the fact that the production of Brazilian leather shoes is not as expressive in 2017, the bovine leather market is still of great importance in exports. Brazilian exports of bovine leather in 2018 totaled 143.1 million m^2 and is divided by the three types of leather processing: wet blue (chrome-tanned, but neither dried, dyed nor finished)—43.8%; crust (dried after tanning but has not yet been dyed)—8.8%; and finished (treated with a top coating)—47.4%; being that these types follow an increasing order of the leather tanning production process [4].

In addition to bovine leather, which is a major player in the leather industry, fish and goat leather stand out in the Brazilian hides' industry. Among the fish species can be highlighted "pirarucu" and "tilapia" [6–8].

In general, the synthetic materials most employed are: polyurethane (PU); polyvinyl chloride (PVC); synthetic rubber of styrene/butadiene (SBR); ethylene/vinyl acetate copolymer (EVA); and thermoplastic rubber (TR)[9]. For shoe uppers are: (i) bovine leather (majority) or from other animals; (ii) textiles from synthetic (nylon, polyester, polypropylene, elastane, etc.) or natural fibers (cotton, jute, linen, etc.) in weave, knit or nonwoven constructions; (iv) synthetic laminates (polyurethane, polyvinyl chloride, etc.) [10].

Aspects of shoe production that can make shoes more environmentally sustainable are pointed mainly: (i) Sustainable shoe material selection; (ii) environmentally friendly footwear production processes; (iii) waste reduction in footwear manufacturing; and (iv) sustainable shoe material selection [10]. Also, for the implementation and management of cleaner production in plastic footwear companies: (i) energy and water consumption; (ii) green technology and innovation: (iii) environmentally friendly materials; (iv) waste and residues generation; (v) environmental efficiency; (vi) human resources; and (vii) environmental management [11].

In addition to these issues, this study is composed by this introduction (Sect. 1 subitem) and eight following main topics. Sects. 2 and 3 detail data on Brazilian footwear production, materials and processes, including the national seal of sustainability of the footwear industries, restriction of harmful substances in the components and use of regulated cruelty-free skins. Sect. 4 reports waste management and prevention in the footwear industry. Sect. 5 outlines the sustainable materials in Brazilian footwear, by means of biodegradability on materials and processes: natural fibers, Amazonian latex, vegetal leather tanning, and the use of synthetic laminates for extending durability and as a vegan option. Sects. 6 and 7 explore sustainable design concepts, Brazilian ecodesign in footwear and recycled raw materials. Sect. 8 indicates the increase in ecofashion fairs and market, besides small brands aiming sustainability, in which their products are engaged in this purpose. At the end, Sect. 9 presents last considerations in conclusion.

2 Panorama of Brazilian Footwear Industry

2.1 A Brief History of the Brazilian Shoe Industry

The origin of Brazilian footwear production is closely related to European immigration in the country. German immigrants settled in the southern region (in the state of Rio Grande do Sul) and inaugurated the first rudimentary site for the manufacture of saddles and shoes in the late nineteenth century. Then, at the beginning of the twentieth century, Italian immigrants who settled near the city of Franca (interior of São Paulo), after the decline of agricultural coffee production in the region, opted to work in the footwear industry, since they already had experience in your home country. In the South, with the advent of the Paraguayan war (1864–1870), it was necessary to expand and improve the manufacturing quality in making footwear, which was still entirely handmade [12, 13].

The advent of the sewing machine was an important event and several other technological improvements were made until 1920, what made possible for the production to become somewhat more industrialized [13]. Only in 1960, after 40 years of stagnation, that production gained volume and also began to export. Exports intensified with investments in productivity, technology and government incentives through rate reductions, so between 1970 and 1990 there was an increase in exports. In 1973, 22 million pairs were sold with a gain of US$93 million, in 1983, 93 million pairs and US$682 million were exported, the largest gains occurred in 1993 in 201 million pairs and US$1.846 million, being that about two thirds were directed to the North American Market [12, 14].

In the period of strong growth in Brazilian exports (1970–90), Vale do Rio dos Sinos region, footwear cluster in the extreme south of the country, had great prominence in exports, with the amount varying between 40/50% of exported volume. Since the Americans were the main buyers, the reasons that contributed to buy decision were as follows: (a) the well-organized and experienced leather and footwear production chain; (b) the good quality of local suppliers; (c) the stable and specialized business environment; and (d) tax incentives for exportation offered by the Brazilian government [15].

During the 1990s, there was a drastic reduction in Brazilian exports, largely because internal issues, as well as to currency overvaluation and market opening, but there was also a great influence of China's entry into the international competitive environment [16]. To the extent that prices from China were lower than the production costs in other footwear-producing countries, there was the advent of the transfer of the production process of major international sports brands of sneakers that started to manufacture in China. Vietnam and Indonesia also have lesser cost of production and along with China are major exporters of shoes worldwide [15].

Following these changes, the Brazilian footwear sector reacted in order to reduce the production cost: producing regions had shifted to gain tax benefits and find a cheaper workforce. Another strategy to reduce the production cost was the

Table 1 Exportation of Brazilian footwear by type of raw material in the years of 2003 and 2011 [2]

	Leather (%)	Synthetic (%)	Textile (%)	Inject (%)	Others (%)
2003	63	26	5	5	1
2011	22	73	4	4	1
2019	15.2	76	9.2	0.3	0.5

replacement of the shoes' raw material, in order to substitute leather [17, 18]. The replacement of material over time can be seen at exportations data (Table 1).

In addition to the substitution of raw materials, there was a geographical change in part of footwear companies in the south and southeast, due to the offer of salary, tax and financial advantages [19].

Until the 1980s and 1990s, the south and southeast regions held about 68% of national production, compared to the Northeast, which was not very active, when in 1991 it held only 3.3% of formal direct jobs in the sector. But, already in 2001, there was a transformation, jumping the amount of jobs in the region, to more than 20% of the national sector [20].

Finally, new relationships were established with China's entry into the world footwear market. The economic importance of Brazilian exports was reduced, but even then, there was an increase in the price of the pair of shoes, an increase in the productivity of factories and new configurations in the national market.

2.2 Manufacturing Clusters

The Brazilian footwear sector was impacted by the reduction of exports since the 1990s. The internal reaction occurred, in addition to the replacement of materials, also by the development of the global value chain and influenced in shorter product life cycles, new configurations in the way of production, in the relationship between suppliers and in geographical locations [21, 22]. Prior to this decade, it was common for Brazilian industries to be vertical, including all stages of the leather footwear production cycle: creation, cutting, pre-sewing, sewing or pre-joining, assembly and finishing; this demanded an intense use of specialized labor and a slow development process [21].

The result of these changes in configurations was the development of regional hubs (clusters) with segmentation of the production stage. Thus, part of the processes is outsourced or suppliers sell finished components, with better costs [17, 23]. In Fig. 1 the main Brazilian footwear clusters are presented.

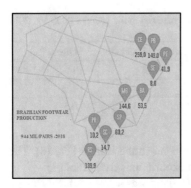

STATE	AMOUNT (%)	MAIN CLUSTER	CLUSTER SECTOR PRODUCTION (%)	
Ceará (CE)	27.4%	Sobral	Female	60.4%
Rio Grande do Sul (RS)	20.1%	Vale dos Sinos	Female	41.6%
Paraíba (PB)	15.8%	Campina Grande	Flip Flop	96.9%
Minas Gerais (MG)	15.3%	Nova Serrana	Sportive	50.9%
São Paulo (SP)	6.4%	Birigui	Children	47.2%
Bahia (BA)	5.6%			
Pernambuco (PE)	4.4%			
Santa Catarina (SC)	1.6%	São João Batista	Female	77.4%
Paraná (PR)	1.1%			
Sergipe (SE)	0.9%			
Others	1.3%			

Fig. 1 Quantity of shoes produced by state in Brazil and the main footwear clusters in 2018 [2]

2.3 Production Data

Brazil is the fourth largest shoe producer in the world, with a total production of 943 million pairs (2017), only behind China (11.4 billion pairs), India (2.8 billion pairs) and Vietnam (1.2 billion pairs). In 2018, the footwear sector added up to 271.1 thousand formal jobs, in a total of 6.6 thousand establishments [2].

In 2017, the Brazilian consumption, classified by type of materials employed in shoes, was: plastic and rubber, 52.3%; synthetic laminates, 25.6%; leather, 18.1%; textiles, 3.1%; and others, 0.9%. This followed the similar trend of the large volume of exports of synthetic footwear. In per capita consumption, the country was only 31st place with an average of 4 pairs/year, while the USA, leader of the ranking, the consumption is 7.2 pairs/year [2, 3].

Brazilian footwear exports have less impact on the sector's market, when compared to previous periods. In 2017, only 13.5% of the Brazilian produced shoes (943 million pairs) was exported [2, 3]. Data show the Brazilian import in 2019 corresponded to US$373.9 million, of which 49.2% of sports shoes and 47.6% made in Vietnam. While exports represented US$967.1 million in 2017, which mainly composed of shoes with synthetic uppers (45.4%), indicating that the country maintains a positive trade balance [3]. One of the justifications for the positive trade balance in recent years is the application of the antidumping fee, correspondent to the amount of US$10.22/pair, established since 2009 and valid until March 2021 [14].

It is important to highlight the consequences of the tax on imported shoes to Brazil, as this occurred at a time of growth in imports of Chinese finished shoes between 2001 and 2009, reaching more than US$160 million in 2008 (Fig. 2). In addition, the market reaction in relation to taxation was the dribbling of Brazilian companies when bringing separate components from China. This can be observed from 2010 to 2019 in Fig. 2, in which there was a substantial increase in the import of components, and, in the same period, reduction of the import of ready shoes [19].

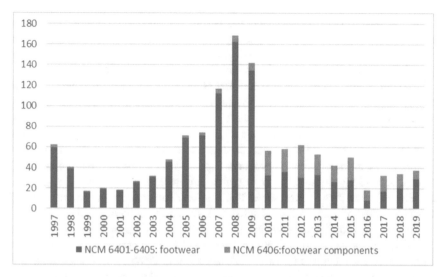

Fig. 2 Brazilian imports of Chinese shoes versus footwear components from 1997 to 2019 (in US$ million), highlighting the difference between the periods 1997–2009 and 2010–2019 [19, 24]

2.4 Brazilian Certification of Footwear—"Sustainable Origin": A Safety to Exported Products

This text was developed based on data collected from scientific articles, websites of the participating entities, the manual of seal implementation "Methodology for Implementing Sustainability Requirements" and an interview with the coordinator of innovation projects at Instituto By Brazil, Mrs. Linda Pienis, who accompanied the development of the seal since the beginning in 2010.

The "Sustainable Origin" seal was developed jointly between Brazilian sectoral footwear associations—the responsible for the component industries, Assintecal (Brazilian Association of Companies of Components for Leather, Footwear and Artifacts), and the responsible for the shoe manufacturing industries, Abicalçados (Brazilian Association of Footwear Industries). The seal came into effect in 2013, and in 2019 (July), a reformulation was launched. The current model of the seal is more demanding, mainly in obtaining the first category.

The certification aimed to increase engagement on sustainability issues, expand export opportunities and align the footwear and components industry with the UN Sustainable Development Goals, together with international programs and seals. Certification also assists in business management by seeing what happens internally in processes and actions for non-conformities, in addition to the involvement of people and leaders [25]. Also, to face international competition, the footwear sector has been investing in design, new technologies and sustainability—aiming to mitigate the industry's main impacts, such as the use of potentially toxic substances and diversity of working conditions [26].

In the new version of the seal, the before so-called pillars now are dimensions, and the sustainability management links all four dimensions of the program (economic, social, environmental and cultural). The dimension of sustainability has been included in all areas in way it is included in the company's business strategy.

According Mrs. Pienis, the seal contributes mainly to the issue of exports, in addition to valuing waste management and including concepts of the circular economy within companies. Other indicators were no less important, but the companies had already solved them well, such as: social (such as: job security and compliance with labor laws) and environmental (companies already have strict licenses). It was pointed out that exportation of products depends on the booklet that the customer brings, depending on the country in which they will be sold, showing which and the allowed quantities of substances, that footwear can contain. In this way, Table 2 shows the most relevant restricted substances in each shoe component.

The certification process begins with the signing of a contract between the company and the association to which it belongs (shoes or components), upon "payment of a fee and receipt of the program's methodology." When the company already feels prepared to be audited, it needs to provide a list of documents, as follows: CNPJ (an identification number issued to Brazilian companies by the Department of Federal Revenue of Brazil), contact person, Environmental License (state/municipal), Federal Environmental License (issued by IBAMA), Operating Permit (city hall), intended level of certification and scope of certification. The obtaining of the seal occurs after the audit finds in the company the data previously presented, according to 124 indicators among the five dimensions, in which there are 32 mandatories. There is a hierarchy between the types of seal, so when reaching 50% of the indicators, the corresponding seal is bronze, 75% is silver, 90% is gold, and 100% is diamond.

One of the main consequences of the seal for the chain, according Mrs. Pienis, was the integration of the sector, since the suppliers had to adapt to the requirements for the fulfillment of the seal and brought more security regarding the identification of the compounds of the final product. Because, in addition to the footwear exporter's duty to present the footwear certificate, in which states not contain or be at a permitted level in relation to the restricted substances mentioned (Table 2), it forced the chain, mainly component suppliers, to declare the raw materials contained in each part of the footwear.

3 Materials and Processes Currently in Brazilian Footwear Production

3.1 Brazilian Leather Production (Bovine, Caprine, Fish, Etc.)

One of the by-products generated and discarded by the meat-producing agribusiness is animal skin. When processed, it can become a raw material with high added value,

Table 2 List of the main restricted substances that can be contained in a shoe, according to the type and which part are found in the components [27]

Test	Substances	Components									
		L	T	ST	PM	CM	MC	CP	T	A	P
1	Alkylphenols (NP, OP) e ethoxylated alkylphenols (APEO, NPEO)	x	x	x	x	x		x	x	x	x
2	Azo dyes	x	x	x				x	x	x	
3	Chlorophenols	x	x								
4	Volatile organic compounds (VOCs)							x			
5	Chromium 6	x									
6	Formaldehyde	x	x			x		x	x	x	x
7	Dimethyl fumarate										x
8	Phthalates	x		x	x			x	x	x	x
9	Heavy metals (Arsenic, cadmium, lead and mercury)	x	x	x	x	x	x		x	x	x
10	Organotins			x	x			x	x	x	
11	Polycyclic aromatic hydrocarbons (PAHs)			x	x			x	x	x	
12	Residual solvents (DMF, DMAC, NMP e formamide)			x	x			x	x	x	
13	Nickel (Ni) release								x		
14	Dispersed dyes		x								

L leather, *T* textiles (natural, synthetic and blend fibers), *ST* synthetic laminates, *PM* polymer materials, *CM* cellulosic materials, *MC* metallic components, *CP* chemical products, *T* trims, *A* adornments, *P* packaging

leather. The skin is no longer waste and becomes a source of income. The process of transition from skin to leather needs to receive a treatment called tanning, where the waste from the food industry is treated with tanning substances to remain immune to the decomposition process [28]. The process of transforming animal skin into leather, in addition to preventing it from rotting, provides softness and flexibility to the material. The final product can be destined to different sectors, such as the textile, automobile, furniture and leather footwear industries. The hides used in the manufacture of leather can be originate from several animals, but bovine leather is the most found due to the size of the food industry related to this animal [29].

Brazil is the second largest cattle creator in the world, behind India only. By highlighting that a large part of the Indian herd is not destined for slaughter, Brazil ranks in the first place as the largest breeder of entirely commercial cattle in the world. As the meat industry, leather production also stands out in the international market, in which Brazil is among the largest leather exporters in the world [4, 5]

Despite the fact that the Brazilian production of leather shoes, the country has always imported leather for this purpose, and national leather has always been focused on exports. In 2013, the country was the second largest producer and third exporter in the world and, in that year, it totaled a record of exported value, US$2.511 billion, corresponding to a growth of 20.8% in relation to 2012 [30]. Since then, there has been a downward trend; in 2017, the exported value was US$1.899 billion and in 2018, US$1.442 billion [4].

Brazilian exports of bovine leather in 2018 totaled 143.1 million m^2 and is divided by the three types of leather processing: wet blue (chrome-tanned, but neither dried, dyed nor finished)—43.8%; crust (dried after tanning but has not yet been dyed)—8.8%; and finished (treated with a top coating)—47.4%; being that these types follow an increasing order of the leather tanning production process. A highlight in the segment was the increase in exports of finished leather in recent years, since in 2000, 57% of the exported product was wet blue, while in 2018, it dropped to 43.8% [4]. Thus, the increase in the use of leather with greater processing corresponds to the use of a product with greater added value [30]. Even though the country having finished leather produced domestically, there was an import of 2.27 million m^2 (2018), with a value of US$36.97 million [4].

The main purchasers of Brazilian leather are China (including Hong Kong), which accounts for 25% of the export in FOB value, followed by the USA, 17.1% and Italy, 16.9% according to 2019 billing data. The main leather exporting Brazilian regions are Rio Grande do Sul, São Paulo, Paraná and Goiás [31]. Approximately 77.6% of the leather produced in Brazil is exported, 40% of which is destined for the footwear, artifacts and clothing segments. In the domestic market, the footwear industry absorbs about 30% of domestic consumption, where most of it is employed by the automobile and upholstery industry [5].

The Brazilian leather industry is predominated by small- and medium-sized tanneries, in addition to small artisanal tanners that do not have formal records, which implies in the lack of accuracy in statistical data. In 2009, 702 sprayed tanneries were accounted for across the country. However, the small portion of large tanneries takes on a large proportion because they have direct link with the slaughterhouses. They hold about 60% of leather production in Brazil, which transfers them great relevance in the leather industry [32].

In addition to bovine leather, which is a major player in the leather industry, fish and goat leather stand out in the Brazilian hides' industry. Among the fish species can be highlighted "pirarucu" and "tilapia." Pirarucu (*Arapaima gigas*) is one of the largest rivers and lake freshwater fish in Brazil. It can reach 3 m and its weight can go up to 200 kg. It is a fish that is usually found in the Amazon basin [6, 7]. Tilapia is the common name for many species of freshwater cichlid fish (Cichlidae family and subfamily Pseudocrenilabrinae subfamily) and in particular from the genus Tilapia. They are native to Africa, but have been introduced in many places in open waters worldwide, including Brazil [8].

The use of fish leather in the footwear sector has stood out mainly for its ecological appeal. The Brazilian fishing industry produces an average of 800,000 tons of fish per year, where 8–12% of this represents the skin, which is mostly discarded because it is considered garbage [33]. The transformation of the waste from the fishing industry into leather is a sustainable alternative for the use of the by-product generated by the cultivated or fished animal, in addition to being an income-generating practice that contributes to the socioeconomic development of Brazilian fishing communities [7]. Considered as exotic and innovative, fish leather is very well received by the leather footwear sector, mainly in the luxury market.

Fish leather is a noble and high-quality product, with resistance as a peculiar characteristic. In addition to this characteristic, for fish species with scales, the protective coverslips at the insertion of the scales results, after tanning, in a leather of typical appearance and difficult to be imitated, guaranteeing exclusive patterning with high visual impact [34].

Tilapia is the main species cultivated in Brazil, and its skin is found in abundance (Fig. 3), but the leather of the pirarucu fish is the most exalted for its exuberance and resistance (Fig. 4) [35].

Garments originating from goatskin are light and soft. Goatskin is a resistant leather, widely used in the manufacture of women's shoes and bags. It can be worked for clothing in the quality of suede or kidskin (smooth), in pieces such as gloves and garments with flesh split. The best parchments and leathers for bookbinding are from goat. The screwed goat flesh skins have the same characteristics as the chamois, which is type of porous leather from sheep and lambs. It is a soft, highly elastic leather that has the ability to easily absorb and eliminate large amounts of water [34].

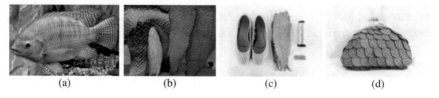

(a) (b) (c) (d)

Fig. 3 Tilapia fish: **a** animal [36]; **b** tanned leather [33]; **c** feminine slipper [37]; **d** women's pouch [38]

(a)   (b)   (c)   (d)

Fig. 4 Pirarucu fish (*Arapaima gigas*) [35]: **a** animal; **b** raw hide; **c** tanned leather; **d** fashion jackets

The goat leather industry is a niche of high importance and very characteristic of the northeast region of Brazil. Its concentration, located in the northeastern states, where 93% of the country's goats are created, is due to the climatic conditions of the region that favor the animal's breeding. Traditionally, there is a culture of vegetable tanning with "angico"—it is the common name of several trees of the genera Piptadenia, Parapiptadenia and Anadenanthera of the subfamily Mimosoideae. They are native to tropical America, mainly from Brazil, and are also explored and or culti-vated due to the good quality of their wood [39]—and other regional tannins, made in tanks and homemade streams for later the leather to be used in typical footwear and artifacts of Northeastern culture [40]. The cultivation of goats in Brazil is very promising for the leather segment; however, it is necessary to invest in improving the quality of the skin, which involves an adequate animal management and the improve-ment of the tanning processes currently employed [41]. More information about goat leather production in Brazil and the employed vegetal tanning is presented in item 5.3 (Leather tanning employing vegetal extracts).

3.2 Main Shoe's Components

The parts that make up a shoe have specific functions and require some characteristics of comfort and resistance, because despite the aesthetics involved by the social dress code and the great influence of fashion, the main function of a shoe is the protection of the feet [42].

In general, the main parts that build a shoe can be composed of upper, lining, stiffener and toe puff, insole, shank, midsole, outsole and heel (Fig. 5a, b).

According to the "Study of Quantification of Materials in the Brazilian Footwear" [44], considering the segments women, men, children, sports and security, the produc-tion of shoes in 2019 in Brazil can be classified as: only assembled (93.10%); injected and assembled (4.93%); and only injected (1.97%). Also according this reference, the employed materials in injected and assembled shoes are shown in Table 3.

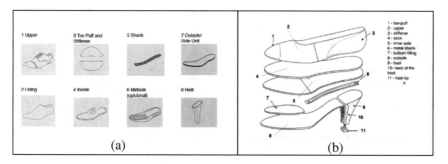

Fig. 5 Shoe [43]: **a** general components; **b** high heel pumps anatomy

Table 3 Materials employed in injected and assembled shoes [9, 44]

Materials of injected shoes	PVC, PU, TR, TPU, Rubber, EVA, others
Types of materials used in assembled shoe uppers	Leather, laminated PVC, laminated PU, textiles, others
Types of materials used in assembled shoe linings	Leather, laminated PVC, laminated PU, textiles, others
Types of materials used in assembled shoe soles	PVC, TR, PU, TPU, expanded PVC, rubber/board/SBR/latex, leather, EVA, others
Types of materials used in assembled shoe ornaments	Metals, plastics, imitation gemstones, textiles, others
Types of adhesive used in prefabrication, preparation and assembly of shoe	Solvent based, water based, hot melt, layers, others

3.3 Shoe Uppers

In general terms, the most common employed materials for shoe uppers are: (i) bovine leather (majority) or from other animals (tanned, suede, etc.); (ii) textiles from synthetic (nylon, polyester, polypropylene, elastane, etc.) or natural fibers (cotton, jute, linen, etc.) in weave, knit or nonwoven constructions; (iv) synthetic laminates (polyurethane, polyvinyl chloride, etc.) [10].

The shoe upper provides characteristics to auxiliary the modify and increase, it can improve the comfort and shoe fitting [45]. Depending on materials can or cannot dissipate heat and wick moisture away from the skin surface and influence of microclimate inside shoes [46].

Leather is the oldest material employed in shoes. This material provides full comfort to the user. This may be attributed to high capacity the evaporation of water into the environment, besides being soft in direct contact with the skin and still adapting to the shape of the foot [47].

Textile uppers are very common done by synthetic fabrics, as polyester and polyamide, due to the properties such as flexibility, air permeability, lower water absorption, quick dry, low weight and low cost [48]. However, fabrics of polyester

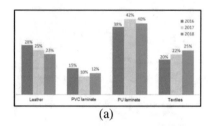 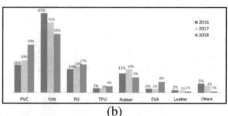

(a)   (b)

Fig. 6 Evolution of materials in the Brazilian footwear market from 2016 to 2018 [51]: **a** shoe uppers; **b** soles

(also called ethylene polyester—PET, which is the most commonly produced form of polyester fiber) are normally not so comfortable to users, because they cannot absorb the sweat properly, but it can be improvement by blends with cellulosic fibers for instance as polyester–bamboo fabric. Day use shoes does not have necessity of performance, and it permits to apply other fibers, as cotton twill, jute yarns, and it also occurs in Brazilian market [18].

Synthetic polyurethane (PU) laminate is generally produced by coating PU resin over a base fabric, forming open cells to maintain air permeability. Due to this molecular structure, PU has both tenacity and elasticity, and its physical properties vary greatly depending on the type and composition of the monomer which is used as the binder [49].

Synthetic polyvinyl chloride (PVC) laminate is used in the shoe upper construction as a substitute of leather because of the great versatility of colors, prints and thicknesses. The laminate is composed of PVC and textile fiber, usually in the mass ratio of 60/40 [50].

In the Brazilian materials' market for upper shoes, there is the tendency to decrease the use of bovine leather, meanwhile the growth of textiles remains and laminates of PU and PVC oscillate (Fig. 6a).

3.4 Sole (Outsole, Insole and Midsole)

Footwear comfort is influenced by several factors such as fitting the foot to the footwear, temperature and humidity in the internal environment, plantar distribution and impact of forces on the soil. The impact forces on the ground are the factors that most contribute to discomfort and can still cause pain or illness and this has to do with the construction of the sole [52, 53]. In general speaking, it can be considered in three parts: outsole, insole and midsole (Fig. 5a, b).

The evolution of sole materials in the Brazilian market from 2016 to 2018 is presented in Fig. 6b.

3.4.1 Outsole

The outsole, also known as the sole, is the bottommost part of a shoe that comes in direct contact with the ground. The outsole can be made out of a variety of materials, including leather and rubber. Certain types of outsoles provide more traction than others by using specific materials or designs. It all depends on the style and intended purpose of the shoe. Shoes are designed with durable outsoles to be as long-lasting as possible, but it will wear out over time, being in fact the part of a shoe that wears out first. Most shoes have soles made out of natural rubber, polyurethane or PVC compounds [54]. These materials are better specified as follows:

Rubber outsole
Rubber materials usually present the features of good shock absorption behavior, a good relationship between the shock absorption and elasticity, and resistance to fatigue besides particular physical properties, such as high tear and tensile strength, lower abrasion, resistance to oils and durability. The main rubbers are natural rubber (NR), synthetic isoprene rubber (IR) and styrene-butadiene rubber (SBR) [52].

Thermoplastics outsole
Thermoplastic rubber (TPR) outsoles can be produced quickly in high volumes and require inexpensive low-pressure injection molds. They do not have the slip resistance or durability, and for this reason, TPR does not used for performance footwear, but it is typically applied for making women shoes of good quality [10]. Even so TPR has positive attributes as water resistant, very flexible, resilient, and it can be subjected to recycling processes [55].

PVC is very popular because the low price, the texture can be adjusted by pre-injection molding, it can be hard or soft. The production of soles presents the advantages of other plastic materials, besides facilities to processing and plasticity [56].

Foam outsole
Ethylene-vinyl acetate (EVA) is a copolymer that can be used as thermoplastic and elastomer depending on vinyl acetate (VA) content in the copolymer. VA also reduces crystallinity and the improvement in flexibility, impact and a reduction in hardness [57].

3.4.2 Midsole

Midsole requires resilience, so foams are usually selected and closed cell ones have better capabilities, by being very dense. The most common closed cell foams include ethyl vinyl acetate (EVA), polyethylene (PE), styrene-butadiene rubber (SBR), polyurethane (PU), latex and neoprene, each one presenting specific properties [10].

Polyurethane or PU foam is a very popular choice for midsoles. Like EVA, it can be formulated into different densities suitable for many various types of footwear midsoles. PU presents superior results in durability and resistance tests when compared to EVA. It is used in sports shoes with high levels of impact absorption where the mass of the footwear is not the main objective [58].

Thermoplastic polyurethane (TPU) is a recent foam material designed by BASF™ [10]. It is one of the most versatile thermoplastics presenting elastomeric properties and higher tensile strength, high abrasion resistance and interfacial adhesivity [57].

3.4.3   Insole or Footbed

Footbeds are made of compression of molded EVA, poured polyurethane foam (PU), latex and cork, sponge rubber, or polyethylene (PE) foam, and also they can be cut from flat materials [10].

The sole materials in the Brazilian footwear market are shown in the Fig. 6, in which PVC consumption grew up significantly, meanwhile thermoplastic rubber declined in recent years [51].

3.5   Heels and Others (Midsole, Buttress, Toecap, Lining, Etc.)

The shape and volume of the fitting area of the toe cap are important, because when a shoe is tight, the toes can cause bruises and calluses and even affect the plantar distribution [59]. One of the oldest materials that has easy molding on the foot is bovine leather, as it has the ability to follow the contour of the foot [47].

Thermal comfort is another parameter that depends on the type of material used at the top and lining, influenced by the performance in the transfer of heat and humidity that is felt by the user's foot [46]. The desirable temperature for the feet is between 27 and 33 °C. Temperatures between 35 and 38 °C already cause irritation, and the humidity varies depending on the person and the region of the foot [60].

Materials in contact with the foot are usually lining fabrics and help control the microclimate inside shoes; it must have several characteristics as water-absorbing behavior, water-holding capacity and drying time [61]. Natural fibers fabrics (cotton, bamboo) have the ability to absorb more moisture [62], and synthetic fabrics such as polyester and polyamide (sports application) dry faster and permits the air permeability [48]; to assist in thermal comfort, the choice of lining material should depend on the type of footwear used.

PU can be processed by open cell foam plastic; it is available in different densities and in almost every thickness and color, and it usually is applied at tongues, collars of shoes and back fabric. When PU foam is reticulated, it can assist for ventilation features [10].

Tighter heel support reduces hindfoot movement and improves comfort [59]. Heel elevated to the 20 mm limit protects the posterior muscles (triceps of the leg, tibia and back) [63].

Women's shoe high heels are usually made of polyvinyl chloride (PVC). Acrylics and acetates appear as transparent raw materials used for thick high heels and wedge heels models (platform shoes) [64].

From this, the main parts of the footwear, their functions, structures and ideal materials were presented to better fit the foot and comfort the user.

4 Environmental Solutions in Waste Management and Prevention in Shoe Manufacturing

The impulse that society promotes by imposing changes for transformation aimed at socio-environmental management within organizations is a very efficient tool for its implementation. Companies direct efforts toward sustainable development based on external impositions such as government, environmental legislation, consumers, as well as internal factors that promote economic gains [65]. Environmental issues have promoted initiatives within the leather footwear sector, and sustainable alternatives are being presented in the market. Less polluting materials have been used by companies concerned with mitigating their negative impacts, seeking to replace dangerous inputs and processes and/or using recycled materials. The increase in consumer demand for cleaner products stimulates the sustainable positioning of companies, combining efforts in search of improvements that reduce negative impacts on the environment [66].

The footwear industry presents many advances in its manufacturing process, innovations in its machinery and production process. However, the wide variety of materials used in its composition remains, generating toxic, flammable, pathogenic and corrosive residues [67]. The production of a pair of shoes can reach 40 different types of materials, of different compositions, which can contain plastic, rubber, metal, PVC, textiles, among others [68, 69]. Data collected in 2017 show the types of materials most used by the footwear sector. Plastic/rubber corresponds to 52.3%; synthetic laminates represent 25.6%; leather, 18.1% and textiles, 3.1% [2].

The disposal process at the end of the footwear's life presents difficulties, mainly caused by the great variety of components and the number of adhesives and seams used, resulting in an amount of hybrid solid waste that ends up being sent to landfills, causing contamination of the soil and waterbodies. The level of disposal of shoes generated after consumption is directly related to the accelerated fashion chain, in which the product's life cycle is getting smaller and its production is getting bigger, consequently increasing consumption and disposal [70]. For this reason, it is possible to realize that disposal at the end of the footwear life is a problem that is growing and difficult to solve, reinforcing the urgency of repositioning the industries in face of this problem. It is also worth noting that the large volumes of low-quality shoes are the most prominent waste generators [71].

The end of the footwear life cycle is the key issue to trigger low environmental impact; however it is the most challenging problem to be faced. Studies in this sector are still scarce and segmented. The development of shoes that include function, aesthetics and the ability to reproduce on a large scale within the guidelines of sustainability is a matter of great complexity [72].

The footwear productive field in Brazil has a self-sufficient productive chair, in which it is possible to find great efficiency in processes, internal production for all inputs, machinery and high training for competent people to act in all segments. However, an analysis of the environmental management practices of a group of medium-sized industrial Brazilian companies that exports, that is, which need to meet international requirements, showed that investments in this regard are insufficient, pointing to timidity in the actions carried out by organizations. The expectation is for greater precautions, mainly because these are companies that are positioned as committed to issues of waste generation, effluents and less polluting processes [65].

Recent changes in demand and new standards for discharge of effluents forced important changes in the production processes of bovine leather tanneries, since it is a widely exported product. As the footwear market is integrated with this whole range of international protocols and controls, due to exports and imports of raw materials and finished products, a continuous change in actions for the preservation of the environment is expected. Brazil was also affected, prevented from salting the skin for conservation and having greater control over the incineration of chromium [73].

Effective management of post-consumer waste is a rather complex issue made up of many components. Based on EU hierarchy, an integrated waste management framework for footwear products has been developed and presented in Fig. 7 [74].

This proposed framework (Fig. 7) sections the waste management options for shoes into two major approaches: proactive and reactive. **Proactive approaches** include all measures that are taken with the aim to reduce or minimize waste at the source. Reduction of waste, also referred to as waste minimization, is a proactive approach because simply, waste which is avoided needs no management and has no environmental impact. These can include design and materials improvements. On the other hand, **reactive approaches** include all the other waste management options which act in response to the waste problem when the useful life of the product has ended. Reactive waste management approach is also referred as end-of-life management, such as reuse, recycling, energy recovering from waste and disposal. The integration of all these actions should be considered in planning sustainable reverse logistics, which will converge to a circular economy approach [74].

According Mia et al. [75], the management of solid waste has become an urgent problem. Product quality means that a product will accompany its producer from cradle to grave; prevention, recycling and disposal of waste are part of a theory of the firm. Lean manufacturing tools are one of the most influential and most effective methodologies for eliminating wastes, controlling quality and improving overall

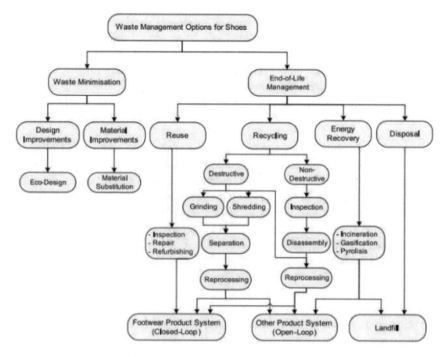

Fig. 7 Waste management framework for shoes [74]

performance of any machine, system or process in any industry with the complete assurance of large annual profit margins.

On the other hand, Shah [76] states that the number of footwear used by each person or by reusing the shoes which are not used further could help. Besides presenting a prototype of furniture (Shoe Rack) made by him with footwear end product, this author presents some modular footwear already existing in market (German shoe retailer Mime et Moi has option of 5 interchangeable heels; Zipz shoe company, with detachable base of shoes; Tanya Heath, Paris, has various detachable heels for one sandal; and Mahabis slip on's, which have a foldable topline, making it convenient to wear without using your hand) and some already existing solutions for reduction of footwear waste, highlighting New Balance, Nike and Adidas cases [76].

Typically, shoes are cut from large sheets into patterns, leaving 30% or more material to be discarded. New balance (among others) have tried to curb this by introducing the "puzzle cut" upper, but the new knit innovations may eliminate this problem altogether. Adidas Primeknit and Nike Flyknit lightweight running shoes (unlike traditional sewn shoes) are knitted from front to back, reducing wasted material by about 80%. Taking this one step further is Adidas. Their collaboration with Parley has resulted in an Adidas Recycled Ocean Waste Shoe. Constructed from fishing nets that were abandoned in the ocean, reused as nets are woven into the shoes upper. With no shortage of available material (there are fishing nets like this clogging oceans and seas around the world), this could be developed into an actual scalable product. Removing these nets that kill and maim hundreds of thousands of ocean animals, including turtles, dolphins, sharks, rays and even whales [76].

According to the Brazilian Association of Shoe Artifact Shopkeepers, 972.6 million pairs of sneakers, shoes and related ones were sold in Brazil only in 2017. What many people do not know, however, is that the product is difficult to recycle [77].

In addition, for the production of a single pair of shoes, an average of 133 grams of waste is generated, that is, near 133 thousand tons of waste are produced annually. Taking worldwide, there are 2.6 million tons of waste generated only from the production of footwear. In addition to the waste generated in this production, the shoes are produced far above the demand they have. And every year, they are incinerated because, simply, the industry did not sell them. In Brazil, the number is 91 million shoes that go unsold every year. Even if added to the waste generated to produce and the waste that old shoes represent, we do not have the exact number of wasted footwear for the Brazilian case [78].

For post-consumption footwear recycling, the disassembly process is manual. The solidity of the whole shoe makes disassembly work difficult. In addition, there is a low utilization of raw materials. Another issue that impacts recycling is the difficulty of adding large volumes of components present in shoes for subsequent sale or reuse. Women's shoes are a huge problem, because they have many ornaments. It is difficult, for example, to obtain volume from the same imitation gemstone that adorns a shoe. It would be necessary to gather many equal pairs, which is not the case. For experts, the result is that almost all shoes will end up in landfills at some moment of their life cycle [77].

An upward line of thought points to reducing consumption. When buying a pair, choose products made from recyclable materials, as natural fabrics or so-called vegan shoes. These have no leather and, consequently, do not contain chrome. The trend for the future will be biodegradable shoes, which decompose within four years after being buried. However, these solutions do not yet exist on the market. Another alternative is to choose to repair shoes and sneakers whenever possible, avoiding disposal. In addition, it is worth donating those which are no longer useful [77, 78].

A Brazilian case is "Rabble Shoes," a pioneer brand in the solution of collecting and managing waste, not only from its own shoes, but also receiving shoes from other companies, also producing vegan shoes. In case of donation of shoes in good conditions, the "eCycle" website can be accessed, which will indicate several places that accept to receive, in addition to the options to customize their own shoes and go to the shoemaker more often [78].

In São José dos Campos city, in the countryside of São Paulo state, "Recicalce" is one of the few NGOs in Brazil that collect used shoes and make small repairs to subsequently donate them to institutions that care people in social vulnerability. In seven years of activities, the NGO received 110,000 used parts, half of which could be repaired and donated. Shoes that do not have conditions for reuse are disassembled to use recyclable materials. The rest is sent to an industrial landfill, in compliance with current environmental standards [77].

Lopes et al. [79] studied the production of natural and synthetic rubber/waste— ethylene-vinyl acetate (EVA) composites using natural rubber, styrene-butadiene rubber or acrylonitrile-butadiene rubber and 10–20 parts per hundred rubber (phr) of waste. Several physical and mechanical properties of the composites were further determined according to the requirements of the footwear sector, confirming the possibility to reintroduce part of raw material recycled from EVA—waste in the footwear industry.

Among other different aspects of shoe production that can make shoes more environmentally sustainable, Motawi [10] highlights mainly: (i) sustainable shoe material selection; (ii) environmentally friendly footwear production processes; (iii) W\waste reduction in footwear manufacturing; and (iv) Sustainable shoe material selection.

Filho et al. [11] characterized and discussed cleaner production (CP) opportunities and challenges for its implementation in a plastic footwear industry. The study approached the case study analysis in the practical context of six Brazilian footwear plants. In order to address these important issues from a managerial perspective, seven guidelines were proposed for the implementation and management of CP in plastic footwear companies related to: (i) energy and water consumption; (ii) green technology and innovation: (iii) environmentally friendly materials; (iv) waste and residues generation; (v) environmental efficiency; (vi) human resources; and (vii) environmental management.

Despite of the related issues, Brazilian footwear industry has technology and a high capacity to project environmental management throughout the footwear life cycle. It is up to government institutions together with companies to join forces to promote more effective changes involving sustainability [71, 72].

5 Sustainable Materials and Processes in Brazilian Footwear

5.1 The Usage of Amazon Latex in Shoes

Natural rubber is a typical product of Amazonian extractivism in Brazil. The states of Acre, Amazonas, Rondônia and Pará were the birthplace of an abundant extraction of latex in the country. Between 1870 and 1920, rubber was the main product responsible for the Brazilian economy, where the Amazon during years was the only source of supply worldwide. After the smuggling of rubber tree seeds opened a market in the English colonies and in the East, the exploitation of latex in the Amazon became economically unfeasible, due to the great fall in the price of rubber [80]. Within 100 years, the country lost space in the market, falling from the position of largest exporter in the world to importing rubber produced in Asia. Since then, its share represents only 1% of the global market, with São Paulo and Bahia states as the main latex producers [81].

Latex extraction in the Amazon is considered very important because it is a resource of its biodiversity, which generates income for the natives, maintains the forest ecosystem and contributes to the climatic balance. In addition, rubber trees have a large storage capacity for carbon extracted from the atmosphere, which not only happens in timber tree species but also in that producers of natural rubber [80].

The Amazonas State Research Support Foundation (FAPEAM) has been supporting scientific studies aimed at the region. Some studies indicate that the Amazon's wild rubber tree is promising for the income generation of the Amazon population and that for the productive growth of natural rubber it is necessary to invest in the genetic improvement of the plant, in addition to providing technical support and adequate training for the process of latex extraction. Currently, 24 cities in Amazonas participate in the project "New technologies to boost the production of natural rubber in the Amazon" developed by Brazilian Agricultural Research Corporation (EMBRAPA) [82].

A practice that can be found among the natives of the Amazon region is the elaboration of latex boards for the development of handicrafts. An experimental study by Garcia [83] analyzed the development of a latex laminate made by rubber tappers from a cooperative in order to apply the material to a shoe production. The result was good adhesion in the production process, involving seams and glue, stating that the material is suitable for the requirements required for this type of product.

In scientific articles, the use of latex, extracted from the *Hevea brasiliensis* tree, is widely studied for use for sustainable soles and orthopedic devices. The intensification of medical research is the ability to adapt to the shape of the body and slowly recover from the pressure caused by the body and has the function of helping to protect orthopedic equipment, especially for diabetics [84, 85]. Sustainable uses in soles use residues in search of improvement of mechanical properties, as the mixture of latex and kenaf [86] and latex and leather [87].

It is worth to mention "Production of FDL and FSA—Training Guide," which is a publication launched by World Wide Fund for Nature (WWF) with the University of Brasília for community and riverside people on how to produce FDL (liquid smoke rubber sheet) and FSA (semi-artifact rubber sheet)—materials that have greater added value and can be sold at more expensive prices. With this, these professionals are able to generate income in a rational and sustainable way, in addition to conserving the resources of the rubber tree, the tree that generates latex/rubber. The launch of this guide is part of the Protecting Forests Initiative (Sky Rainforest Rescue project), a partnership between WWF-Brazil, WWF United Kingdom, the State Government of Acre and the British TV network Sky—these institutions seek to maintain a billion trees standing in that state [88].

In Brazil, latex in shoes also presents an appeal of sustainable product, but values the design-craftsmanship of the traditions of local Amazon communities. Native Amazonian latex is a viscous yellow-white liquid produced by nature. It is collected from native rubber trees through various incisions in the tree, according the following steps: **cutting**—it occurs at dawn, from 20° to 30°, and at height from 1.20 to 1.50 m, on days without rain and a bowl is placed at the bottom; **bleeding**—from three to four hours, after cutting, the latex slowly leaves and accumulates in the bowls; **collection**—with a bucket of zinc, a daily harvest ranges from 10 to 20 L [89]. A pre-treatment is still done in the forest, depending on the type of application that will be made with the latex, so there are: to produce the vegetable laminate, it is necessary to take it immediately to the concentration units where the vulcanizing agents are mixed and the smoking coagulation is performed and; in rubber for soles and slippers, half a liter of ash water and 600 ml of a solution of vulcanizing agents are mixed, it is heated for one hour until reaching the temperature of 90 °C, and finally, it is cooled and stored [90].

The Cazumbá-Iracema Extractive Reserve (state of Acre, northern Brazil) is located in the Amazon rainforest domain (GPS 09° 01'–10° 12' S and 68° 50'–70° 11' W) has an area of 748,817 ha with more than 190 families of rubber tappers [91] and other regions in communities in the state of Pará (northern Brazil) participate in the latex supply network for the company "Encauchados de Vegetais da Amazônia." It has been producing decorative objects and shoes for over 10 years, such as the flip-flops (Fig. 8) without the use of petroleum products, it was replaced by vegetable oil, carnauba wax, vegetable stearin and other vegetable fillers from agribusiness residues—"açaí" (*Euterpe oleracea*), "murumuru" (*Astrocaryum murumuru*), "castanha do Brasil" (*Bertholletia excelsa*), "andiroba" (*Carapa guianensis*) and "pracaxi" (*Pentaclethra macroloba*) (native fruits of the Amazon) [90, 92].

Another local production in Acre state, working with vegetable laminate, is represented by "Doutor da Borracha" brand (Fig. 9).

In addition to artisanal processes using latex in Brazil, there is an industrial scale of vegetable laminates, as an example by the company "Ecologica" that transforms the compound into natural latex, cured in a vulcanization system with a protective layer with water-based varnish made from a latex layer on cotton and natural fibers blanket basis (Fig. 10) [96].

The French footwear brand VEJA, which in Brazil is called VERT, presents guidelines for sustainable development and uses the production of Amazonian latex for the production of its shoe soles. The company claims that the Amazon is the only place where rubber trees grow in the wild. Their cultivation generates income for the natives who once left the extractive practices to turn to cattle breeding and logging, activities that contribute to the deforestation of the forest [97].

(a)   (b)

Fig. 8 Flip-flops by Encauchados made from latex and açai residues [93]

(a)   (b)

Fig. 9 **a** Latex laminates [94]; **b** sandals made from latex by "Doutor da Borracha" [95]

Fig. 10   Latex and cotton laminate for shoe uppers by Ecologica [96]

5.2   *Natural Fibers Employment*

In the Brazilian footwear market, only 4.5% was constituted by fabrics [2]. This value does not include linings and nonwoven textiles that complement the footwear construction. The kinds of fabrics are not classified according their fibers (natural, artificial or synthetics), so it is known from this number for certain use of natural fibers, but two most produced natural fibers in the world are cotton and jute, which are also produced in Brazil [18].

The materials of vegetal basis are traditionally used in the footwear industry, such as: raw cotton twill fabrics, jeans and canvas [98], braided textile fabrics such as cords and jute braids that are used to cover heels [99], buttress/stiffeners of virgin cardboard or pressed wastes [100]. The Brazilian footwear market follows the same trend of materials and uses them in its national production (Fig. 11).

Fabrics made from 100% bamboo fabrics were tested and the results compared to the properties of cotton, presenting better water vapor absorption and good elongation, suitable for the upper and shoe lining [103, 104].

(a) (b)

Fig. 11 a Sneaker: stamped cotton canvas and jute covering sole (Fam Rio) [101]; **b** sandal: natural fiber straw (unspecified) (Arezzo) [102]

Despite of the huge abundance and variety of natural fibers found in nature and cultivated in Brazil [105], the literature found on vegetable fibers in elastomeric composites present sisal fiber as one of the most used fibers in the studies, due to the tensile properties and because Brazil is one of the main world producers of this fiber.

Li and Ye [106] review developments involving composites reinforced with sisal fibers, as well as their characteristics. They mention the use of rubber matrices as the second most used for composites reinforced with sisal fiber. Also, that the main areas of study take into account the effects of fiber size, orientation, concentration, types of bonding agent and the interaction between fiber and matrix on mechanical properties, rheological behavior, thermal aging, resistance to gamma radiation and ozone of the composite. Among the experimental results, elastomeric composites (NR and SBR) reinforced with fibers of approximately 6 mm in size present the best balance of their properties.

Martins and Mattoso [107] characterized the mechanical and dynamic properties of NR and SBR elastomeric matrix composites reinforced with sisal fibers, in order to investigate the main effects of the chemical treatments used on the fiber (alkaline treatment at 5% solution of NaOH and alkaline treatment with acetylation), fiber size (2, 5, 10, 15, 20 and 25 mm) and amount of fiber used as reinforcement (5 and 10% of fibers without and with treatment). The results showed that composites reinforced with 10 mm fibers presented the best performance, in which their Young's modulus increased 180% and the rupture stress 25% when compared to the matrix without fiber. Composites with fibers treated by mercerization and acetylation improved the mechanical properties when compared to untreated fibers, through better adhesion between the fiber and the matrix.

Iozzi et al. [108] studied the influence of the length of the sisal fibers as reinforcement in a matrix of nitrile rubber/calcium carbonate. Composites of NBR/sisal obtained through calendering, with 6 mm fibers showed the best values for tensile strength. Larger fibers showed a loss in Young's modulus, due to the difficulty in dispersing the fiber in the elastomeric matrix.

Iozzi et al. [109] studied the influence of different treatments of sisal fibers on the properties of nitrile rubber composites reinforced with sisal fibers and nitrile rubber and calcium carbonate composites reinforced with sisal fibers, processed in calenders. The influences of the treatments (untreated fibers, washed and mercerized for 3, 5 and 10 h in a 10% NaOH solution) on 6 mm long fibers and the influence of the fiber content (0; 5.5; 11; 22; 33 and 44 phr—parts per hundred parts of resin) were analyzed. The results indicated that the used fiber washing methods show similar results and that there were no major changes in relation to their morphology. The different times of treatment by mercerization did not show significant changes in the mechanical properties of the composite. Regarding the tensile strength at rupture, there was a tendency to increase this property with the increase in the calcium carbonate content for the composites. The addition of fibers reduced this property and composites with untreated fibers showed the lowest performance, due to their low adhesion in relation to the matrix and calcium carbonate. The Young's modulus of the composite presented an increase with the addition of the fibers, and similarly to the breaking strength, the mercerized fibers performed better than those with the washed fibers and those with untreated fibers, thus the mercerization treatment of the fibers contributed to increase the mechanical properties of nitrile composites reinforced with sisal fibers.

There are also articles showing the use of vegetable fibers as a raw material for the development of polyurethanes for soles and uppers, such as polyurethane based on castor oil and palm fiber for insoles [110], epoxidized soybean oil and palm oil lauric acid methacrylate for sole and, this oil mixture was also used on cotton base for the development of vegetable laminates [111].

In conclusion, the Brazilian employment of natural fibers in footwear corresponds to a small part in comparison of synthetic fibers. The footwear use of cotton, raffia and jute fabrics is especially present in summer collections. Cellulosic fibers have been researched to reinforce rubbers and they are applied to footwear soles. These fibers are being researched as biodegradable materials, showing innovations for instance, the use of leather waste, coconut fiber and latex, multilayers fabrics and finishing process adding bactericide properties to cotton. Brazilian footwear industry is finding solutions to reduce production costs in order to compete in the international market. One of the strategies is the gradual replacement of leather by fabrics, synthetic polymers and rubbers. Thus, the increasing of researches about recycling footwear components and the development of biodegradable materials points a sustainability trend in this sector [18].

5.3   Leather Tanning Employing Vegetal Extracts

The leather industries could be considered recyclers, because their raw material is waste from the meat industry, but their process with heavy metals implies several types of environmental contamination [112]. Among tanners around the world, on average 80% of them use chromium salts to tan their skins, while the use of plant extracts is between 10 and 15%. The others use oxazolidine, glutaraldehyde, among others [113].

The tanning process made with tannin consists in the use of organic compounds that have a tanning substance in their properties, tannin, generally found in bark of trees, leaves, wood and branches of plants. This method of delaying the deterioration and conservation of leather skin was developed in the Mediterranean region, which for centuries was the dominant process in the transformation of hides into leather. Its replacement occurred gradually by tanning with chromium minerals during the nineteenth century [114].

Environmental improvements aimed at the tanning process are currently advancing due to several factors such as legislative requirements, corporate environmental and social responsibility, technological availability and demand from consumers. Vegetal tanned leather has been used by footwear and clothing companies as a sustainable alternative in the development of greener products and using terms such as "bio-,", "organic," "without heavy metals," "biodegradable" and "without metal" and "natural" in their marketing campaigns [115].

Environmental concerns are forcing tanneries to seek more sustainable means in leather production, and the vegetable tanning system has been approached as a possible substitute for the chromium process, promoting a resurgence of market interest in leather with vegetal tanning. The advantages of the leather resulting from this vegetal process are its compatibility with human skin, comfort and high dimensional stability, in addition to viable treatment and reuse of waste generated in the process. However, an issue to be addressed in the process is the salinity, which generates serious concerns [116].

Historically, in the northeast region of Brazil, the vegetable tanning process is made with the bark powder of some tree species, which acts to close the pores of the leather and increases its resistance, making it less spongy and more malleable. A type of leather resulting from this process is known as "atanado," which is typically used in soles of shoes, and for this purpose the leather is prepared to have more rigidity. In this Brazilian region, there is a very strong tradition in the production of vegetal tanned leather, mainly in the hinterland locals. In Pernambuco state, there was a period when the barks from mangrove' species was widely used for tanning hides, but it was banned due to devastating exploitation that affected the mangrove biome. Of northeastern origin, barks from "cajueiro" or cashew tree (*Anacardium occidentale*), "jurema" (*Mimosa tenuiflora* or *Acacia jurema*) and "angico" (common name of several trees of the genera Piptadenia, Parapiptadenia and Anadenanthera of the subfamily Mimosoideae) also can be used, in addition to the "açoita-cavalo" (*Luehea divaricata*); species that vegetates on the banks of the São Francisco river and other regions of the hinterland [117].

According Paes et al. [118], vegetable tannins are found in several forest species. Actually, the Northeast Brazilian tanners have, on angico-vermelho (*Anadenanthera colubrina* var. *cebil*), their only source of vegetable tannins. This study aimed to evaluate the tanning capacity of tannins extracted from four vegetable species of Brazilian semiarid region, seeking to make possible the diversification of species to be used in tannings in the region. Thus, physical and mechanical characteristics of caprine (goat) treated skins with extract tannic of "angico" (*Anadenanthera colubrina* var. cebil), "cajueiro" (*Anacardium occidentale*), "jurema-preta" (*Mimosa tenuiflora*) e de "jurema-vermelha" (*Mimosa arenosa*) were analyzed. The caprine (goat) skins were tanned with extracted tannins of those species and compared to tanned skins by tannins of "acácia negra" or *Acacia mearnsii* ("Seta Natur" commercial tannin). The samples of tanned skins were submitted to tension, elongation and progressive tear resistance tests. Good results were verified to tanned skins by *Mimosa tenuiflora* and *Mimosa arenosa*. At last, according these authors, due to the abundance of these species in Brazilian semiarid region, they showed potential to exploration of tannins; however, researches are necessaries to indicate the best forms of application of obtained tannins.

Among the various regions of Brazilian Northeast, which have the tradition of vegetable tanning, the municipality of Cabaceiras, in Paraíba state, Cariri hinterland, represents a large production of goat leather and handmade shoes. A documentary made by JPB ("Jornal da Paraíba") and released by the G1 website in 2017 [119] showed that in the last 10 years the "Arteza cooperative" has contributed to improvements in the artisanal vegetable tanning process, expanding the production of leather that is destined for 26 workshops that produce shoes in this municipality. Together they produce 12 thousand pairs of shoes per month and the cooperative moves more than R$ 1 million (near US$ 250 thousand) per month in a community that has only near 600 inhabitants (Fig. 12a–c).

In the cultural history of leather production in the Brazilian northeast, there is the "vaqueiro" ("cowboy"), a character who explored the hinterland taking care of cattle dressed totally of leather garments, all of them made by artisans. The tradition of leather clothing influenced the regional culture, and today some pieces are still in use. Leather is part of the identity of Northeastern handicrafts that represent a great

Fig. 12 Artisanal goat leather production in Cabaceiras (Paraíba state—Brazil): **a** goat creation [119]; **b** vegetal tanning [119]; **c** leather sun drying [119]; **d** tanned goat leather [120]; **e** goat leather accessories [121]; and **f** Northeastern cowboys' clothing [122, 123]

source of income in the region (Fig. 12d–f) About 64.3% of cities in Brazil have handicraft production in which handicraft techniques are inherited [122, 123].

It is interesting to emphasize the leather festival that takes place in the municipality of Cabaceiras, in Paraíba state, Cariri hinterland. The event, called "ExpoCouro-Bode," aims to highlight goat farming and expose the work of artisans who use goat leather as raw material, the main means of subsistence and economy in the city [121].

5.4 Natural Leather Dyeing

Color is one of the most important parameters of leather, as it is the first property of leather to be evaluated by consumers. The leather is widely used (approximately 60–70%) for the construction of the upper parts of the shoe and synthetic dyes are used abundantly in dyeing it. Natural dyes have been known for a long time and are derived from natural sources such as plants, insects, animals and minerals. On the other hand, synthetic dyes are widely available at an economical price and produce a wide variety of colors; these dyes, however, produce skin allergies, toxic residues and other damages to the human body [124, 125].

Different types of natural dyes have been used for leather dyeing from ancient times until middle of nineteenth century. Since then, a number of synthetic dyes are used for leather dyeing purposes that are continuously released into the environment and caused great damage to biodiversity due to the release of large volumes of pollutant effluents. A recent crucial issue of the leather industry is to reduce environmental pollution caused by leather processing. In this way, the use of eco-friendly and non-toxic natural dyes has become a matter of significant importance due to the increased environmental awareness to avoid some risky synthetic dyes [125].

In order to illustrate the importance growing of this issue, a preliminary search in recent literature about natural dyes employed in leather dyeing is presented in Table 4.

Taking in account the potentialities from Brazilian Amazonian biome, the use of natural dyes from wood residues, fruit peels, seeds, flowers and other materials that preserve the Amazonian flora and do not damage health and the environment can be used as an alternative for dyeing fish leather from skins that previously were discarded [126].

"Matrinxã" (*Brycon amazonicus*) is a medium-sized fish, presenting scales, silver color, elongated body, capable of reaching 80 cm in length and five kilograms in weight, widely distributed in Central and South America, found in the Amazon, Araguaia-Tocantins, Orinoco, Prata and São Francisco basins. In order to evaluate the vegetal dyeing efficiency on the leather of this fish, Melo [126] employed the following species and respective parts: "Açaí" (*Euterpe precatoriia* Mart.) (draff and fruit); "Bacuripari" (*Rheedia macrophylla*) (fruit peel); "Bananeira"—banana tree (*Musa* spp) (inflorescence); "Cacauí" (*Theobroma speciosum*) (flower); "Cajueiro"—cashew tree (*Anacardium occidentale*) (bast fiber); "Crajiru" (*Arrabidaea chica* Verlot) (leaves); "Cupuaçu" (*Theobroma grandiflorum* Schum) (seeds);

Table 4 Recent literature on natural dyes employed in leather dyeing (elaborated by authors)

Origin of natural dye and processing	Type of leather	Result	References
Coreopsis tinctoria flower petals	Conventional chromed wet blue goat leather	The extraction used ultrasound. Leather dyeing was optimized with the aid of ultrasound and magnetic stirring. Dyed leather samples showed good fastness against washing, light, and dry and wet rubbing. Bulk properties, such as softness, were found to be improved by the use of *Coreopsis tinctoria* yellow and brown dyes using an ultrasonic and magnetic stirring dyeing process	Velmurugan et al. [127]
Indigenous plants of Central and Northern Punjab—*Arbutus unedo* L., *Arucaria angustifolia, Callistemon citrinus* L., *Ficus benghalensis* Linn., *Iresine paniculata* L., *Mangifera indica* L., *Moringa oleifera* Lam., *Morus alba* Linn., *Musa acuminate* L., *Nerium oleander* L., *Polyalthea longifolia* Sonn., *Punica granatum* L. and *Syzygium cuminii* L. were selected for dye extraction	Goat crust leather	The pre-mordanting method was employed with oxalic acid to improve the color. Results of the study revealed that *C. citrinus* produced high dye yield. *A. unedo* and *F. benghalensis* have developed dark shades	Pervaiz et al. [128]

(continued)

Table 4 (continued)

Origin of natural dye and processing	Type of leather	Result	References
Flowers of *Bidens macroptera* Mesfin and leaves of *Eucalyptus Camaldulensis* Dehnh	Sheep crust	Color fastness from very good to excellent and free of heavy metals. *E. camaldulensis* leaf and *B. macroptera* flowers extract have potential applications for leather dyeing	Mengistu [129]
Insect (*Laccifer lacca*). The lac dye was applied on the leather samples with and without using mordants	Wet blue cow hide	The results of color fastness of the mordanted and unmordanted leather samples showed excellent (5) and best (4–5) gray scale rating, respectively	Deb et al. [125]
Microbial colorant—*Penicillium minioluteum*. Dyeing at pH-2.0 altered after 30 min of dyeing, temperature-80 °C and time- 60 min	Wet blue goat nappa skin leather	Dyed samples showed good rub fastness with moderate to remarkable fading to light. No allergy on human skin	Sudha and Aggarwal [130]
Tagetes erecta L. (Marigold flower)	Goat leather	Twenty various shades were obtained with pre-mordanting and post-mordanting methods. Color fastness (daylight) results have been recorded fair to good	Pervaiz et al [131]
Bark of *Cassia singueana* (Hambo Hambo)	Chrome-tanned sheep skin crust leather	*Aloe vera* juice and mango bark extract use as natural mordants for leather coloration was investigated. Fastness good-excellent. Better color fastness using aloe vera as premordant	Berhanu and Ratnapandian [132]

(continued)

Table 4 (continued)

Origin of natural dye and processing	Type of leather	Result	References
Barks of *Mangifera indica* L., *Syzygium cumini* L. and *Eucalyptus camaldulensis* Dehn	Crust blue leather of goat	Mordants $CuSO_4$, $FeSO_4$, $KMnO_4$ and Potash Alum were used. *S. cumini* extract showed more variation in colors. *M. indica* showed good fastness properties as compared to others. The formation of light and soft colors with different mordants was observed	Shamsheer et al. [133]
Dried fruit of *Terminalia chebula* retzius (T. chebula)	Commercial vegetable-tanned cow hide	The effect of enzymatic post-tanning process on dye affinity is evaluated by using different type of proteases such as flavourzyme, alcalase and bromelain. The fastness properties against the rubbing and dry cleaning of the dyed leather were improved by the enzymatic post-tanning process	Song et al. [134]
Two varieties of saffron (*Crocus sativus*)	Goat leather	Color fastness to mild washing evaluation showed almost no color change in red *Crocus sativus* and same with sodium dichromate (4) and slight color change (3-4) in all other samples which showed the retaining capacity of colorant at leather surface	Gull et al. [135]
Fungi *Monascus purpureus*	–	Review on recent literature about improvements of leather dyeing techniques and the search of natural dyes for industrial uses	Fuck et al. [136]

(continued)

Table 4 (continued)

Origin of natural dye and processing	Type of leather	Result	References
Aronia melanocarpa (black chokeberry) Extracted red natural dye was used alone or together with commercial silver nanoparticles	Wet blue goat leather	The dry and wet rubbing fastness values for dye alone and dye with nanoparticles were grade 4–5 and 4, respectively	Velmurugan et al. [137]

"Mangostão" (*Garcinia mangostana* L.) (fruit peel); "Murici" (*Byrsonima crassi-ifolia* L.) (fruit peel); "Patauá" (*Oenocarpus bataua*) (fruit); "Rambutã" (*Nephelium lappaceum* L.) (fruit peel); "Tangerina"—tangerine (*Citrus reticulata*) (fruit peel); "Tucumã" (*Astrocaryum aculeatum*) (fruit peel); and Urucum (*Bixa orellana* L.) (seeds). All the obtained extracts were chemically evaluated and selected for dyeing the "matrinxã" fish leather. "Crajiru" dyeing presented higher resistance to the action of the light and water. "Crajiru" presented the coloration changing from the red-purple to the wine and the "cacauí" of the light and dark lilac, taking into consideration to the ratio of dyeing. Despite of there be no statistics significance among the of 10 and 15% contents this last resulted at more intense coloration. The conditions and proportions of both were shown appropriate. Their uses will depend of the tonality, coloration and availability of the source of coloring wished. Therefore, according this author, "crajiru" and the "cacauí" can be excellent sources of natural on the staining of leather matrinxã.

Other study, conducted by Karnarski [7], aimed to identify the best concentration of vegetable tannins in the treatment of the pirarucu skin; and, to compare vegetable tannins with chromium salts, to identify sustainable alternatives for leather produc-tion, replacing chemical dyes by natural ones. Pirarucu (*Arapaima gigas*) is one of the largest rivers and lake freshwater fish in Brazil and a potential species for the production of ecological leather in the Amazon. Scaled leather accounts for 10–20% of body weight, is exotic and innovative with specific strengths and is widely accepted in various market segments. The use of vegetable tanneries to replace chromium is an alternative to ecological production and aggregation of values in the fish chain and the use of leather in the Amazon. This author concludes that when comparing tanners (chromium salts and vegetable tannins) with coloring ("urucum"—*Bixa orellana* L.—and chemical), the vegetable tannin provides leather thicker, more resistant to the application of force; more resistant to traction tension and more elastic compared to leather tanned with chromium salts. In this way, the use of natural dyes and tanners is effective in replacing chemical tanners and dyes in the Pirarucu ecological leather production process.

5.5 Synthetic Materials

As far as plastic laminates are concerned in Brazilian market, these materials play an important role as a good option to leather and textiles for shoes, bags and accessories. These Brazilian sectors majority employ PVC and PU [44].

5.5.1 PU Laminates

Most of the volumes of PU offered in the Brazilian market are imported from China, Taiwan and Vietnam [19]. Its main applications are for women and sport shoes uppers.

Women shoes uppers have been normally traded as a material of short lifetime because of speed of lady shoes fashion. Based on this idea, China has been using (Brazil and all the world also) lower quality and lower price PU, based in polyester diol type (explanation as following) that has a maximum lifetime of two years taking in account the time of production of the laminated material [138–141].

One important point is that this "guarantee" is based in an average temperature of 25 °C, therefore, if the average temperature is higher than that, faster degradation will occur.

This degradation is called hydrolysis, the reaction of PU with water (just air humidity is already enough), that has as final result the total damage of material because the polymers chains start to break. It starts to lose mechanical properties very fast and in the final stage, is not possible even to get a good recyclability.

There are basically three types of polyurethane that behave very differently in terms of resistance to hydrolysis: (i) polyester—2 years resistance; (ii) polyether—5 years resistance; and (iii) polycarbonate—10 years resistance. These three different classes are named based on the type of polyol that reacts with isocyanate to produce the polyurethane polymer) [138–142].

As a commentary, highlighting the differences between male and female market, during mid of 2000s, shoes makers tried to use polyester PU in lining for male shoes and it did not work at all since a lot of complains and recalls emerged: men are slower to exchange shoes and fashion changes for male shoes are not so fast and sharp as women, allowing higher time of usage and consequently, more quality problems!

For sport shoes, the most used type is PU polyether type, as sport shoes are more quality and time usage demanding. Therefore, a better and more expensive PU type was offered and an important volume of this material comes from Taiwan and Vietnam [19].

All PU products offered in all these cases and until some years ago have been only produced by coagulation process or also called wet process. Wet process materials in almost cases used to be post-finished in dry coating machines. The coagulation process (wet process) is showed briefly in Fig. 13 [143]. Besides the coagulation line also is needed one distillation column to recover the DMF (dimethyl formamide

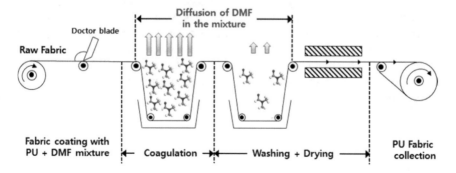

Fig. 13 Scheme of the procedure for manufacturing PU products based on wet process [143]

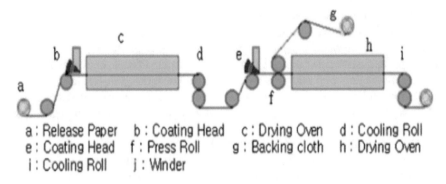

a : Release Paper   b : Coating Head   c : Drying Oven   d : Cooling Roll
e : Coating Head   f : Press Roll   g : Backing cloth   h : Drying Oven
i : Cooling Roll   j : Winder

Fig. 14 Finishing of PU laminates employing dry coating process [144]

used as solvent in the process, along with water) and to reuse it again—it is a closed system. Anyway, a lot of energy is used to proceed this operation. Besides this, the handling of large amounts of DMF to produce these laminated materials is a critical operation in terms of safety and environmental aspects.

After the coagulation process, normally at least one finishing stage is required to give the proper quality and the looking to the material to be sold. Most of the time, this finishing is done using dry coating process. A brief description of the process is presented in Fig. 14 [144].

A type of release paper (special paper properly treated for this operation) is the carrier in the dry coating system, and it also gives the embossing of the final product. Sometimes a flat paper is used, and then, a post-process of embossing is applied to give also same embossed material compared with embossed paper. This release paper is separated in the final of process to be reused again, since your cost is an important part of the whole cost of the product. This paper can be reused sometimes, depending on the quality and type of paper [145].

In the last years, new options of PU with very low solvent levels have emerged PU HS (called high solids PU) [138–141]. These types of products have high resistance to hydrolysis and no use of coagulation process, being produced by dry coating process. It is becoming a very good option in terms of softness and visual patterns to substitute regular coagulated products, and it is being introduced and developed in Europe and China. Just it is usually a higher price technology, but with a higher durability compared with the polyester conventional coagulated products. It is being introduced in higher aggregated value markets. In shoes market, it is starting to be used in top application (e.g. sport shoes) and high-quality brands, even for female shoes market.

5.5.2 PVC Laminates

Brazilian shoes market uses widely PVC plasticized laminated material for shoes upper. All chemical supply chain needed to cover this area is present within the country and complemented by international suppliers well-structured in Brazil [44].

In this case because of flexibility of PVC technology and dry coating/finishing processes, it is possible to get a huge number of visual patterns, thicknesses (0.5– 8 mm—from linings to insoles) and different levels of softness.

The main production process is the dry coating; it is similar to presented in Fig. 14. Also, like PU, there are many finishing processes: printing, embossing, sanding that enables producer to have a real leather looking. Normally, many types of woven/nonwoven substrates are used to give the proper mechanical properties for the different applications in shoes, bags and accessories [145].

PVC is well accepted in Brazilian market, supplied mostly, in the case of PVC, by Brazilian suppliers with a very good cost benefit. PVC now, is starting to be realized as a high-technology material with a good cost/benefit. Its technology allows to have a wide range of products in all technical aspects (from a very low cost/lower quality to a very high aggregated value/top quality material) [145].

Taking in account synthetic materials more sustainable from the point of view of increasing the useful life, Cipatex® has launched Vynyltech®, laminated material for shoe uppers and shoe lining, also bags and accessories based on most modern technology vinyl based. The main property of this brand product is durability, leading to less disposal, an opposite situation when compared to low cost PU (Fig. 15a, b). This meets to a more conscientious consumption. Besides this, the percentage of raw material coming from renewable raw material source is remarkable. Vegetal plasticizer is already widely used, and the possibility of supply PVC resin from sugar cane source is also possible in the future [146], being already used for polyethylene.

An restricted substances list (RSL) is being prepared in Brazil to be applied to shoes and shoes components Brazilian chain [150]. At the moment of writing this book chapter (February 2020), the list was going to be submitted to public voting— this is very important to improve the quality of the shoes and to enable Brazilian shoes chain to get closer to worldwide standards, since it is being based on REACH [151–153] (EU standard to control safety of articles in terms of chemical risks). This work is being leaded by IBTeC® (Brazilian Institute of Leather, Shoes and

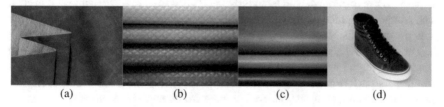

(a)   (b)   (c)   (d)

Fig. 15 Cipatex® special PVC laminates examples: **a** e, **b** Vynyltech® Ecologic Jeans [147, 148]; **c** Vegan® [147, 148]; **d** sneaker made with Vegan® laminate material [149]

Articles), one of the most important footwear technology centers in Brazil, with cooperation of Assintecal (Brazilian Association of Companies of Components for Leather, Footwear and Artifacts) and Abicalçados (Brazilian Footwear Industries Association).

PVC mostly during the 90s was blamed as a no sustainable polymer. Many big brands had banned all PVC usage in their chain. Part of automotive industry had been planning to phase out PVC completely until beginning of twenty-first century.

The production chain of PVC in Europe was able to change the image of this polymer through scientific effort and coordinated work. It includes a movement coordinated by the chain, initially called Vinyl 2010 and later renamed Vinyl Plus. This movement is aimed to develop the control of the gas emissions, implementing a significant recycling of the polymer in the last 15 years (from almost zero in the late 1990s to 640,000 tons in 2017); sustainable use of additives, energy and raw materials; as well as the diffusion of sustainable awareness along the chain.

Also, part of this new scenario is the important role of regulatory activity (for example in the case of Europe REACH), which helped to clarify doubts that the population had about substances used in this chain. In this way, from damn material in the late 1990s, ready to be banned from the automotive industry and avoided at the Sydney Olympics in 2000; became today the premium material used by major automotive brands inside cars and widely used at the London Olympics in 2012 [154].

5.6 The Case of Vegan Plastic Laminate Coverings

The vegan movement is on the rise, not only because of the numbers of followers, but also because of the contribution it has made to raise awareness in dealing with animals. Considering the fact that this public uses PVC laminates as an alternative to replace products of animal origin, the company researched in depth and identified that it had materials from animal source in its composition [155, 156].

Based on this information, work had started to replace these raw materials. The process involved not only laboratory tests, but also performance and field tests, since some substances had to be developed to meet this demand. Therefore, alternatives of vegetal origin or of another source had to be developed. The partnership of suppliers with extensive experience and technology in PVC was essential to achieve this goal. The company counted on the active participation of suppliers in the development of new raw materials, since the substances commonly used come from animal origin.

Partner clients also participated in the design and manufacture of prototypes in order to study the various designer possibilities offered by the product, as well as confirm its performance in the field. The product also has in its composition an important percentage of raw material coming from renewable source.

It is worth remembering that the laminate is not only free of toxic heavy metals, but also meets REACH, a regulation on the registration, evaluation, authorization and restriction of chemicals in the European Community [151, 153, 154].

The product launched by CIPATEX® for this line is VEGAN® (Fig. 15c, d). Another important issue that the vegan laminate contemplates is the aspect of leather availability, as a "by-product" of the meat market, which may generate a deficit of leather, in relation to meat. Such a global imbalance could create complicated situations from livestock to logistics, with its large volume of transport demand and its consequences.

In addition to involving the issue of veganism, the company aimed to produce a laminate that would provide comfort and beauty to the feet, with an appearance and texture similar to leather. The result was extremely positive, presenting the footwear industry with a noble material, with a selected velvety touch substrate that provides excellent conformation, in addition to the advantage of being versatile, with the possibility of making women's, men's, children's shoes, bags, among others. This product opens a new vision of serving markets and trends of the twenty-first century, with the strengthening of the industry of products free of raw material from animal source.

6 Eco- and Sustainable Design in Brazilian Footwear

6.1 Concepts of Eco- and Sustainable Design in Footwear

Ecodesign, as a projection of objects in their functional complexity, not only has the possibility to design their shape, but also to renew production processes and behavioral habits, with a view to greater environmental sustainability. Ecodesign aims to design to reduce the use of non-renewable resources or to minimize their environmental impact during their life cycle, reducing waste generation and saving final disposal costs [157].

Rech and Souza [158] describe in the article "Eco luxury and Sustainability: a new consumer behavior," about the challenges of contemporary society and address the possibility of contemplating an ecological vision concept, through the preservation of natural resources and the preservation of the environment. Therefore, a reflection and profound change in the sectors of production and consumption. Resources can be minimized in production through the reduction of material used in the manufacture of products, waste, the use of energy and the use of resources applied in the production process. The factors related to the adoption of sustainable measures and processes are related to the entire life cycle and processes applied to the production and development of products, because the impacts that a process or product can cause to the environment, are related to the extraction of resources, production of materials, transformation of raw materials, distribution, use and disposal [159]. Thus, it comprises the entire production chain of a segment.

The footwear sector has the opportunity to align growing production with the sustainability agenda. The broad concept called sustainability, which includes the environmental, social, economic, cultural and political pillars, presents itself as one of the most complex challenges for industry and society. To try to achieve environmentally oriented solutions, industry, companies and organizations need to adapt and search for research that encompasses environmental management and production management, fostering new consumption dynamics [160]. The consumption of resources refers to inputs of materials (i) that impact the exposure of human beings and environmental systems and energy (ii) that requires the use of fossil fuels, biomass, nuclear energy, hydroelectric, geothermal, solar or wind [72, 161].

Since the 2000s, Ecodesign, centered on environmental aspects, grew to more comprehensive concepts and a holistic approach to sustainable development, which also starts to consider social issues and economic aspects [162]. Currently, sustainability is researched with greater emphasis on its social and environmental approaches. In the first, there are visions and practices centered on people, collaborative processes and communities. In the second, there are processes and practices more focused on reducing environmental impacts [163].

Sharing resources becomes one of the lines of sustainable fashions products, when sharing data between designers and producers to use the same resources in search of ecodesign. There is a proposed program called "Cradle to Cradle Fashion Design (C2CAD)" which shares data in other supply chains for selecting fabric types, for example [164].

The footwear industry produces about 22 billion pairs of shoes worldwide [165, 166], and this volume of production creates enormous environmental challenges, such as the treatment of manufacturing waste and the ample volume of products that reach the end of their useful life, which are not always disposed of correctly, these being the sensitive points for incorporating sustainable development into the chain [72, 167].

Products should be designed with the aim of reducing the ecological impacts they cause and contributing to the deceleration of replacement cycles, thus avoiding the disposable, which means operating in a system opposite to that of accelerated consumption [168]. In addition, gradually, there is an increase in demand for products of ethical brands that include environmental preservation in the development and production of their products, by a portion of consumers of footwear, the same ones who seek information on how harmful the product is and impacts in the environment, which characterizes a conscious consumption movement [72, 169].

It is perceived that it is necessary to look for alternatives that guarantee, simultaneously, the maintenance of business and the environment. In this sense, the first step for change would be to alter the sequence of activities that typically occurs in industry [170], to include sustainable development strategies in the design and production processes, redefining the actions of each stage [162], so that they contemplate a systemic view of environmental precautions in the development, production and at the end of the product's useful life [72]. In this way, the concept of life cycle "refers to the exchanges (input and output) between the environment and the set of processes that accompany the birth, life and death of a product" [171]. Thus, to develop projects that contemplate this path of sustainability, one must consider the life cycle of the products, or the design for the life cycle (Life Cycle Design) [172].

This consideration needs to be taken into account from the beginning of the project, so that plans are developed that adapt to the premises of the sustainability path. According to Manzini et al. [171], sustainable design must act in several spheres such as [172]:

- DfA (Design for Assembly)—Design for Assembly: uses an easier assembly with lower manufacturing cost, reducing expenses and improving the qualities of the products;
- Design for Manufacture (DfM)—Design for manufacturing makes a selection of materials, having processes and modulated projects, using standardized components, multipurpose quick couplings and assembly directed to minimization;
- Design for Service (DfS)—Design for service provides a longer service life, greater product reliability, easy maintenance and repair, classic design related to the user's style and zeal. It offers a maintenance service during the life of the product and its reconditioning when necessary;
- Design for Disassembly (DfD)—Design for disassembly maximizes recycling sources and minimizes the potential for product pollution. It has an easy dismantling project.

During pre-production, semi-manufactured raw materials are produced. At this stage, resources are acquired, which can be virgin, secondary or recycled. Virgin resources, in turn, can be from renewable or non-renewable sources. In this phase, there is also the transport and transformation of resources into materials and energy. The form of assembly is relevant in the development of products, as well as the raw material used and the disassembly of the product both for maintenance and after use for disposal. In this way, an eco-efficient product must incorporate, through design, special requirements that present a differentiation from other products, such as in manufacturing, transportation, disposal and recycling processes [172].

Motawi [10] suggests design simplification for sustainable shoe design and production, guiding by the concept of less. The simplification of shoe design can lead to fewer materials, less waste and less energy to produce. For instance, the number of components can be reduced through digital fabrication (zero waste cutting, 3D printing production, etc.). Also, footwear products with a long lifespan will help keep shoes out of landfills. In this way, design and material choices are the deciding factors for products' lifespan. At last, this author list features about ethical shoe production, some of them that can incorporated in shoe's design, such as: ethical and low-impact production; made from and/or of recycled, organic and upcycled materials; zero or low waste production; eco-friendly production, etc.

6.2 Brazilian Ecodesign in Footwear

In research on the global footwear market and sustainability, it is shown that the adoption of sustainability in companies has two objectives: the first to demonstrate to the consumer the perception of the issue in the footwear market; and second, to

(a)   (b)

Fig. 16 Shoes made with Amni Soul® fabric: **a** Arezzo ZZBio sneakers [178], **b** Top Shoes Brasil sneakers [177]

show the opportunities for companies in order to maximize the benefits and minimize the risks of implementation in the business [173]. The benefits for the company, such as reputation and the reduction of supply chain risks, are encouraging the inclusion of sustainability in industries [174]. In addition to the media, it also disseminated products and brands that use renewable or recyclable materials [175].

The principal national footwear producers, mainly in the female segment, are launching sustainable footwear lines, which are usually more sporty models, following this model's trend in the female market. These lines can employ materials made from natural sources or incorporating additives that modify the intrinsic non-biodegradable properties, accelerating the degradation capacity. One example is the polyamide 6.6 yarn (Fig. 16a, b), developed by Rhodia Solvay group (launched in 2014), which is able to decompose in less than 3 years when properly disposed of in landfills [176]. The footwear proposed by the company Top Shoes Brazil (Fig. 16b) was a joint project, which in addition to the biodegradable upper, uses 3D printing ("Fiber Corporation"), recycled buttress ("Ambiente Verde") and polyurethane midsole based on polyol, recyclable and biodegradable (company Basf) [177].

In this way, the development of ecological and ethical technologies and models in footwear is continuous. As cited before, in item 2.4, in Brazil, the "Sustainable Origin" seal was developed jointly between Brazilian sectoral footwear associations—the responsible for the component industries, Assintecal (Brazilian Association of Companies of Components for Leather, Footwear and Artifacts), and the responsible for the shoe manufacturing industries, Abicalçados (Brazilian Association of Footwear Industries). The seal came into effect in 2013, and in 2019 (July), a reformulation was launched. The current model of the seal is more demanding, mainly in obtaining the first category.

In 2018, the Brazilian footwear sector added up 6.6 thousand establishments [2]. To face international competition, the footwear sector has been investing in design, new technologies and sustainability—aiming to mitigate the industry's main impacts, such as the use of potentially toxic substances and diversity of working conditions [26]. On the other hand, the national competition in the nowadays economic crisis (especially for the medium and small-sized companies producing conventional shoes), in which they compete for the same resources and market, negatively impacts the environment in due the necessity of increasingly lower production costs. In this scenario, the interest in sustainable issues for this part of enterprises, including ecodesign and certification, is still secondary.

However, at the same time, small and micro-enterprises working in niche markets, including that sustainable ones, which incorporate ecodesign principles in their production, are arising and growing in Brazil. Some examples are presented in the next items.

In this way, the Brazilian footwear sector, which presents a huge potential to green sustainable products and processes, has been growing in terms of the development of technological materials, as well as their application in the market, and new guidelines are already focusing on strategies for improve the manufacture of conventional and sustainable footwear. The society's demands in the search for technologically sustainable products, with less impact on the environment, promotes the competitive differential to reach the new consumer profile concerned with finding sustainable solutions to everyday problems. Thus, although the nowadays non-favorable economic situation, it is expected that issues related to sustainability and ecodesign will develop further within the national and global context.

7 Sustainable Brazilian Footwear Components

7.1 Recent Initiatives of Recycled Footwear Raw Materials

In Brazil, there is an event, the "Inspiramais" that greatly encourages the exhibition of sustainable brands of national components for footwear. This is the largest trade show for footwear and fashion components in Latin America. It is held twice a year in the largest city of the country (São Paulo, Brazil). It started in 2009, and the last edition "Inspiramais 2021" (January 14–15, 2020) was visited by 7.8 thousand people, and 190 brands offered solutions in soles, accessories, fabrics, laminates, leather, among others, for all footwear, fashion, accessories and furniture sectors [179].

With this, the "Brazilian Association of Companies of Components for Leather, Footwear and Manufactured Goods" (ASSINTECAL), responsible for organizing the Inspiramais fair, together with the national footwear industries, boost the development of new companies focused on Brazilian sustainable components.

The company "Ambiente Verde" (medium-sized company) uses the thermoplastic residue from the shoe factories to develop stiffeners/buttresses and reaches a product with competitive cost and adequate mechanical properties. Operating in the market since 2012, in 2018, it produced about 69.5 million insoles (Fig. 17), corresponding to 7.36% of the share of this component in the national market and recycled 3.4 million kg of industrial waste [180].

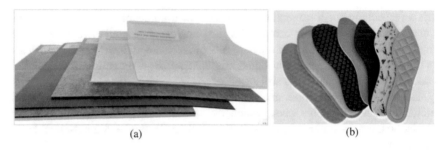

(a)   (b)

Fig. 17 **a** Boards made with thermoplastic waste from the footwear industry [180]; and **b** insoles made from these boards with a superior finishing [180]

Ethylene-vinyl acetate (EVA) from sugar cane for soles was developed in 2018 in a joint project enrolling the London footwear brand "All Birds" and Braskem, the sixth largest producer of plastic resins and the world leader in the production of biopolymers. Braskem is also headquartered in Brazil and the largest production of biodegradable inputs is the green polyethylene (the industrial plant is in the city of Triunfo, state of Pernambuco, northeastern region), whose raw material comes from the residue of sugar cane production [181]. In this way, Braskem supplies resin to national companies that develop the components, such as Cofrag (textiles and insoles) (Fig. 18).

Brazil is one of the main producers of vegetable textile fibers in the world, mainly cotton and sisal, as well as jute, curauá, among others. In addition, national textile production is structured with spinning and weaving/knitting and various biodegradable fabrics are produced nationally, including for the national footwear market, such as braided and jute, raffia and sisal fabrics (Fig. 19) [18].

Less than 12% of the potentially recyclable materials are collected in the Brazilian. In South and Southeast regions, mainly in São Paulo state, there are some chemical

Fig. 18 Sugarcane EVA insole (Product slogan: "I'm green, Braskem") [182]

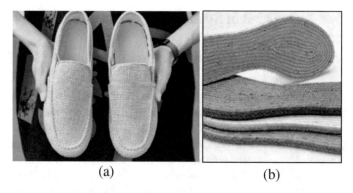

(a)   (b)

Fig. 19 Natural fibers produced in Brazil applied to the footwear market: **a** shoes employing organic jute from company "Castanhal" [183]; **b** soles and insoles in sisal produced by the company "Bertex" [184]

and mechanical textile waste recycling companies, some of them closer to the principles of solidarity and circular economy [185]. Since 2010, all the production of "Ecosimple" weaving company is made of 100% recycled materials from plastic bottles, discarded clothing or fabric scraps from clothing factories and waste from spinning processes, and this company also produces fabrics to upper shoes (Fig. 20a, b) [186]. Another article pointed out more than 20 companies in Brazil that use post-industrial waste of natural, fibers, chemical fibers, synthetic fibers and carpet waste, as "Cotton Move" company that recycles denim to produce yarns (Fig. 20c) [187].

(a)   (b)   (c)

Fig. 20 a Sneakers from the French footwear brand VEJA, which in Brazil is called VERT, employing Ecosimple fabric for uppers [188]; **b** recycled fabrics produced by "Ecosimple" [188]; **c** recycled denim yarn produced by "Cotton Move" [189]

8 Market and Design of Sustainable Brazilian Shoes

8.1 The Growth of the Brazilian Sustainable Fashion Market

Fashion Revolution is a non-profit global movement (from UK) that its mission brings unite people and organizations to work together toward radically changing the way of clothes are sourced, produced and consumed. It is a global movement (92 countries) that runs to celebrate fashion as a positive influence while also scrutinizing industry practices and raising awareness of the fashion industry's most pressing issues [190]. It was established in 2013, to respond Rana Plaza disaster in Bangladesh, a catastrophic incident, which major fast fashion brands produced their clothes and the building collapsed leaving 1127 dead and about 2500 were injured [191].

In Brazil, the movement has taken place on the internet and in presential networks since 2014. In 2017, 37 cities joined the movement with 225 events, mainly in the south and southeast of the country with lectures, reports, research, debates and mini courses [192]. The national movement, from the beginning, mobilized many students and colleges and became the largest in the world, but with little participation from fashion brands. So in 2018, the Fashion Transparency Index was created, based on six criteria clarify policy and commitments; governance (social and employee impacts); traceability (suppliers and outsourcing); knowledge and communication (supplier evaluation); featured topics (gender equality SDGs, decent work, responsible consumption and production, global climate change) [190].

In the participation of the Fashion Transparency Index in Brazil (2019), 30 companies were selected, among them 6 (Arezzo, Dumond, Havaianas, Melissa, Moleca and Olympikus) in the footwear industry and another, Osklen, which commercializes clothing and footwear. They were chosen for their turnover, diversity of performance in the market segments and positioning. The result obtained occurred after filling out a questionnaire based on the six criteria previously presented, and two more meetings in person. Havaianas and Osklen achieved the highest marks (over 40%). This index was scored from 0 to 100%, among the initially 30 companies, 50% answered, 20% did not answer and 30% answered the first questionnaire, but did not follow up, showing that the Brazilian clothing and footwear segments are still well closed to expose their information to the market [190].

Since 2017, an alternative fashion circuit to the country's official fashion shows has started, focusing on sustainability, called "Brasil Ecofashion Week" (Fig. 21). In that year, it attracted around 3 thousand people and, in the last, in 2019, it reached the visit of about 7 thousand people. This event has strengthened the culture of sustainability in fashion and gives visibility to those who do differently through exhibition of brands, fashion shows, mini courses, lectures, on themes of: business models and circularity, certifications, upcycling, textile recycling, biodesign, transparency, fair trade, etc., [193] and is expanding in the country. Previously, some topics had already been addressed in the Brazilian official fashion circuit, including issues environmental awareness, sustainability, environmental protection, water, forest peoples major fashion events such as "São Paulo Fashion Week" [194].

Fig. 21 Brasil Ecofashion 2019—photographs of a fashion show and exhibition of products and materials at the event [193]

In the specific footwear area, there is the Inspiramais fair, as mentioned above, which promotes the Brazilian footwear chain from sustainable materials, still proposing two projects: "Inovamais" (materials developing) and "Conexão Criativa" (combining sustainability, plurality, technology and handicraft work) to involve the entire industry and create an authentic fashion that values the national identity (Fig. 22). The "Inovamais" project develops products on the themes of composites, prototyping, additives, functional fabrics, among others [195].

8.2 Sustainable Product Design Solutions: Small and Micro-Footwear Companies

Sustainability in fashion business models is divided into 5 pillars, which are the main ones: circular economy (Upcycling, Recycling and Vegan), sharing economy and collaborative consumption (collaboration, second hand and fashion library), technological innovation (sustainable raw materials, zero waste, wearable technology) and consumer awareness (capsule wardrobe, "lowsumerism" and slow fashion) [197]. Brazilian sustainable footwear niche is predominant in micro- and small companies, which mainly focus on the circular economy issue, and in addition, they also use the theme of material innovation (previously described), and most of them use it as a concept consumer awareness and slow fashion [198]. Examples of these Brazilian companies are presented as follows.

Insecta Shoes
Focused on upcycling, recycling, vegan and slow fashion (Fig. 23). The main features are [199]:

- Sole is made by recycled rubber;

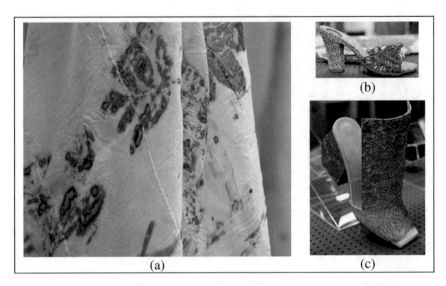

Fig. 22 Photographs of materials and products from the "Conexão Criativa": **a** natural dyeing with aroeira (*Schinus terebinthifolia*) in Brazilian silk (designer Luciano Pinheiro) [195]; **b**: shoe developed with waste from the local footwear industry by the company Trammô (state of Santa Catarina) [196]; **c**: upper and heel shoe covering in flounder fish leather by "Cooperativa Rio e Mar" (state of Paraná) (designer Luciano Pinheiro) [196]

Fig. 23 Main features of Insecta Shoes (shoe upper of recycled cotton, recycled rubber sole and insole made from fabric waste) [199]

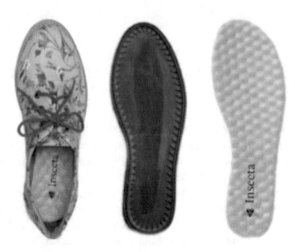

- Fabric waste from the company are reused to insole;
- Shoe uppers utilize second hand fabrics, PET fabrics and recycled cotton;
- Certifications "B System Certification" and "PETA—Approved Vegan."

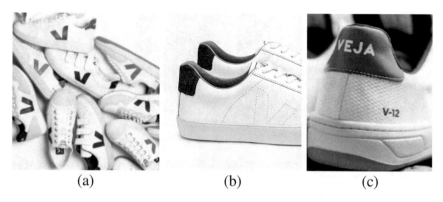

(a)   (b)   (c)

Fig. 24 Veja (Vert) brand [200]: **a** sneakers; **b** tilapia leather detail; and **c** B-mesh application

Veja or Vert shoes (named in Brazil)

Focused on technological innovation, upcycling and slow fashion. This French footwear brand started in 2004; it produces shoes in Brazil and utilizes almost everything materials from there [94, 97, 200]:

- The organic cotton (canvas and lining) used by VEJA is made by farmer associations in Brazil (northwest) and Peru, but Brazilian companies produces yarn, fabric and the assembly shoes in a factory near Porto Alegre city (south). It is important to note that the Vert company encourages local organic cotton producers, together with other entities such as Esplar with productive technical knowledge (Fig. 24a);
- VEJA soles are made from 18 to 22% natural rubber (latex). Since 2004, the company purchased 195 tons of wild rubber from Brazilian Amazonia;
- The employed leather comes from southern Brazil. From 2008 to 2015, 100% VEJA leather was vegetable-tanned leather, but it changed due to the prohibitive costs and quality level. Now, vegetable tanning is used on 10% of models, but all of leathers are now tested and meet REACH standards and are without chrome III;
- Upcycling employing tilapia fish leather. Every pair made with Tilapia is unique and vegetable tanning is used exclusively (Fig. 24b);
- B-Mesh: from recycled plastic bottles to innovative sneaker (Fig. 24c).

Vegano shoes

Focused on recycling and vegan products. Vegano Shoes was born in 2013 in the city of Franca, one of the largest footwear clusters in Brazil [201]:

- Use of recycled raw materials in its products, such as reused canvas fabric truck and tire chamber, combined with foams of mineral and vegetable origins.

Ahimsa

Focused on vegan products and slow fashion. Founded in 2013, Ahimsa is a vegan brand that prioritizes love [202].

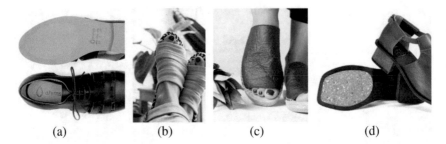

(a)   (b)   (c)   (d)

Fig. 25 **a** Example of Ahimsa woman footwear; Kasulo shoes brand [202]; **b** fabric reused upper [203]; and **c** Piñatex upper [203]; **d** Sandal Ciao Mao made form goat leather and recycled rubber sole [205]

- In 2016, vegan leather was introduced to the collection, made of polyurethane (PU), a very multipurpose material on the market that is becoming increasingly sustainable (Fig. 25a).

Kasulo
Focused on upcycling, sustainable raw materials and slow fashion, the main features are [203]:

- Fabrics purchased from the fashion industry surplus (Fig. 25b);
- Using "Piñatex" [204] ananas ("pineapple") material for shoe uppers. It is sustainable, ecological, vegan and biodegradable fiber collected by farmers in communities in the Philippines and then sent to Spain where the fabric is finished (Fig. 25c);
- Employing jute in soles;
- Cotton and polyester recycled fabrics used to upper shoes.

Ciao Mao
Focused on technological innovation, collaboration and slow fashion. This brand was created in 2007, with the purpose of original design focuses on four main attributes that express our way of being and doing: Brazilian culture; curatorship and co-creation; quality and comfort; and sustainable network (Fig. 25d). The main features are [205]:

- The vast majority of models are made of goat or bovine leather, from companies that respect quality and all environmental standards;
- The lining of almost all models is made of semi-finished leather (little pigment) to better absorb perspiration and provide pleasant skin-to-skin contact;
- Recycled rubber (natural latex with rubber residue) in the soles.

9 Conclusion

This study aimed to present a brief panorama of Brazilian footwear industry, leather production (bovine and fish and goat origins), environmental issues related to this sector, employed shoe's manufacture materials (natural and synthetic) and processes, including synthetic vegan options, Amazonian latex, natural fibers, vegetal leather tanning and natural dyeing. In addition, it is pointed the eco- and sustainable design in Brazilian footwear, indicating recent initiatives of recycled footwear raw material and sustainable product design solutions proposed by Brazilian small and micro-footwear companies.

In addition to bovine leather, which is a major player in the leather industry, fish and goat leather stand out in the Brazilian hides' industry, especially when associated to the use of vegetal tanning and natural dyes. There is a huge potentiality worldwide employing wood residues, fruit peels, seeds, flowers and other materials, and taking in account the Brazilian Amazonian and other biomes, their responsible use could preserve the environmental and do not damage human health.

Latex extraction in the Amazon is considered very important because it is a resource of its biodiversity, which generates income for the natives, maintains the forest ecosystem and contributes to the climatic balance. In addition, rubber trees have a large storage capacity for carbon extracted from the atmosphere, which not only happens in timber tree species but also in those producers of natural rubber. Latex in shoes also presents an appeal of sustainable product, but values the design-craftsmanship of the traditions of local Amazon communities.

The increasing of consumption awareness, which is aligned in many points to the vegan movement, is on the rise. In the case of veganism, not only because of the numbers of followers, but also because of the contribution it has made to raise awareness in dealing with animals. This public uses shoes made from natural/recycled fabrics, PU and PVC laminates, natural/recycled rubber, as an alternative to replace products of animal origin.

Nowadays, without recycling options and with toxic products in its composition, the footwear has only one solution to avoid further harming the environment: reducing consumption. When buying a pair, choose products made from recyclable materials, as natural fabrics or so-called vegan shoes. These have no leather and, consequently, do not contain chrome. The trend for the future will be biodegradable shoes, which decompose within four years after being buried. However, these solutions do not yet exist on the market. Another alternative is to choose to repair shoes and sneakers whenever possible, avoiding disposal. In addition, it is worth donating those which are no longer useful.

The simplification of shoe design can lead to fewer materials, less waste and less energy to produce. For instance, the number of components can be reduced through digital fabrication (zero waste cutting, 3D printing production, etc.). Also, footwear products with a long lifespan will help keep shoes out of landfills. In this way, design and material choices are the deciding factors for products' lifespan.

Product quality means that a product will accompany its producer from cradle to grave; prevention, recycling and disposal of waste. Improvements can be related to: (i) energy and water consumption; (ii) green technology and innovation: (iii) environmentally friendly materials; (iv) waste and residues generation; (v) environmental efficiency; (vi) human resources; and (vii) environmental management. In addition, the lean manufacturing tools implementation applied on these issues after all improves the waste reduction, control quality and the overall performance of footwear industries.

In Brazil, the "Sustainable Origin" seal was developed jointly between Brazilian sectoral footwear associations—the responsible for the component industries, Assintecal (Brazilian Association of Companies of Components for Leather, Footwear and Artifacts), and the responsible for the shoe manufacturing industries, Abicalçados (Brazilian Association of Footwear Industries). The seal came into effect in 2013 and in 2019 (July) a reformulation was launched. The current model of the seal is more demanding, mainly in obtaining the first category.

To face international competition, the Brazilian footwear sector has been investing in design, new technologies and sustainability—aiming to mitigate the industry's main impacts, such as the use of potentially toxic substances and diversity of working conditions. On the other hand, the national competition in the nowadays economic crisis (especially for the medium- and small-sized companies producing conventional shoes), in which they compete for the same resources and market, negatively impacts the environment in due the necessity of increasingly lower production costs. In this scenario, the interest in sustainable issues for this part of enterprises, including ecodesign and certification, is still secondary.

In Brazil, there is an event, the "Inspiramais" that greatly encourages the exhibition of sustainable brands of national components for footwear. This is the largest trade show for footwear and fashion components in Latin America. It is held twice a year in the largest city of the country (São Paulo, Brazil). With this, the "Brazilian Association of Companies of Components for Leather, Footwear and Manufactured Goods" (ASSINTECAL), responsible for organizing the Inspiramais fair, together with the national footwear industries, boost the development of new companies focused on Brazilian sustainable components.

Sustainability in fashion business models is divided into 5 pillars, which are the main ones: circular economy (Upcycling, Recycling and Vegan), sharing economy and collaborative consumption (collaboration, second hand and fashion library), technological innovation (sustainable raw materials, zero waste, wearable technology) and consumer awareness (capsule wardrobe, "lowsumerism" and slow fashion). According the examples presented, Brazilian sustainable footwear niche is predominant in micro- and small companies, which mainly focus on the circular economy issue, in addition they also use the theme of material innovation and most of them use it as a concept consumer awareness and slow fashion.

In conclusion, the Brazilian footwear sector, which presents a huge potential to green sustainable products and processes, has been growing in terms of the development of technological materials, as well as their application in the market, and new guidelines are already focusing on strategies for improve the manufacture of conventional and sustainable footwear. The society's demands in the search for technologically sustainable products, with less impact on the environment, promotes the competitive differential to reach the new consumer profile concerned with finding sustainable solutions to everyday problems. Thus, although the nowadays non-favorable economic situation, it is expected that issues related to sustainability and ecodesign will develop further within the national and global context.

Acknowledgements We gratefully acknowledge Coordination for the Improvement of Higher Education Personnel of Brazilian Education Ministry (**CAPES**).

References

1. Villela C (2020) Setor calçadista espera crescer até 2,5% no ano. Jornal do comércio – O jornal de economia e negócios RS (2020). Available at: https://www.jornaldocomercio. com/_conteudo/economia/2020/02/724671-setor-calcadista-espera-crescer-ate-2-5-no-ano. html Accessed on: 11 Feb 2020
2. ABICALÇADOS (2019) A. B. da I. do C. Relatório Setorial da Indústria de Calçados - Brasil 2019
3. ABICALÇADOS (2019) A. B. da I. do C. Relatório Comércio Exterior - Dezembro 2019
4. CICB (2018) Exportações brasileiras de couros e peles
5. Caldas R (2007) O Brasil e o Mercado Mundial de Couro. (CICB—Centro das Indústrias de Curtume do Brail)
6. Arantes CC, Garcez DS, Castello L (2008) Densidade de pirarucu (*Arapaima gigas*, Teleostei, Osteoglossidae) em lagos e reserva de desenvolvimento sustentável Mamirauá e Amanã. Amazonas, Brazil. Sci. Mag. UAKARI 2, 37–44. https://doi.org/10.31420/uakari.v2i1.13
7. Kanarski PS de O (2018) Extratos de origem vegetal em substituição aos produtos químicos no curtimento e tingimento da pele do pirarucu na Amazônia. (Fundação Universidade Federal de Rondônia)
8. De Azevedo Santos VM, Costa Neto EM, Stripari NDL (2010) Concepção dos pescadores artesanais que utilizam o reservatório de Furnas, Estado de Minas Gerais, acerca dos recursos pesqueiros: um estudo etnoictiológico. Biotemas 135–145. https://doi.org/10.5007/2175-7925.2010v23n4p135
9. Villar WD (1998) Química e Tecnologia dos Poliuretanos. Rhodia Solvay Group
10. Motawi W (2018) Shoe material design guide: the shoe designers complete guide to selecting and specifying footwear materials, volume 2 of How Shoes Are Made. (Wade Motawi)
11. Carvalho Filho J, Nunhes TV, Oliveira OJ (2019) Guidelines for cleaner production implementation and management in the plastic footwear industry. J Clean Prod 232:822–838. https://doi.org/10.1016/j.jclepro.2019.05.343
12. Kayser PAB (2008) The Brazilian shoe industry and the Chinese competition in international markets. Ohio University
13. Flores GD (2018) The men footwear industry in Brazil: challenges and opportunities. Fundação Getúlio Vargas
14. ABICALÇADOS (2016) A. B. da I. do C. Relatório Setorial da Indústria de Calçados - Brasil 2016
15. Schreiber D (2015) Reflective analysis of the organizational culture underlyning innovation by footwear manufactures. Bus Manag Rev 4:1–12
16. Guidolin SM, Costa ACR da, Rocha ÉRPda (2010) Indústria calçadista e estratégias de fortalecimento da competitividade. BNDES Setorial 31:147–148

17. Cavalheiro GM do C, Brandao M (2017) Assessing the IP portfolio of industrial clusters: the case of the Brazilian footwear industry. J Manuf Technol Manag 28:994–1010. doi:10.1108/JMTM-10-2016-0137
18. Kohan L, Martins CR, Oliveira Duarte L, Pinheiro L, Baruque-Ramos J (2019) Panorama of natural fibers applied in Brazilian footwear: materials and market. SN Appl Sci 1:895. https://doi.org/10.1007/s42452-019-0927-0
19. de Santana SKS (2015) O impacto da reconfiguração internacional do mercado calçadista sobre o segmento brasileiro de couro e calçados. IPEA—Inst Pesqui Econ Apl 2114:24
20. Castro I, Moreira C (2009) Reestruturação da indústria de calçados na Região Nordeste nas décadas 1990/2000. Rev Econ Nordeste 40:851–868
21. Belso-Martínez JA (2008) Differences in survival strategies among footwear industrial districts: the role of international outsourcing. Eur Plan Stud 16:1229–1248. https://doi.org/10.1080/09654310802401649
22. Dunoff JL, Moore MO (2014) Footloose and duty-free? Reflections on European union-Antidumping measures on certain footwear from China. World Trade Rev. 13:149–178. https://doi.org/10.1017/S1474745614000056
23. Gunawan T, Jacob J, Duysters G (2016) Network ties and entrepreneurial orientation: innovative performance of SMEs in a developing country. Int Entrep Manag J 12:575–599. https://doi.org/10.1007/s11365-014-0355-y
24. MDCI (2020) Base de dados do Comex Stat
25. Brasil IB (2020) Origem Sustentável. 2020 Available at: https://www.institutobybrasil.org.br/origem-sustentavel. Accessed on: 20 Feb 2020
26. Carvalho TCMB, Riekstin AC, Francisco GA, Guimarães I (2014) Sustainable origin seal—increasing the Brazilian footwear sector international competitiveness. Int J Sustain Energy Dev 3:136–146. https://doi.org/10.20533/ijsed.2046.3707.2014.0019
27. SENAI (2020) PPSR—Programa de Proficiência em Susbstâncias Restritas. (2020). Available at: https://www.senairs.org.br/PP-Substancias-Restritas. Accessed on: 05 Feb 2020
28. Maluf MLF, Hilbig CC (2010) Curtimento ecológico de peles de animais para agregação de valor através da confecção de artesanato. Varia Sci 9:75–79
29. Menda M (2012) Tratamento químico de couros e peles. Consellho Regional de Química - IV Região. Available at: https://crq4.org.br/couros_e_peles. Accessed on: 21 Feb 2020
30. CICB (2014) Brazilian Leather
31. CICB (2019) Exportações Brasileiras de Couros e Peles - Janeiro de 2019. CICB - Centro das Indústrias de Curtume do Brasil 7. Available at: http://cicb.org.br/storage/files/repositories/phpKi072B-total-exp-jan19-vr.pdf. Accessed on: 31 Jan 2020
32. ABDI (2011) Relatório de acompanhamento setorial—Indústria do couro
33. SEBRAE (2014) O mercado de curtume do couro de peixe. SEBRAE Mercados—Agronegócio, Alimentos e Bebidas, Indústria, Pesca e Aquicultura. Available at: https://respostas.sebrae.com.br/o-mercado-de-curtume-do-couro-de-peixe/. Accessed on: 31 Jan 2020
34. Nobrega LCOO (2016) couro de peixe e seus benefícios na indústria têxtil e de confecção. Universidade de São Paulo. https://doi.org/10.11606/d.100.2016.tde-08092015-131941
35. CICB (2017) Luxo Exótico. CICB—Centro das Indústrias de Curtume do Brasil. Available at: http://www.cicb.org.br/cicb/noticias/luxo-exotico. Accessed on: 26 Feb 2020
36. SEBRAE (2014) Informações técnicas para manter um viveiro de Tilápia. SEBRAE—Serviço Brasileiro de Apoio às Micro e Pequenas Empresas. Available at: https://respostas.sebrae.com.br/informacoes-tecnicas-para-manter-um-viveiro-de-tilapia/. Accessed on: 05 Feb 2020
37. Heliconia (2015) Sapatilha de couro de peixe tilápia. Heliconia, Brasil. Sapatilha nude—coleção verão 2015
38. Heliconia, Clutch de couro de peixe tilápia. Heliconia, Brasil
39. Torres CM, Repke DB (2006) Anadenanthera: visionary plant of ancient South America. Haworth Herbal Press
40. Rey S, Acosta JM, Carvalho FFR, Camacho ME, Costa RG (2007) O couro: contribuição na caprinocultura sustentavel. Arch Zootec 56:731–736

41. Ximenes LJF, da Cunha AM (2012) Setor de peles e de couros de caprinos e de ovinos no Nordeste. *Inf. Rural Etene—Banco do Nord.* Ano VI, 22
42. Tyrrell W (2016) Footwear assessment Chapter 9. Musculoskeletal key—fastest musculoskeletal insight engine. Available at: https://musculoskeletalkey.com/footwear-assessment/. Accessed on: 26 Feb 2020
43. Fashionary Shoe Design (2015) Fashionary
44. ASSINTECAL (2019) Estudo de Quantificação dos Materiais na Indústria Calçadista Edição 2019
45. Melvin JMA, Price C, Preece S, Nester C, Howard D (2019) An investigation into the effects of, and interaction between, heel height and shoe upper stiffness on plantar pressure and comfort. Footwear Sci 11:25–34. https://doi.org/10.1080/19424280.2018.1555862
46. Yick K, Yu A, Li P (2019) Insights into footwear preferences and insole design to improve thermal environment of footwear. Int J Fash Des Technol Educ 12:325–334. https://doi.org/10.1080/17543266.2019.1629028
47. Williams A (2007) Footwear assessment and management. Pod Manag 26:165–178
48. Uttam D (2013) Active sportswear fabrics. Int J IT Eng Appl Sci Res 2:2319–4413
49. Roh EK, Oh KW, Kim SH (2013) Classification of synthetic polyurethane leather by mechanical properties according to consumers' preference for fashion items. Fibers Polym 14:1731–1738. https://doi.org/10.1007/s12221-013-1731-x
50. Dreger AA, Barbosa LA, Santana RMC, Schneider EL, Morisso FDP (2018) Caracterização mecânica e morfológica de solados produzidos com resíduos de laminados de PVC da indústria calçadista. Matéria (Rio Janeiro) 23. https://doi.org/10.1590/s1517-707620170001.0288
51. Guimarães IMB (2018) Panorama do complexo calçadista e coureiro: contexto e perspectivas. ASSINTECAL—Associação Brasileira das Empresas de Componentes para Couro, Calçados e Artefatos 33. Available at: http://feiplar.com.br/pos_feira/apresentacoes/dia07/calcadista/painel_calcadista_assintecal.pdf. Accessed on: 31 Jan 2020
52. Silva RM et al (2009) Evaluation of shock absorption properties of rubber materials regarding footwear applications. Polym Test 28:642–647. https://doi.org/10.1016/j.polymertesting.2009.05.007
53. Yick KL, Tse CY (2013) Textiles and other materials for orthopaedic footwear insoles. In: Handbook of footwear design and manufacture. Elsevier, pp 341–371. https://doi.org/10.1533/9780857098795.4.341
54. Stimpert D (2019) Shoe glossary: outsole every outsole is different. Liveaboutdotcom
55. Paiva RMM, Marques EAS, da Silva LFM, António CAC (2015) Surface treatment effect in thermoplastic rubber and natural leather for the footwear industry. Materwiss Werksttech 46:632–643. https://doi.org/10.1002/mawe.201500403
56. Chen D-C, Jin H-W, Su J-Z (2019) Investigation on optimization of baking in sole bonding process. IOP Conf Ser Mater Sci Eng 644:012002. https://doi.org/10.1088/1757-899X/644/1/012002
57. Ma J, Shao L, Xue C, Deng F, Duan Z (2014) Compatibilization and properties of ethylene vinyl acetate copolymer (EVA) and thermoplastic polyurethane (TPU) blend based foam. Polym Bull 71:2219–2234. https://doi.org/10.1007/s00289-014-1183-5
58. Brückner K, Odenwald S, Schwanitz S, Heidenfelder J, Milani T (2010) Polyurethane-foam midsoles in running shoes—impact energy and damping. Procedia Eng 2:2789–2793. https://doi.org/10.1016/j.proeng.2010.04.067
59. Branthwaite H, Chockalingam N, Greenhalgh A (2013) The effect of shoe toe box shape and volume on forefoot interdigital and plantar pressures in healthy females. J Foot Ankle Res 6:28. https://doi.org/10.1186/1757-1146-6-28
60. Smith CJ et al (2013) Design data for footwear: sweating distribution on the human foot. Int J Cloth Sci Technol 25:43–58. https://doi.org/10.1108/09556221311292200
61. Yuan Y, Lee TR (2013) Contact angle and wetting properties. In: Bracco G, Holst B (eds) Surface science techniques, 3–34 (Springer Berlin Heidelberg). https://doi.org/10.1007/978-3-642-34243-1_1

62. Irzmańska E, Brochocka A (2014) Influence of the physical and chemical properties of composite insoles on the microclimate in protective footwear. Fibres Text East Eur 5:89–95
63. Hessas S, Behr M, Rachedi M, Belaidi I (2018) Heel lifts stiffness of sports shoes could influence posture and gait patterns. Sci Sports 33:e43–e50. https://doi.org/10.1016/j.scispo.2017.04.015
64. de Paula VB, Paschoarelli LC (2007) Aplicação do design sustentável no desenvolvimento de calçado. In: IV CIPED—4o Congresso Internacional de Pesquisa em design
65. Scherer FL, Gomes CM, Madruga LRG, Crespam CC (2009) Estratégia e práticas de gestão socioambiental: o caso das empresas brasileiras exportadoras do setor de calçados. Rev Adm FACES J 8:116–136. https://doi.org/10.21714/1984-6975FACES2009V8N4ART167
66. da Silva AH, Moraes CAM, Modolo RCE (2015) Avaliação ambiental do setor calçadista e a aplicação da análise de ciclo de vida: uma abordagem geral. In: 6o Forum Internacional de Resíduos Sólidos (Instituto Venturi para Estudos Ambientais)
67. Robinson LC (2009) Estudo sobre o nível de evolução da indústria calçadista para o desenvolvimento de calçados ecológicos. Centro Universitário Feevale
68. Lee M, Rahimifard S (2010) Development of an economically sustainable recycling process for the footwear sector. In: Press H (ed) Proceedings of the 17th CIRP life cycle engineering conference (LCE2010) (U. of T.), pp 334–339
69. Lee MJ, Rahimifard S (2012) A novel separation process for recycling of post-consumer products. CIRP Ann 61:35–38. https://doi.org/10.1016/j.cirp.2012.03.026
70. Francisco GA, Dias SLFG (2015) Design para a Sustentabilidade e Resíduos: Reflexões a partir da Prática na Indústria de Calçados. In: 7o. Encontro Nacional da Associação Nacional de Pós Graduação e Pesquisa em Ambiente e Sociedade. Universidade de Brasília
71. Francisco GA (2017) Prevenção de resíduos: um estudo de caso na indústria calçadista brasileira. Universidade de São Paulo. https://doi.org/10.11606/d.106.2017.tde-10012017-132000
72. Guarienti GRO (2018) cenário calçadista ambientalmente orientado e as práticas de design que reduzem o impacto do fim de vida útil dos calçados. Universidade Federal do Rio Grande do Sul
73. Rajamani SG (2014) Growth of leather sector in asian countries and recent environmental developments in world leather sector. In: ICAMS 2014—5th international conference on advanced materials and systems, 491–496
74. Staikos T, Rahimifard S (2007) End-of-life management considerations in the footwear industry. Glob Footwear Ind Emerg Trends 16
75. Mia MAS, Nur-E-Alam M, Murad ABMW, Ahmad F, Uddin MK (2017) Waste management & quality assessment of footwear manufacturing industry in Bangladesh: an innovative approach. Int J Eng Manag Res 7:402–407
76. Shah B (2018) To study the waste caused by discarded footwear in India and finding a solution for the reduction of the same. National Institute of Fashion Technology, Mumbai
77. Valle L (2018) Prejudicial ao meio ambiente, calçados devem ser consumidos sem excesso Comprar sapatos de materiais recicláveis e consertá-los pode ajudar a reduzir impactos negativos. Cidadania. Available at: https://www.institutonetclaroembratel.org.br/cidadania/nossas-novidades/noticias/prejudicial-ao-meio-ambiente-calcados-devem-ser-consumidos-sem-excesso/. Accessed on: 05 Feb 2020
78. Brasil M (2018) Calçados: dos resíduos da produção aos resíduos dos sapatos velhos. Moda sem Crise—Jornalismo Consciente. Available at: http://modasemcrise.com.br/calcados-dos-residuos-da-producao-aos-residuos-dos-sapatos-velhos/. Accessed on: 26 Feb 2020
79. Lopes D, Ferreira MJ, Russo R, Dias JM (2015) Natural and synthetic rubber/waste e Ethylene-Vinyl Acetate composites for sustainable application in the footwear industry. J Clean Prod 92:230–236. https://doi.org/10.1016/j.jclepro.2014.12.063
80. Sarmento F (2014) Design para a sociobiodiversidade: perspectivas para o uso sustentável da borracha na Floresta Nacional do Tapajós. Universidade de São Paulo. https://doi.org/10.11606/t.16.2014.tde-28072014-111246

81. Globo Rural (2013) Depois de 100 anos, látex ainda é importante fonte de renda na Amazônia. G1—Globo Rural. Available at: http://g1.globo.com/economia/agronegocios/noticia/2013/01/extracao-do-latex-ainda-e-importante-fonte-de-renda-na-amazonia.html. Accessed on: 21 Feb 2020

82. de Melo H (2019) Novas tecnologias aplicadas para a seringueira da Amazônia podem fazer a borracha natural voltar a ter importância econômica para a região. FAPEAM—Fundação de Amparo à Pesquisa do Estado do Amazonas. Available at: http://www.fapeam.am.gov.br/novas-tecnologias-aplicadas-para-a-seringueira-da-amazonia-podem-fazer-a-borracha-natural-voltar-a-ter-importancia-economica-para-a-regiao/. Accessed on: 31 Jan 2020

83. Garcia EA (2012) Aplicação do látex da Hevea brasiliensis em produtos têxteis sustentáveis, como material alternativo no design de moda. Universidade da Beira Interior

84. Roslim R et al (2018) Novel deproteinised natural rubber latex slow-recovery foam for health care and therapeutic foam product applications. J Rubber Res 21:277–292. https://doi.org/10.1007/BF03449175

85. De Azevedo Cunha J, Carneiro M, De Souza K, Santos Júnior F, Ceccatto V (2019) Biomaterials characterization for orthopedic orthoses: a systematic review. J Mater Sci Nanotechnol 7:101

86. Kudori SNI, Ismail H, Shuib RK (2019) Kenaf core and bast loading versus properties of natural rubber latex foam (NRLF). Bioresources 14:1765–1780

87. Teklay A et al (2018) Preparation of value added composite sheet from solid waste leather—a prototype design. Sci Res Essays 13:11–13. https://doi.org/10.5897/SRE2017.6551

88. do Nascimento KR, Pastore Júnior F, Peres Júnior JBR (2015) Produção de borracha FDL e FSA: guia de treinamento. WWF, Brasil

89. Emperaire LAMB (2002) Seringueiros e Seringas. In: Cunha C, Almeida MB (eds) Enciclopédia da Floresta. Cia. das Letras, pp 1–7

90. Samonek FA (2006) borracha vegetal extrativa na amazônia: um estudo de caso dos novos encauchados de vegetais no estado do Acre. Universidade Federal do Acre

91. MMA—Ministério do Meio Ambiente & ICMBio—Instituto Chico Mendes de Conservação da Biodiversidade. Plano de Manejo da Reserva Extrativista do Cazumbá-Iracema

92. PPA—Programa de Aceleração. Encauchados de vegetais da Amazônia. PPA—Programa de Aceleração. Available at: http://aceleracao.ppa.org.br/portfolio-de-negocios/encauchados/. Accessed on: 21 Feb 2020

93. Rosa M (2018) Empresa social cria produtos com látex orgânico da Amazônia. Ciclovivo. Available at: https://ciclovivo.com.br/inovacao/negocios/empresa-social-cria-produtos-com-latex-organico-da-amazonia/. Accessed on: 21 Feb 2020

94. Jardim A (2016) Florestas Plantadas cultivam tradição acreana na economia. Fotos públicas. Available at: https://fotospublicas.com/florestas-plantadas-cultivam-tradicao-acreana-na-economia/. Accessed on: 24 Feb 2020

95. ARTESOL - Artesanato Solidário. Doutor da Borracha. ARTESOL—Artesanato Solidário Available at: https://www.artesol.org.br/rede/membro/doutor_da_borracha. Accessed on: 26 Feb 2020

96. ECOLOGICA LATEX (2020) Laminado vegetal estampado. Ecologica. Available at: http://ecologicalatex.com.br/category/laminados/estampado/. Accessed on: 05 Feb 2020

97. VERT (2020) Borracha Nativa. VERT. Available at: https://www.vert-shoes.com.br/content/36-borracha. Accessed on: 24 Feb 2020

98. Zhai JJ (2011) The Denim's characteristics as upper material of footwear. Adv Mater Res 332–334:1643–1646. https://doi.org/10.4028/www.scientific.net/AMR.332-334.1643

99. Santos HM, Razza BM, dos Santos JEG (2015) História da alpargatas: um modelo resistente ao tempo e ao modismo. In: VII Congresso International de História 2043–2055. https://doi.org/10.4025/7cih.pphuem.1428

100. Tatàno F et al (2012) Shoe manufacturing wastes: characterisation of properties and recovery options. Resour Conserv Recycl 66:66–75. https://doi.org/10.1016/j.resconrec.2012.06.007

101. FARM RIO (2019) Moda feminina (Feminine fashion). Farm Rio. Available at: https://busca.farmrio.com.br/busca?q=sapato+feminino&p=sapato&ranking=2&typeclick=1&ac_pos=header. Accessed on: 26 Feb 2020

102. AREZZO (2019) Sandálias (Sandals). Arezzo. Available at: https://www.arezzo.com.br/c/sap atos/sandalias. Accessed on: 21 Feb 2020

103. Ławińska K, Serweta W, Gendaszewska D (2018) Applications of bamboo textiles in individualised children's footwear. Fibres Text East Eur 26:87–92. https://doi.org/10.5604/01.3001. 0012.2537

104. Ławińska K, Serweta W, Jaruga I, Popovych N (2019) Examination of selected upper shoe materials based on bamboo fabrics. Fibres Text East Eur 27:85–90. https://doi.org/10.5604/01.3001.0013.4472

105. Satyanarayana KG, Guimarães JL, Wypych F (2007) Studies on lignocellulosic fibers of Brazil. Part I: source, production, morphology, properties and applications. Compos Part A Appl Sci Manuf 38:1694–1709. https://doi.org/10.1016/j.compositesa.2007.02.006

106. Li Y, Mai Y-W, Ye L (2000) Sisal fibre and its composites: a review of recent developments. Compos Sci Technol 60:2037–2055. https://doi.org/10.1016/S0266-3538(00)00101-9

107. Martins MA, Mattoso LHC (2004) Short sisal fiber-reinforced tire rubber composites: dynamic and mechanical properties. J Appl Polym Sci 91:670–677. https://doi.org/10.1002/app.13210

108. Iozzi MA, Martins MA, Mattoso LHC (2004) Propriedades de compósitos híbridos de borracha nitrílica, fibras de sisal e carbonato de cálcio. Polímeros 14:93–98. https://doi.org/10.1590/S0104-14282004000200012

109. Iozzi MA et al (2010) Estudo da influência de tratamentos químicos da fibra de sisal nas propriedades de compósitos com borracha nitrílica. Polímeros 20:25–32. https://doi.org/10.1590/S0104-14282010005000003

110. Vidal AC, Mulinari DR (2014) Preparação e caracterização de biocompósitos de poliuretano reforçados com fibras da palmeira para aplicação em palmilhas. Cad UniFOA 1:89–93

111. Cao H et al (2014) Development and evaluation of apparel and footwear made from renewable bio-based materials. Int J Fash Des Technol Educ 7:21–30. https://doi.org/10.1080/17543266.2013.859744

112. Riba MTL, Miró EPi (2007) O couro : as técnicas para criar objectos de couro explicadas com rigor e clareza. Estampa

113. de Paiva RM, Morisso FDP (2009) Uma breve revisão sobre processos de curtimentos alternativos e considerações sobre a aplicação de cálculos químicos na previsão de propriedades desses sistemas. Rev Tecnol e Tendências 8:89–96

114. Falcão L, Araújo MEM (2011) Tannins characterisation in new and historic vegetable tanned leathers fibres by spot tests. J Cult Herit 12:149–156. https://doi.org/10.1016/j.culher.2010.10.005

115. Laurenti R, Redwood M, Puig R, Frostell B (2017) Measuring the environmental footprint of leather processing technologies. J Ind Ecol 21:1180–1187. https://doi.org/10.1111/jiec.12504

116. Sundar VJ, Muralidharan C (2017) Salinity free high tannin fixation vegetable tanning: commercial success through new approach. J Clean Prod 142:2556–2561. https://doi.org/10.1016/j.jclepro.2016.11.021

117. de Moura AS (2014) O beneficiamento do couro e seus agentes na capitania de Pernambuco. Universidade Federal de Pernambuco, pp 1710–1760

118. Paes JB, Marinho IV, de Lima RA, de Lima CR, de Azevedo TKB (2006) Viabilidade técnica dos taninos de quatro espécies florestais de ocorrência no semi-árido brasileiro no curtimento de peles. Ciência Florest 16:453. https://doi.org/10.5902/198050981927

119. Fernandes B et al (2017) Série sobre produção de calçados mostra a produção artesanal em Cabaceiras)

120. Duarte T, Cabaceiras PB (2015) Inovação tecnológica e capacitação na produção associada ao turismo. Oconcierge. Available at: https://oconciergeonline.com.br/coluna/cabaceiras-pb-inovacao-tecnologica-e-capacitacao-na-producao-associada-ao-turismo/. Accessed on: 21 Feb 2020

121. Prefeitura Municipal de Cabaceiras (2019) Veja programação completa do Festival do Couro 2019, no distrito da Ribeira de Cabaceiras. Prefeitura Municipal de Cabaceiras. Available at: http://cabaceiras.pb.gov.br/veja-programacao-completa-do-festival-do-couro-2019-no-distrito-da-ribeira-de-cabaceiras/. Accessed on: 21 Feb 2020

122. de Zuim VA, Farias ACS, de Vasconcelos AFP, de Held MSB, Kanamaru AT (2014) As transformações do couro no trabalho de Espedito Seleiro como alternativa de superação para as adversidades do sertão. Rev. Labor 1:58–72

123. de Zuim VAS (2016) Espedito Seleiro: análise dos métodos e processos produtivos artesanais como possibilidades criativas no Design de Moda. Universidade de São Paulo. https://doi.org/10.11606/d.100.2016.tde-23052016-111839

124. Rungruangkitkrai N, Mongkholrattanasit R (2012) Eco-friendly of textiles dyeing and printing with natural dyes. In: RMUTP international conference: textiles & fashion 2012, vol 17

125. Deb AK, Shaikh MAA, Sultan MZ, Rafi MIH (2017) Application of lac dye in shoe upper leather dyeing. Leather Footwear J 17:97–106. https://doi.org/10.24264/lfj.17.2.4

126. de Melo KSG (2007) Extração e uso de corantes vegetais da Amazônia no tingimento do couro de Matrinxã (Brycon amazonicu Spix & Agassiz, 1819). Instituto Nacional de Pesquisas da Amazônia—Universidade Federal do Amazonas

127. Velmurugan P, Shim J, Seo S-K, Oh B-T (2016) Extraction of natural dye from *Coreopsis tinctoria* flower petals for leather dyeing—an eco-friendly approach. Fibers Polym 17:1875–1883. https://doi.org/10.1007/s12221-016-6226-0

128. Pervaiz S, Mughal TA, Khan FZ (2016) Green fashion colours: a potential value for punjab leather industry to promote sustainable development. Pak J Contemp Sci 1:28–36

129. Mengistu M (2017) Natural dyes from (*Bidens macroptera* Mesfin and *Eucalyptus camaldulensis* Dehnh.) leaves and their application for leather dyeing. Addis Ababa University

130. Sudha CG, Aggarwal S (2017) Eco-benign wet processing of leather: from dyeing to after treatment. Int J Home Sci 3:693–697

131. Pervaiz S et al (2017) Environmental friendly leather dyeing using *Tagetes erecta* L. (Marigold) waste flowers. Int J Biosci 10:382–390)

132. Berhanu T, Ratnapandian S (2017) Extraction and optimization of natural dye from Hambo Hambo (*Cassia singueana*) plant used for coloration of tanned leather materials. Adv Mater Sci Eng 2017:1–5. https://doi.org/10.1155/2017/7516409

133. Shamsheer HB, Mughal TA, Ishaq A, Zaheer S, Zahid K (2017) Extraction of ecofriendly leather dyes from plants bark. Pak J Sci Ind Res Ser A Phys Sci 60:96–100

134. Song JE, Kim SM, Kim HR (2017) Improvement of dye affinity in natural dyeing using Terminalia chebula retzius (*T. chebula*) applied to leather. Int J Cloth Sci Technol 29:610–626. https://doi.org/10.1108/ijcst-03-2017-0029

135. Gull S, Mahmood MH-U-R, Ishfaq A, Ramzan S, Iqbal Z (2018) Extraction of natural colorants from two varieties of saffron (*Crocus sativus*) and their dyeing evaluation on goat leather. World J Pharm Res 7:45–52. https://doi.org/10.20959/wjpr201818-12861

136. Fuck WF, Brandellib A, Gutterresa M (2018) Special review paper: leather dyeing with biodyes from filamentous fungi. J Am Leather Chem Assoc 113:299–310

137. Velmurugan P, Vedhanayakisri KA, Park Y-J, Jin J-S, Oh B-T (2019) Use of *Aronia melanocarpa* fruit dye combined with silver nanoparticles to dye fabrics and leather and assessment of its antibacterial potential against skin bacteria. Fibers Polym 20:302–311. https://doi.org/10.1007/s12221-019-8875-2

138. Stahl, Mild-Solvent PERMUTHANE® PU Coatings. Stahl. Available at: https://www.stahl.com/performance-coatings/permuthane. Accessed on: 11 Feb 2020

139. Stahl. Reactive High Solids PERMAQURE® PU Coatings. Stahl. Available at: https://www.stahl.com/performance-coatings/permaqure. Accessed on: 11 Feb 2020

140. Stahl. Water-Borne PERMUTEX® PU Coatings. Stahl. Available at: https://www.stahl.com/performance-coatings/permutex. Accessed on: 11 Feb 2020

141. Stoye D, Freitag W, Beuschel G (1997) Resins for coatings: chemistry, properties and applications. Choice Reviews Online vol 34. Hanser Verlag

142. Blesius J (2018) An overview of polyurethane fabric. Mitchell—performance Faux leathers. Available at: https://mitchellfauxleathers.com/Default/ViewPoint/Read/faux-leather-viewpoint/2018/03/16/an-overview-of-polyurethane-fabric. Accessed on: 24 Feb 2020

143. Park YG, Ji W, Han KS, Jee MH (2017) A survey and studies on the residual content of dimethylformamide and its reduction in polyurethane-based consumer products. J Korean Soc Qual Manag 45:769–780. https://doi.org/10.7469/JKSQM.2017.45.4.769

144. Trade Korea (2020) Process synthetic leather and PU coated spilt leather. Trade Korea. Available at: https://www.tradekorea.com/product/detail/P301994/Process-Synthetic-Leather-and-PU-Coated-Spilt-Leather.html. Accessed on: 05 Feb 2020
145. Rodolfo Junior A, Tsukamoto CT (2019) Tecnologia do PVC. Instituto Brasileiro do PVC. Instituto Brasileiro do PVC
146. Martinz D, Quadros J (2008) Compounding PVC with renewable materials. Plast Rubber Compos 37:459–464. https://doi.org/10.1179/174328908X362917
147. Textilia.net (2019) Cipatex® lança linha ecológica com tecnologia Vinyl Tech. Textilia.net— O maior conteúdo da cadeia têxtil. Available at: http://www.textilia.net/materias/ler/moda/Cal cados/cipatex-lanca-linha-ecologica-com-tecnologia-vinyl-tech. Accessed on: 05 Feb 2020
148. Cipatex (2020) VynilTech—Cipatex. Cipatex. Available at: https://www.catalogocipatex. com.br/pt/moda/vinyl-tech/
149. Grupo A (2018) Cipatex anuncia sua primeira linha vegana. TM - Tecnologia de Materiais. Available at: http://tecnologiademateriais.com.br/portaltm/cipatex-anuncia-sua-primeira-linha-vegana/. 05 Feb 2020
150. Brandão F (2018) Novos Tempos para o Calçado. Revista Tecnicouro 80–81. Available at: https://issuu.com/marcelachavesdasilva/docs/ed._308_-_completa. Accessed on: 11 Feb 2020
151. ECHA (2013) European Chemicals Agency. Evaluation of new scientific evidence concerning DINP and DIDP In relation to entry 52 of Annex XVII to REACH Regulation (EC) No 1907/2006
152. European Union (2017) Commission implementing decision (EU) 2017/1210 of 4 July 2017 on the identification of bis(2-ethylhexyl) phthalate (DEHP), dibutyl phthalate (DBP), benzyl butyl phthalate (BBP) and diisobutyl phthalate (DIBP) as substances of very high concern according
153. ECHA (2018) European Chemicals Agency. Committee for risk assessment RAC opinion proposing harmonised classification and labelling at EU level of 1,2-Benzenedicarboxylic acid, di-C8-10-branched alkylesters, C9- rich; [1] di-"isononyl" phthalate; [2] [DINP]. 35
154. Mottram R (2018) Welcoming change: a decade of innovations and developments - helping to shape the future of the PVC industry. In: PVC formulation 2018 4. Applied Market Information Ltd. UK
155. Wikipedia (2020) Veganism. Wikipedia. Available at: https://en.wikipedia.org/wiki/Veg anism. Accessed on: 11 Feb 2020
156. PETA—People for the Ethical Treatment of Animals (2012) 'PETA-Approved Vegan' Logo. PETA. Available at: https://www.peta.org/living/personal-care-fashion/peta-approved-vegan-logo/. Accessed on: 11 Feb 2020
157. Barbero S, Cozzo B, Tamborrini P (2009) Ecodesign. HF Ullmann
158. Rech SR, de Souza RKR (2009) Ecoluxo e sustentabilidade: um novo comportamento do consumidor. DAPesquisa 4:602–608. doi:10.5965/1808312904062009602
159. Santolaria M, Oliver-Solà J, Gasol CM, Morales-Pinzón T, Rieradevall J (2011) Eco-design in innovation driven companies: perception, predictions and the main drivers of integration. The Spanish example. J Clean Prod 19:1315–1323. https://doi.org/10.1016/j.jclepro.2011.03.009
160. Jørgensen MS, Jensen CL (2012) The shaping of environmental impacts from Danish production and consumption of clothing. Ecol Econ 83:164–173. https://doi.org/10.1016/j.ecolecon. 2012.04.002
161. ISO/TR 14062:2002(en). ISO/TR 14062:2002(en) Environmental management—integrating environmental aspects into product design and development
162. Gwilt A, Longarço M (2015) Moda sustentável: Um guia prático. Editora Gustavo Gili
163. da Silveira ALM, Franzato C, van der Linden J (2014) Caminhos para a Sustentabilidade através do Design. Uniritter
164. Jin Gam H, Cao H, Farr C, Heine L (2009) C2CAD: a sustainable apparel design and production model. Int J Cloth Sci Technol 21:166–179. https://doi.org/10.1108/095562209 10959954
165. APICCAPS—Portuguese Footwear C, LGMA (2014) The world footwear 2014 yearbook

166. CouroModa (2015) O cenário mundial do calçado e as oportunidades para o Brasil. CouroModa. Available at: https://www.couromoda.com/noticias/ler/o-cenario-mundial-do-calcado-e-as-oportunidades-para-o-brasil/. Accessed on: 20 Feb 2020

167. Francisco GA, Dias SLFG, de Carvalho TCMB (2013) A cadeia reversa do calçado: uma revisão da literatura com foco no resíduo. In: XVI Simpósio de Administração da Produção, Logística e Operações Internacionais (SIMPOI)

168. Braga JC (2010) O design industrial como ferramenta para a sustentabilidade: estudo de caso do couro de peixe. Rev Espaço Acadêmico 114:110–117

169. Herva M, Álvarez A, Roca E (2011) Sustainable and safe design of footwear integrating ecological footprint and risk criteria. J Hazard Mater 192:1876–1881. https://doi.org/10.1016/j.jhazmat.2011.07.028

170. Fletcher K, Grose L (2012) Moda & sustentabilidade: design para mudança. Senac São Paulo

171. Manzini E, Vezzoli C, de Carvalho A (2011) O Desenvolvimento de Produtos Sustentáveis. Os Requisitos Ambientais dos Produtos Industriais. EDUSP

172. Ventura FC (2014) Aplicabilidade da metodologia ecodesign à produção de calçados femininos. Universidade Estadual Paulista

173. Polese F, Ciasullo MV, Troisi O, Maione G (2019) Sustainability in footwear industry: a big data analysis. Sinergie Ital J Manag 37:149–170. https://doi.org/10.7433/s108.2019.09

174. Perry P (2012) Exploring the influence of national cultural context on CSR implementation. J Fash Mark Manag 16:141–160. https://doi.org/10.1108/13612021211222806

175. Caniato F, Caridi M, Crippa L, Moretto A (2012) Environmental sustainability in fashion supply chains: An exploratory case based research. Int J Prod Econ 135:659–670. https://doi.org/10.1016/j.ijpe.2011.06.001

176. Bonaldi RR (2018) Functional finishes for high-performance apparel. In: High-performance apparel. Elsevier, pp 129–156. https://doi.org/10.1016/b978-0-08-100904-8.00006-7

177. Cunha R (2020) Rhodia e grupo de empresas lançam novo conceito de calçado sustentável e confortável. Stylo Urbano. Available at: https://www.stylourbano.com.br/rhodia-e-grupo-de-empresas-lancam-novo-conceito-de-calcado-sustentavel-e-confortavel/. Accessed on: 26 Feb 2020

178. AREZZO (2020) ZZBio Arezzo. Arezzo. Available at: https://www.arezzo.com.br/c/zzbio. Accessed on: 26 Feb 2020

179. INSPIRAMAIS (2020) Inspiramais 2021_I: sincronia, um conceito que linka mundo high tech às relações humanas e sustentabilidade. Inspiramais. Available at: https://www.inspiramais.com.br/conteudo/1991/inspiramais-2021-i-sincronia-um-conceito-que-linka-mundo-https://www.inspiramais.com.br/conteudo/1991/inspiramais-2021-i-sincronia-um-conceito-que-linka-mundo-high-tech-as-relacoes-humanas-sustentabilidade. Accessed on: 26 Feb 2020

180. Ambiente Verde (2019) Ambiente Verde—Apresentação Institucional. Ambiente Verde. Available at: https://d7ae7404-52ff-4157-9544-63fc5059bd89.filesusr.com/ugd/813307_24d323732a1f4d73b26efab7d6a24f3d.pdf. Accessed on: 21 Feb 2020

181. Pagliaro M (2019) An industry in transition: the chemical industry and the megatrends driving its forthcoming transformation. Angew Chem Int 58:11154–11159. https://doi.org/10.1002/anie.201905032

182. COFRAG (2019) Nossa palmilha TECHGREEN - Cofrag. COFRAG. Available at: https://www.cofrag.com.br/nossa-palmilha-techgreen/. Accessed on: 21 Feb 2020

183. Cunha R (2020) Castanhal investe no setor calçadista e lança artigos em juta no Inspiramais 2021. Available at: https://www.stylourbano.com.br/castanhal-investe-no-setor-calcadista-e-lanca-artigos-em-juta-no-inspiramais-2021/. Accessed on: 26 Feb 2020

184. BERTEX (2015) Bertex Componentes para Calçados. Bertex. Available at: http://www.bertex.com.br/. Accessed on: 26 Feb 2020

185. Baruque-Ramos J et al (2017) Social and economic importance of textile reuse and recycling in Brazil. IOP Conf Ser Mater Sci Eng 254. https://doi.org/10.1088/1757-899x/254/19/192003

186. Leal W et al (2019) A review of the socio-economic advantages of textile recycling. J Clean Prod 218:10–20. https://doi.org/10.1016/j.jclepro.2019.01.210

187. do Amaral MC et al (2018) Industrial textile recycling and reuse in Brazil: case study and considerations concerning the circular economy. Gestão & Produção 25:431–443. doi:10.1590/0104-530x3305

188. ECOSIMPLE. Ecosimple - Segmentos. Ecosimple. Available at: http://ecosimple.com.br/seg mentos/. Accessed on: 26 Feb 2020

189. Maturo J (2019) Cotton move amplia canais de venda. GBL Jeans. Available at: https://www. gbljeans.com.br/mercado/producao-limpa/cotton-move-amplia-canais-de-venda/. Accessed on: 26 Feb 2020

190. Simon F (2019) Índice de Transparência da Moda Brasil. Fashion Revolution

191. Tan K (2019) Audience concern of eco-fashion by fashion revolution Indonesia through marketing communication in Jakarta. J Commun Stud 5:81–94

192. Golfetto IF, Fialho FAP, Pellizzoni RC (2019) Inovação e design na trajetória de indústrias ultracentenárias. e-Rev LOGO 8:21–37. https://doi.org/10.26771/e-revista.logo/2019.1.02

193. Brasil Ecofashion Week (2019) Brasil Ecofashion Week—O Evento. Brasil ecofashion week. Available at: https://befw.com.br/o-evento/. Accessed on: 26 Feb 2020

194. Berlim LG (2009) Moda, a possibilidade da leveza sustentável: tendências, surgimento de mercados justos e criadores responsáveis. Universidade Federal Fluminense

195. Redação Jornal Exclusivo (2019) Inspiramais promove a sustentabilidade na indústria. Jornal Exclusivo. Available at: http://exclusivo.com.br/_conteudo/2019/06/negocios/2426183-ins piramais-promove-a-sustentabilidade-na-industria.html. Accessed on: 26 Feb 2020

196. Tolipan H (2019) Inspiramais 2020_II: projeto Conexão Criativa e comercial é a síntese da diversidade do país + sustentabilidade. Available at: https://heloisatolipan.com.br/moda/ins piramais-2020_ii-projeto-conexao-criativa-e-comercial-e-a-sintese-da-diversidade-do-pais-sustentabilidade/. Accessed on: 26 Feb 2020

197. Todeschini BV, Cortimiglia MN, Callegaro-de-Menezes D, Ghezzi A (2017) Innovative and sustainable business models in the fashion industry: entrepreneurial drivers, opportunities, and challenges. Bus Horiz 60:759–770. https://doi.org/10.1016/j.bushor.2017.07.003

198. Sutter MB, Galleli B, MacLennan MLF, Polo EF, Correa HL (2015) Brazil's fashion and clothing industry: sustainability, competitiveness and differentiation. Lat Am J Manag Sustain Dev 2:280

199. INSECTA SHOES (2020) Insecta Shoes—About us and Collection. Available at: https://ins ectashoes.com/collections/sapatos?gclid=EAIaIQobChMItpaq1_fs5wIVygeRCh3U4wPp EAAYASAAEgJ3l_D_BwE. Accessed on: 26 Feb 2020

200. VERT (2020) Project and Collection. Vert. Available at: https://www.vert-shoes.com.br/

201. Vegano Shoes (2019) The company and collection. Vegano Shoes. Available at: https:// www.veganoshoes.com.br/?gclid=EAIaIQobChMItoXs5_3s5wIVyQeRCh1IUwONEAA YASAAEgKT-PD_BwE. Accessed on: 26 Feb 2020

202. Ahimsa (2020) Ahimsa—brand and collection. Ahimsa. Available at: https://useahimsa. com/?gclid=EAIaIQobChMIzaaW-Y_t5wIVxoCRCh3p-wJXEAAYASAAEgLkpPD_BwE. Accessed on: 26 Feb 2020

203. Kasulo. Kasulo—Brand and Collection. Kasulo. Available at: http://www.kasulo.com.br/. Accessed on: 26 Feb 2020

204. Ananas Anam. Piñatex. Ananas Anam. Available at: https://www.ananas-anam.com/. Accessed on: 26 Feb 2020

205. Ciao Mao. Ciao Mao—Brand and collection. Ciao Mao. Available at: https://www.ciaomao. com/. Accessed on: 26 Feb 2020

Waste and 3R's in Footwear and Leather Sectors

Nilesh C. Jadhav and Akshay C. Jadhav

Abstract In every aspect of human life, there are many various wanted and unwanted materials manufactured or generated. Then later, they are discarded in the environment just because they are considered as waste. This chapter deals with the various processes that are involved in waste generation from the footwear and leather industry sectors. It also enlists the most conventional types of wastes generated from the footwear manufacturing plant and leather industry. The waste generated can be from a specific product or a group of products, and can also be from a particular production technology applied or from the whole production industry. The nature of the wastes are produced daily mostly dependable on the manufacturing process and the materials and methods used for footwear production. Leather is majorly consumed by the footwear sector, which is around 60–65%. Footwear industries are producing the most significant quantity of leather wastes such as leather trimmings, shavings and leather dust. In today's world, there are more than 21 billion pairs of footwear manufactured annually globally. This creates a humungous amount of waste generation from footwear and leather sectors. The maximum percentage of footwear waste is generated from the post-consumer footwear waste, i.e. end of life of footwear which mainly goes into the landfills. In this chapter, the waste hierarchy from the footwear and leather industry is discussed and how it can be implemented to reduce, reuse and recycle the wastes that are generated from these footwear and leather sectors.

Keywords Footwear · Leather · Waste · Reduce · Reuse · Recycle

1 Introduction

The leather sector and the footwear sector are one of the world's largest sectors, which is mostly based upon a by-product. Leather industries majorly comprised of footwear industries, the fashion industry and home furnishing [1, 2]. The leather is

N. C. Jadhav · A. C. Jadhav (✉)
Department of Fibres and Textile Processing Technology, Institute of Chemical Technology, University Under Section-3 of UGC Act 1956, Mumbai 400019, India
e-mail: akshayjadhav488@gmail.com

261

the raw material and the by-product obtained from the slaughter industry. The raw leather can be heavy and light depending on the breed of bovine animals. In contrast, light leather is derived from the sheep and goats that are produced together at the meat industries or slaughterhouses in respective major leather-producing countries across the world [3–6]. The countries which manufacture or produce the largest quantity of leather worldwide are listed in Table 1 and Fig. 1. Here we have listed the top eight countries in the world which produce the highest amount of leather. These rankings are derived from the reports which were published by the World Statistical Compendium for leather, leather footwear 1993–2012. First on the list comes mighty

Table 1 List of countries producing maximum hides or leather in the world

Rank	Country	Production (millions Sq. Ft.)
1	China	2365
2	Brazil	1834
3	Italy	1500
4	Russia	1400
5	India	1400
6	South Korea	1085
7	Argentina	716
8	USA	670

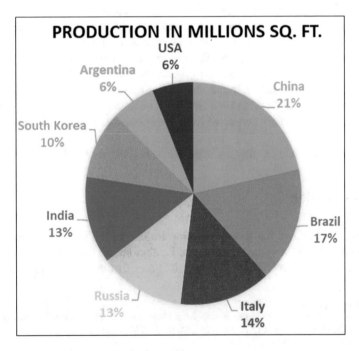

Fig. 1 Top leather/bovine producers in the world

China. They rule the world, in almost each and everything. When it comes to mass production, no one can match them. Everything we use or get comes with a tag of Made in China. They are also the mass producers of leather in the world. China can do what no one else can do. Therefore, the largest leather producer in the world is China. The overall leather production is more or less than 4 million square feet per year. The overall production rate is so significant that even the country which ranks second in the list does not also produce half of the amount of China's overall production. The production majorly constitutes of the light bovine leather, which is almost around 2365 million square feet annually. Global production is about 202 million square feet of bovine skin annually. They are the biggest producer of leather than any other country. Next in the list is Brazil, they are the second-largest producer of the leather. Leather industries of Brazil produce 1833 million square feet of leather. Italy is the third biggest manufacturer of leather on the list. The annual production of leather in Italy is around 1.5 million square feet. Mighty Russia comes fourth in the list of largest producers of leather. It was one of the largest leather producers earlier, but the leather industry of Russia declined drastically. They produce around 1.4 million square feet of leather per year. The fifth country in terms of leather production is India, the leather production rate was continually growing and blooming day by day, and it will very soon replace Russia in the leather-producing countries ranking list. The overall production of leather is about 1.4 million square feet which are almost similar to Russia. South Korea sits sixth on the list, and it produces about 1084 million square feet annually. Argentina holds the seventh rank in the production of leather in the world with 716 million square feet annually. The USA produces about 670 million square feet of leather, and it takes the eighth spot on the list of top leather-producing countries. All the top eight leather-producing countries in the list focus mostly on the light-coloured skin leather from bovine animals [7–12].

In the last two decades, the footwear industry has played a significant role in improvising material efficiency at the time of production. It also focusses on eliminating or reduction in the usage of hazardous materials in footwear production. Footwear sector "consumes" the significant percentage of leather (60%) which is produced annually. The leather sector has even improved the overall economy of the country. The increasing demand for footwear products has considerably increased the production rate immensely, which even overtook the environmental gains achieved. This leather and footwear industries are generating the most significant quantity of solid leather wastes. Relatively, the overall life of the footwear is concise, and it is progressively decreasing due to the rapid market changes and continuous development in the consumer fashion trends. These changes generate a tremendous quantity of footwear waste at the end of life. This waste is generally being disposed off in landfills or even incinerated, which creates air pollution. Unsustainable product consumption, new techniques and new production patterns in the footwear industries are responsible for the generation of enormous waste leather over the decades. This situation alerted the governmental agencies, including the local and national agencies, and has even warned the footwear producers and the people, and they have started taking necessary measures to control the waste at the source. However, it is understood that overall waste cannot be eliminated from the society in any situation it is almost impossible [13–15].

There is always going to be some amount of waste generated at the time of footwear production which cannot be prevented and so it needs to be taken care off when its functional life has ended. The overall waste management is a significant concern, and it can be achieved by looking at the quantity of leather waste generated annually due to the end-of-life products and taking preventive measures to minimize the waste generation by understanding the cause and developing new techniques for the end-of-life (EOL) waste management [16]. The world is very competitive, and the primary concern that should be seriously considered by the footwear manufacturing factories is to improve their customer's satisfaction by continuously improving the quality and fulfilling their needs and demands which can be done by improving the quality of the product at its best level [17]. Nowadays, the customer demands a large variety of specialized or customized footwear collection due to seasonal fashion trends. Therefore, to satisfy the general needs of the clients and to be dangerous in the market, footwear organizations need to take care of two fundamental difficulties: first, they should be responsive with the market changes, and secondly, they have to develop an efficient product to establish themselves according to the new consumer demands. This responsive behaviour of footwear manufactures towards customer demands leads to excessive production of footwears which reduces the life of the footwear, and even decreases the overall product development cycle ultimately leading to the generation of higher amount of post-consumer waste [5, 7]. Producer responsibilities and environmental laws, as well as consumers increasing ecological demands, are on the verge to challenge footwear industry and the way they usually deal with its production rate and managing the end of the production line waste [18]. The overall production of footwears globally has increased by almost 70%, which is around 18 billion pairs of footwears and is expected to reach till 22 billion pairs according to the experts in the footwear sector. Earlier, a consumer used to own a few pairs of footwears according to their daily activities like some for running and gym purpose and other for office work. This trend has changed now; every consumer has a minimum of 9–10 pairs of footwear. For every special occasion, people have different footwears [17–20].

The footwear consumption habits differ according to the country. China has the largest footwear production units and the highest usage of footwear in the world due to its humongous population. The USA has the most noteworthy per capita footwear utilization globally, with each person using up to 7–8 pairs of footwear annually. Similarly, the European footwear industries produce more than 1000 million pairs of footwear annually, out of which 70% consists of upper leather material. This pace of production and consumption produces around 100,000 tons of leather waste, the more significant part of which is dumped in landfills or incinerated. The solid wastes generated during leather processing are significant since leather industry makes use of only 20–25% of the raw material in the finished leather, and rest 75–80% ends up as wastes in the environment. Massive amount of waste is discarded from the leather and footwear manufacturing industries; therefore, essential efforts are made to utilize this waste [2, 21].

However, due to the increasing concerns over environmental legislation, leather waste disposal methods are becoming more demanding and extensively expensive. The European Footwear market is continually under cost-effective pressure from the South American producers and the Asian producers. So due to increasing prices on waste disposal, it will radically hamper the ability of the organization to challenge in the world market. Therefore, a broad industrial approach would help the companies to take essential measures and identify the feasible ways of recycling, reducing and reusing leather, particularly in terms of lowering leather waste. The waste leather can be used in making different products, and their application should be demonstrated in the footwear and other industries. This will not only primarily benefit the footwear manufacturers but also other industrial sectors like automobiles, furniture, machinery manufacturing, etc. This will decrease the impact on the environment. The footwear industries are improving their competitive edge by reducing the overall waste disposal costs. Even though the leather business is a public sector in the worldwide leather trade, but due to its waste generation, it is held accountable for its adverse effect on the environment. Below is the list of top ten exporters of leather footwear [22–24] (Table 2).

The raw material that is used for the production of the upper shoe leathers, upholstery and leather garment is from bovine hides; heavy leather is always obtained from bovine hides obtained from bovine animals. Leather is mainly used for the manufacturing of soles for shoes and different variety of industrial leathers, and sheep skin leather is also utilized in manufacturing of leather garments, bookbinding and gloves production. In contrast, goat leather is used in manufacturing gloves, upper shoe and lining leather. These are some end uses of leather obtained from different animal sources [25–27].

This chapter deals with the current measures to be taken regarding waste generated by the footwear industry mainly focussing on waste management framework. So, the main objective is to recycle, reuse and reduce waste leather by using minimum

Table 2 Top ten exporters of leather footwear

Rank	Country	USD million	World share (%)	Pairs millions	World share (%)	Average price ($)
1	China	11,000	21	730	33.5	15.08
2	Italy	8000	15	135	5.8	62.00
3	Vietnam	6000	12	300	13.6	20.65
4	Hong Kong	2500	4.7	95	4.3	28.00
5	Germany	2400	4.6	69	3.2	34.96
6	Indonesia	2350	4.3	95	4.5	24.10
7	Spain	2200	4.1	60	2.8	38.12
8	Belgium	1988	3.8	75	3.5	28.15
9	India	2000	3.7	120	5.6	17.25
10	Portugal	1900	3.5	65	2.9	32.50

pretreatment to produce a commercial novel product that could be used by the industries which generate waste, or other leather sectors. As of now, endeavours are made to recoup and extricate valuable materials from the leather waste, for example, chromium compounds utilized during tanning procedure can be recuperated by thermal incineration or pyrolysis methods, and proteins can be extracted through chemical digestion method or enzymatic method of leather waste, or it can be used as a reinforcement in the preparation of composites, or as a filler in bricks and concrete. Still, these methods have not significantly applied to industrial processes. The first approach of leather waste management from footwear industry and leather sectors was focussed on the 3 "R's", i.e. Reduce, Reuse and Recycle, to minimize the overall leather waste generated from the footwear industries and the after-life of footwears. During the start of the Millennium, Michael Braungart and William McDonough proposed another "R": Re-thinking or (Re)Designing. This new "R" depended on the behavioural restructuring of the civilization, which focusses on the eco-design procedures, mainly based on sustainability environmentally concerned consumer [25–28].

2 Leather Industry Waste Management

Leather industries are considered as the most heavily polluting industries around the world. It generates a humungous amount of waste in terms of solid waste from the leather/hide processing house and generation of waste in the form of sludge which is discharged from the tannery wastewater treatment. It also releases harmful gases and odour in the atmosphere which is responsible for air pollution. Tannery process house deals in the transformation of raw hides/skins into finished and durable leather. This process consists of mechanical and chemical processes. Tannery process is divided into three stages, i.e. preparatory stages, tanning and crusting. The first stage deals with the preservation, soaking, liming, hair removal, fleshing, splitting, bating, degreasing, bleaching and pickling of the rawhide and skins. In the next step, the processed leather or skin is tanned. In the tanning process, the proteins in the rawhide are transformed into a very stable material which will stop it from rotting, and it could be further implemented in extensive variety of end applications. The third step is crusting; here, the treated skin or hide is thinned, re-tanned and lubricated. Sometimes even colouring is carried out in this step. Lastly, finishing of tanned leather is done, which gives it the waterproofing qualities and the pattern sheen. It includes lubricating, brushing, impregnation, polishing, embossing, coating and plummeting of leather. To get the finished leather, the hides or skins go through various operations which generate a massive amount of solid waste and wastewater which contains a high concentration of harmful pollutants [28]. Tanning process also emits gas and odour due to the protein decomposition and also due to the presence of sulphides, ammonia and other volatile substances. Out of the 1000 kg (100%) of raw hide or skin, nearly about 730 kg, i.e. 73%, of the rawhide, goes into waste generation and the only 270 kg, i.e. 27%, of the rawhide, is converted into a finished

Table 3 Waste generation from 1 metric ton of raw leather

Solid waste	Quantity (kg)
Salt from desalting	82
Salt from solar evaporation pans	208
Hair	108
Lime sludge	64
Raw trimmings	50
Fleshing	120
Wet blue trimmings	25
Chrome tanned unusable splits	60
Chrome shaving	95
Buffing dust	65
Crust trimmings	40
Dry sludge from ETP	130

Table 4 Waste generated from footwear manufacturing industry

Type of waste	Tons/million pairs
Upper	134
Sole	120
Adhesives, oils, solvents	4.8
Household-type waste	11

leather product. The wastes are generated in the mechanical and chemical operations, which mainly contributes to skin trimmings, fleshings, shavings, buffing dust and keratin waste [29]. The primary separation of these leather wastes has been shown in Tables 3 and 4 and Figs. 2 and 3. In India alone, the overall production of hides from the cow is almost 24 million pieces; for buffalo, it is around 22 million, goat skins are approximately 106 million and sheepskins around 37 million pieces. These hide/skins are processed in more than 1600 tanneries. More than 2 lakh tons of solid leather waste are created annually from the leather industries [30–35]. Modern trim pattern of hide is shown in Fig. 4.

2.1 Raw Hide Waste Composition

The chemical composition of the raw hide/skin waste relies upon the type of animal, race, sex and living conditions and the treatment and process conditions applied to the rawhide. The waste consists of fleshing, trimmings, splits and scourings. The main component of the waste constitutes of protein. It consists of 70–80% moisture content. The amount of fat is also very high. The waste also contains around 5% of mineral contents [35–38].

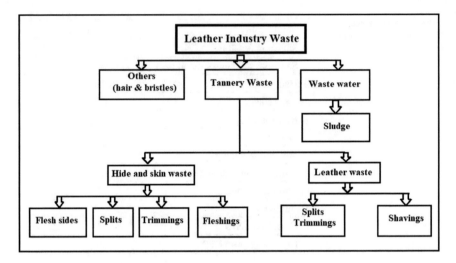

Fig. 2 Sequential process of leather industry waste generation

2.2 *Tanned Leather Waste*

This waste usually consists of chrome shavings and trimmings. Although they come from the same chemical processing, the chemical composition of the waste remains the same. However, the size and state of the individual components vary. The water content of the waste is around 30–60% (w/w), fat substance of 3–6%(w/w) and mineral components around 16% (w/w) having 3–5% (w/w) chromium as a Cr_2O_3 [35–38].

2.3 *Sludge from Wastewater*

Due to the tanning process involved in leather production, a vast amount of wastewater is generated. The wastewater usually consists of leached proteins, products obtained from the skin degradation and chemicals used. In the treatment of wastewater, sludge and sediments are formed, which makes the disposal a severe and difficult problem. It contains 45–65% (w/w) of moisture, around 30% (w/w) of organic components and 2.5% (w/w) of Cr (III) complexes. Preventive measures are taken to store these wastes properly on active landfill sites [35–38].

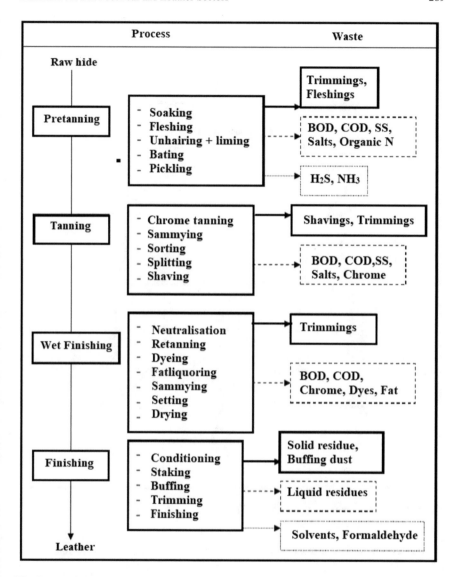

Fig. 3 Description of the tanning process and outputs [35]

2.4 Gases and Odours

The processing of hides or skins in the tanneries goes through various operations. These operations result in the biological decomposition of the hides or skins. Due to this decomposition, the tanneries emit the foul odour of volatile compounds and different gaseous chemical substances. The pollutants that are released from the tanneries have shown traces of ammonia, hydrogen sulphide and other volatile

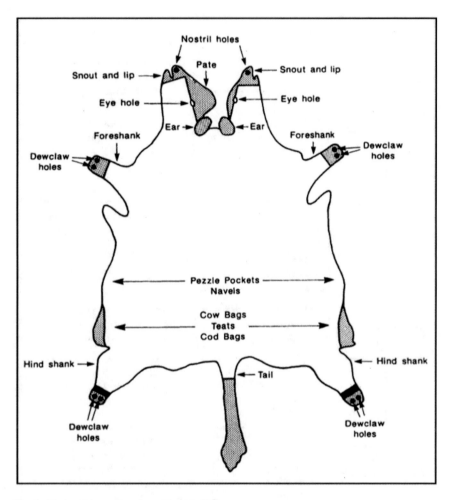

Fig. 4 Modern trim pattern for cattle hide [93]

organic compounds like aldehydes, amines, etc. All these gaseous substances are released in the atmosphere. The type and quantity of pollutants generated differ with the mode of operation implemented and technology implemented for the process [35–38].

3 Waste from the Footwear Manufacturing Industry

The footwear sector is the large and manifold manufacturing sector, which hires a large number of raw materials in the production of footwear and shoes of different styles and fashion according to the current market trends. Even though the shoes are made in various forms and divided into different categories, the essential components of the shoe generally remain common to all types of shoes. Around 40 different kinds of materials are used for shoe manufacturing. The vital pieces of a shoe are comprehensively arranged into three sections, viz. the upper, the lower and the grindery. The upper parts are over the sole, the vamp is sewed together to frame a single unit, and it is appended to the insole and outsole of the shoe. Some lower portion of the shoe, which alludes to the entire base of the shoe, includes the sole and insole of the shoe. Lastly, the grindery consists of the items that do not belong to the upper or the lower part but are incorporated into the shoe such as the stiffener materials eyelets and toe puff. The necessary materials that are used in manufacturing and construction of shoes are leather, synthetic materials, rubber and textile materials. Every element has its unique characteristics and is essential for shoe preparation. These materials are not only crucial for the improved lifespan of the footwear/shoe but are also crucial for the end-of-life waste management of the product [7–9].

The highest amount of leather waste is produced during the cutting procedure, i.e. waste generation from the cutting of leather (e.g. cowhide) 25–35% a leather skin is rarely standardized and rectangular, and the quality of hiding/skin at the corner portions of the skin is usually inferior. The shapes of the leather pieces to be cut are hardly the same. Reducing waste is generally lower because the material is more or less homogenous, i.e. 20–25%. The waste that is produced during shoe fabricating is upper material cutting waste, insole and sole cutting waste, wastes generated during injection moulding, dust and trimmings made due to different procedures [29, 35, 36, 38] (Table 4).

Most of the waste generated during footwear/shoe production is mostly from the cutting procedure. And this depends on the sort of footwear that is being made. Besides this, there will be packing waste, vacant yarn cones waste, and moulding waste will be generated. The last is exceptionally just important to thermoset shaping, e.g. reaction injection moulding polyurethane, like any procedure, waste generated through thermoplastic moulding can be granulated and further recycled again. The most amount of waste is generated from leather. The amount of residue is almost for the cowhide cutting procedure. It is regularly acknowledged that the leather waste is usually generated around 20–45% depending upon the nature and type of leather used and the proficiency of cutting. The overall average waste generated is around 30%. It is regularly acknowledged that the calfskin cutting waste is typically created around 20–45% depending upon the idea of the cowhide and the capability of cutting. The general normal is considered around 30%. It would most likely be increasingly significant to compute ton amounts. For leather, these can be evaluated depending on the accompanying suppositions. In essence, two sq/ft leather for every pair, 30% of wastage is generated per pair, the leather thickness of around 2 mm consists of 0.8 g/cm^3. In this way, the average leather waste generated per pair is around 90 g. The overall leather pairs around the world annually are near to 4600 million pairs of footwears using approximately 6300 million square feet of leather sheets, bringing about 0.5 million tons of wastage. In the developed nations, most of this leather waste that is produced from the footwear sector is disposed of. In developed nations, the waste is now and again used to make leather merchandise. At times, the leather pieces are sewn out to frame a large sheet which is recoloured to make it look even and a while later used to make customized leather accessories. Most of this leather waste is from the leather coating industries. The presence of this polymeric coating material causes issues of reusing and recycling the leather board manufacture [37–39].

4   Waste Management in the Footwear Sector

Footwear industries are one of the largest producers of solid waste. The necessary resources that are used for the footwear manufacturing are leather, synthetic materials, textile materials and rubber. The current fashion trends in the footwear sector have led to relatively decreasing the life of the shoes and footwear, which resulted in rapid market changes. To fulfil the needs of customers and be competitive in the global market, footwear manufacturers should follow and face two essential challenges: they should be quick to adapt the market changes and stay updated and relevant to establish new consumer trends. Mass production of footwear for different occasions and activities has started as per consumers demands. This led the consumers to buy more pairs of shoes at different times. These factors lead to the shorter life cycle of the footwear or shoes, also shorter product development cycle. This action generates a large amount of damaged and rejected shoes or footwears when their useful life has ended, leading to a higher level of the end of life. Post-consumer waste management is an exceptionally perplexing issue which is comprised of various segments. Effective end-of-life waste management should be taken care of by the producer and environmental legislations and by increasing environmental concerned consumer demands. These measures will undoubtedly change the way the footwear industry deals with post-consumer goods. To manage the waste in the footwear sector, European Commission has designed a waste management hierarchy framework which determines the best possible order in which the waste management options and methodology should be considered, with respect to its effect on ecology. This framework will decrease the common practice of waste disposal in landfills or by thermal incineration. This framework segregates the waste management of shoes into a proactive approach and reactive approach (Table 5; Fig. 5).

There are extensive varieties of proactive applications for footwear waste management. The two major advance techniques which should be implemented to minimize the waste generation at the source of footwear industries are design and material improvement. In design improvement, the activities are performed during the start of the product's life cycle, i.e. during product planning and designing stage. The main aim of this phase is to develop a product which is ecological and environmentally friendly and which requires less amount of raw material needed for the development of shoes. This can be achieved by creating the eco-design concept. This concept helps in reducing the waste generated in the footwear sector, which is required to be handled after the end of product life cycle. Design improvement also helps in easy disassembling the footwear and will help in easy reuse and recycling of the different components and parts. The implementation of different designs, styles and materials

Table 5 Footwear manufacturing steps

Process stages	Steps taken
Storage	Raw materials
Design and development	Cutting procedure
Upper footwear manufacture	Cutting of upper material Cutting of lining material Skiving Cementing/stitching
Manufacturing of different shoe parts	Cutting of insole, insock and sole Sole/heel preparation
Assembly	All stages
Finishing	All stages
Storing of finished shoes	Warehouse
Maintenance of manufacturing apparatus	Tasks generating wastes

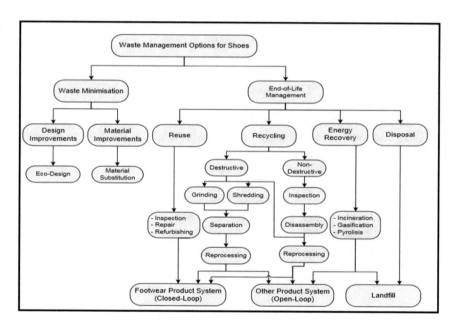

Fig. 5 Waste management framework [7]

during footwear production will help in end-of-life processing of shoe/footwear with the inclusion of environmental and economic implications will help in determining feasible methods to manage this hastily generating footwear waste. This will reduce disposal of waste materials into landfills.

Similarly, material improvement is the second primary improvement method that can be implemented to reduce waste in footwear industries. Innovative materials should be implemented in the manufacturing of footwear which is environmentally friendly. Natural fibres or fabrics and natural rubber can be used to develop eco-friendly footwear which could be easily recycled after the post-consumer life of footwear are ended. Biodegradable materials can also be used to substitute conventional materials like polyurethane, polyester, nylon, etc. because of their biodegradability and compostability at the end of the useful life of footwears. Therefore, material improvement emphasizes the use of renewable resources and environmentally friendly raw materials in the manufacturing of footwear.

On the other hand, reactive approaches manage all the waste management techniques which act in the light of the waste management issue when the functional life of the footwear has finished. Responsive waste management is likewise alluded to as end-of-life waste management of footwears. The reactive approach deals in minimizing waste generated from the footwear industries and also avail the option of reusing the shoes with minimal processing. Recycling is also an act of a reactive approach. The recycling manages the reprocessing of the used and discarded footwear items and parts, either into a similar product framework, i.e. closed-loop framework or into various items, i.e. open-loop framework. These reprocessed footwear waste products can be reintroduced into the markets or can be reused directly. This can be achieved by implementing two significant methods, viz. destructive method and non-destructive methods. Destructive ways are usually made with the shredding process, where the shredded waste footwears materials can be transformed into further useful stuff. These shredded waste materials can be used as fillers in the surfacing of road, reinforcement in composites, sound insulation, playgrounds for kids, shockproof floors, electrical insulation, etc. In contrast, non-destructive recycling methods involve dismantling of waste footwears and further recycling the waste materials into a high calibre of recycled materials that can be utilized in different applications. These are called secondary applications. Non-destructive techniques usually comprised of arranging, examination, dismantling of footwears and afterwards shredding of disassembled footwear waste materials.

These wastes are further used to generate energy and electricity, which can be achieved by thermal incineration method, gasification method and pyrolysis method. Disposal of leather or footwear waste obtained after a reactive approach in landfills if regarded as the last option in a waste management hierarchy [7–14]. Some waste management practices are mentioned in Table 6.

5   Standard Practices for Leather and Footwear Waste Disposal

5.1   Landfill Disposal

The most widely recognized technique for managing wastes is by dumping them on identified landfill sites. Currently, the waste from leather and footwear industries

Table 6 Common waste management practices used

Waste management practice	Code
Reuse as it is	A
Recycle within or outside the industries	B
Incinerate to recover energy	C
Specialized destruction methods	D
Optimized landfill disposal	E
Incinerate without energy recovery	F
Others	G

are usually disposed of in landfills which include keratin wastes, skin trimmings, waste from fleshings, chrome shaving wastes and finally the buffing wastes. The greater part of the post-consumer shoes is likewise discarded in landfills when their useful life has finished. There are very fewer landfill sites available for dumping waste [7]. They all have reached their maximum capacity limits, and government is facing difficulties to get authorization for new landfill sites. The leather waste was also improperly manhandled, and the transportation of leather waste in open dumping trucks creates very poor insanitary conditions. The major problem was the disposal of tanned leather waste in low-lying areas which did not have proper liners; this allowed the leaching of Cr^{3+} mix with the natural groundwaters and rivers which causes water contamination and environmental pollution. Due to the scarcity of vacant land, disposal of a large amount of solid waste in landfills is near to impossible. Also waste disposal in landfills involves a lot of transport facilities as the waste needs to be carried over long distances through dumpers and other vehicles and also requires good infrastructure. These facilities require additional finances which makes the landfilling very costlier operation leading to environmental pollution. Landfill sites also release various harmful gases, out of which 50–60% is methane gas which is majorly responsible for global warming. Hence, due to these reasons, the landfill disposal technique has been strongly discouraged by ecological specialists and researchers. The landfill locations in the European Union nations and Asia have considerably lessened over the last few years due to lack of landfill space [22]. The EU Landfill Directive has promoted recycling of certain types of waste materials from the landfills using a different technique (Council Directive 1999). In Germany, the landfill sites will only take the municipal waste that has undergone mechanical and biological treatment and which is biodegradable, whereas in Austria government has introduced strict rules of landfilling organic wastes. The UK Landfill Allowances and Trading Scheme (LATS) Regulations has provided certain percentage to various biodegradable wastes, e.g. for paper and vegetable oils are possibly (100% biodegradable in nature), furniture, footwear and waste generated from textiles are regarded as (40–50% of biodegradable), and, glass, batteries and waste produced from metals are not at all biodegradable. The products having 50% biodegradability can be used such as natural textile fibres/fabrics, leather products and natural rubber, which are majorly found to be consumed by the footwear industries should either reuse, recycle before disposing of them in landfill areas [7–9, 40].

5.2   Vermicomposting

Vermicomposting of leather solid waste produced from the leather manufacturing industry is a substitute method of waste disposal. This is done with adding cow dung and agriculture residue and *Eisenia foetida* which is an epigeic earthworm species. This earthworm requires the right amount of moisture content, a good quantity of organic matter and dark conditions for the proper growth and development. Vermicomposting is termed as a straightforward and economical process through which leather wastes can be efficiently converted into organic manure which would be very helpful for the farmers. The composition of vermicompost bed consists of solid waste leather 75%, cow dung 17.5% and litter leaves around 7.5%. Earthworm plays a very important role in vermicomposting of solid waste management. There are different species of earthworms such as Perionyx excavatus, *E. foetida*, Eudrilus Eugeniae, *Lumbricus terrestris* and *Eisenia andrei* used for vermicomposting. The action of earthworms on the leather wastes is due to the physical and biochemical reaction. The physical activities incorporate the aeration, blending, grinding and processing leather waste with the assistance of some aerobic and some anaerobic consuming microflora. At the same time, the microorganisms which are in the guts of the worms are entirely liable for the biochemical degradation of leather waste converting and stabilizing them into a finer, humified, microbially active and nutrient-rich humus like material called vermicompost. Apart from vermicomposting application in industries producing leather waste, it also has a vast application in other organic waste-producing industries such as guar gum industrial waste, petrochemical sludge, sewage sludge, municipal solid waste, food waste, textile mill sludge, aquaculture effluent solids, sugar industry wastes and paper waste. Vermicomposting helps in the biotransformation of organic industrial wastes into nutrient-rich manure which is used for farming. The benefit of vermicomposting is that it decreases the toxicity level of organic industrial waste. This process is known as an eco-biotechnological process which helps in transforming a complex organic industrial waste into a stabilized product called vermicompost, and it can be further used as natural fertilizers or soil conditioners which can be used in agricultural/farming applications. Therefore, leather waste is subjected to vermicomposting, along with cow dung and agricultural residues. There are also some disadvantages of vermicomposting which restricts its application, e.g. this process requires more space and labour as the earthworms cannot sustain in the waste heap if it is more than a metre in depth. The worms are very sensitive in terms of temperature, pH, toxic substances and also to waterlogging in the pits. A large number of worms are required to start a vermicompost plant. It also needs to be situated away from the city as it emits a foul smell which can be unbearable. So, vermicomposting is one of the methods used for disposal of industrial leather waste. This converted product is utilized as a manure or as a source of nitrogen for microbial populaces which can be useful to crop development [41, 42].

5.3 Anaerobic Digestion

The present environmental guidelines mandate energy recuperation from solid leather wastes obtained from leather and footwear industries, which is done by digesting the leather waste through anaerobic digestion (AD) process to produce chemical and fuel energy. Anaerobic digestion breakdowns the leather waste material by living microorganisms in the absence of oxygen. Anaerobic digestion is termed as a non-conventional source of energy as it produces methane gas, and carbon dioxide-rich biogas could be utilized in the production of energy, therefore, replacing fossil fuels. Anaerobic digestion system was originated in the tenth century BC and was used by the Assyrians and by the Persians in the sixteenth century. The anaerobic digestion system was carried out in the anaerobic ponds around 1920. The first AD system set up in the industries was started in the year 1859 in Mumbai, India. The development of AD systems for methane gas production was started due to the energy crisis which occurred in the years 1973 and 1979. Most of the leather industries are using the AD systems to treat biodegradable leather waste. The leather waste is co-digested with the secondary biological sludge. This decreases the emanation of greenhouse gases in the environment. The methane gas produced can also be utilized to produce heat and electricity. The two principle preferences of this procedure initially are the production of biogas by decontaminating the leather waste, and the second advantage is that the sludge that is left over after the anaerobic digestion could be utilized as natural manure in agriculture. There are also some disadvantages of this anaerobic digester system. This system generates hydrogen sulphide gas which is responsible for the dry corrosion of the burners; hence, it is required to scrub before considering as fuel gas, which ultimately increases the capital cost, and therefore, this process has limited industrial application [43–47].

5.4 Thermal Incineration

For the disposal of the leather waste, thermal incineration is carried out which includes gasification and pyrolysis and also generating energy from the thrash [48]. The tannery wastes like sludge, shavings and buffing dust from leather industries can be subjected to thermal incineration and later can be disposed off. This is considered as the economical process for the energy generation and reducing the voluminous mass of leather waste. Incineration is the last option used to dispose waste in an incorporated waste management approach [49], and after waste exclusion method, reduction, reusing, recycling and composting have all being implemented and further no more techniques are left for waste disposal. Incineration of wastes is done in an enclosed structure under combustion at controlled temperature of 850 °C. The residue after burning is converted into CO_2, water and non-combustible materials and incinerator bottom ash which always contains a little quantity of residual carbon. Incineration process requires a sufficient amount of oxygen to oxidize the fuel [36, 50, 51]. Special attention is required while incinerating the solid leather waste from tanneries as incineration leads to a discharge of harmful chemicals like, polyaromatic hydrocarbons, chromium (VI) halogenated organic compounds, etc. into the atmosphere. Thermal incineration of solid tanneries waste leads to the formation Cr (III) which further forms $Cr_2(SO_4)_{3(s)}$ and $CrOCl_2(g)$ and $Cr_2O_{3(s)}$ which later creates a way for the formation of Cr (VI) [52–55].

5.5 Pyrolysis and Gasification

For the generation of gas, the thermal degradation of waste leather carried out in the absence of oxygen is called pyrolysis, frequently called as syngas. This procedure is carried out in the range of 400–1000 °C. Gasification is usually carried out at higher temperature than pyrolysis, i.e. in the range of 1000–1400 °C under controlled oxygen. The final product obtained from both the processes of pyrolysis and gasification is called as syngas, whose main composition consists of carbon monoxide (CO) and hydrogen (85%), with small traces of carbon dioxide (CO_2), nitrogen (N_2), methane ($-CH_4$) and various other hydrocarbon gases. Syngas has a calorific esteem and accordingly can be utilized as a fuel to create power or steam or as an essential chemical feedstock in the petrochemical industries and oil refineries [36, 50–55].

6 The Waste Management Hierarchy and the 3 R's

The waste management hierarchy comprises 3R's. These 3R's stand for Reduce, Reuse and Recycle. Waste minimization is a widely used technique applied worldwide to lessen the amount of waste generated from the industries, and this can be achieved mainly via reduction at the source, but along with reduction, it also includes reuse and recycling of waste materials. In the waste management hierarchy, reduction occupies first place in the hierarchy and is the most favoured by the leather and footwear industries. The second option in the waste hierarchy module is reuse. Recycling process occupies third place in the hierarchy module and is a very significant component of the sustainable waste management system. This process converts waste into valuable materials or products that would have become waste if not recycled. The principle of the waste hierarchy is to reduce, reuse and recycle the waste which has been disposed of by the leather industries, footwear industries and also waste generated by post-life of footwear by recovering the useful materials from the waste, which would otherwise have been ended up in the landfill sites. Recycling consists of

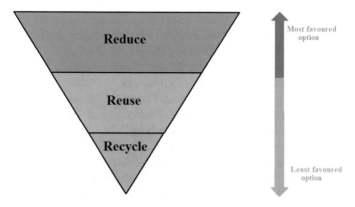

Fig. 6 3 R's of waste management hierarchy

a sequence of different activities which involves assembly, sorting and processing. It mainly deals with the conversion of discarded waste materials into valuable products. Solid leather waste materials are collected from the leather and footwear industries and used as raw materials in generating a new product rather than used as a source for energy generation. It does have many useful benefits, but at last, it should have a market for the end product produced from the waste. Otherwise, this process will not be economically sustainable. Lots of waste materials discarded from leather and footwear industries can be reduced, reused and recycled, which have market value. This will generate an economy source as well as environment is additionally bene- fited from waste reduction technique [7–10, 56]. The waste management hierarchy and 3R's used in this module is shown in Fig. 6.

6.1 Reduction

Waste reduction is at the top of the hierarchy. Several studies and research have been carried out by the scientist who is aimed at waste reduction and the environmental impacts that are caused by leather and footwear production factories. Considering the outcome of leather waste on the health of human beings and its impact on the environment, it is necessary in the first place that there should be a limit on waste generation and it should be stopped immediately. Raw materials and energy are required in the first place to develop or manufacture a new product. All of the raw materials are extracted from the Earth and later processed; later these raw materials are used for the manufacturing of products, they are packaged, and then transported to the markets where it will be finally sold. Every step used for product manufac- turing will produce solid waste and liquid wastes. Every stage will generate waste. The main aim should be to make a product without creating waste or provide less waste during the process. This will effectively help in protecting the environment and natural resources and save money. Industries will have to play an essential part in waste reduction. It has become necessary to adapt and implement new methods in the industries for the product manufacturing processes, which would produce a large number of products without increasing the use of raw materials. The industry can also work to incorporate less content into its products—so for example, a shoe can be manufactured by using very less raw material. The waste reduction should be made at the manufacturing level. Make use of reusable materials rather than using disposable materials. Sustainable design and application should be implemented in the development and production of footwear and shoes. The production is usually based on traditional factors and emerging fashion trends. These factors forced the manufacturers to produce millions of pairs annually. This has led to the waste generation of leather waste in smaller pieces during the manufacturing process. To reduce these waste generations, the design and style team needs to create an innovative model for the usage of leather scraps which would instead have been discarded in waste. This scrap leather which is generated from the central production unit is minimal to cut a piece for shoe out of it. These small leather pieces can be used to produce decorative items like little flowers, ornaments, ribbons, etc. which can be used for designing shoes and jackets. This technique acts as a value addition to the products. Before applying this procedure by the manufactures, details of all the little ornaments are cut into essential material, which at ultimately leads to the production of more waste, which will eventually stop the reuse of waste leather pieces from the decorations because of the small size. This practice requires additional time from the development unit to create unique models and will require extra time and cost from the creative team. The price spent on the acquisition of crude materials was higher, with regard to the cost spent on the final waste disposal. Concerning a decrease of solid waste production and its effect on the environment, the adoption of this new development can reduce around 15–20% of leather waste disposal in a week. With some more improvement in concepts and principles for the design sustainability, progressively definite investigation and identification of improved openings and potential arrangements should be applied to reduce the leather waste. The initial phase in the adoption of sustainable methodologies is the distinguishing proof of the issue, and afterwards, endeavour to imagine new approaches to solve it as the principal objective is to lessen the generation of leather waste and subsequently decrease the environmental impact and the expense that will be occurred due to landfill disposal. The identification of the problem occurs more effectively if the life cycle of the product is known profoundly. The stages of preproduction and production demonstrated a critical point regarding waste generation. So, to identify the waste in preproduction and production stages, there is a need to understand the production cycle process structure and then restructure a new process which will minimize waste from the footwear industries. It is not an easy task to restructure a production cycle process [48]. The rebuilding comprises turning down the order for the activities from the development and planning units, i.e. there is the adoption of a reverse logistics flow. Rather the customary manner by which the development team sends data about how much material is essential for the creation of a new product. Instead, it is the planning sector that sends the message to the development and information department about the material required for the production of footwear. These circumstances create the need for the new design requirement and a challenge for the design creators based on the confinement of available resources, which did not exist before. Adaptation to environmental requirements without loss of competitiveness. These measures result in the reduction of waste generation. The adoption of these strategies to reduce waste generation from the industries requires investments in some departments, especially in the research and development department, cutting department and assembling department. This will require time and labour to stitch and attach the small details. This new assembly for the production of footwear and shoes will increase the costs. However, it is offset by the reduction in waste leather since adopting the new system [7–10, 56].

6.2   Reuse

Reuse can be defined as a waste product which can be used further without changing its original design, i.e. use waste footwear or shoe as it is. Reuse is the second option in the waste hierarchy. Reuse means that less amount of footwear waste is generated. Reuse situation arises when the functional life of the footwear is finished. Every shoe/footwear that is purchased is disposed of when their useful life has ended. This may be because the footwear has become old, worn and torn or maybe just because it is no longer in fashion. Footwear waste is becoming a bigger problem in developed cities because of the emerging fashion trends and lavish lifestyle. On average, one person could own around 4–5 pairs of shoes or footwear for different functions like gym, office, casual wear, etc. Reuse can be very beneficial for disadvantaged people who cannot afford to buy new footwear. By reusing footwears and shoes rather than manufacturing new footwear products from raw materials like leather, rubber, etc., reusing footwear helps in the reduction of raw materials and product imports which will help in stabilizing the economy. Reusing helps many underprivileged people to acquire the shoe or footwear they need at a very low price. It is better to buy used footwear rather than buy a new one. Reusing of footwears or any material which can cause pollution, water consumption and energy consumption will help in reducing pollution. Reusing eliminates the risk of pollution, which would have been created if the footwear waste had been disposed of in the environment. It was observed that almost 60% of the people disposed of their shoes in the garbage, only 20% of the people kept their shoes at home, but no longer wear them. And the remaining 20% of the people said that they donated their shoes to charity shops or directly donated to the orphanages. Some of the shoes which are good condition are sold by the charity shops at a low price so that the poor and needy person can buy them at a very reasonable price. Majority of the shoes are sold to the developing countries.

During the 1990s, Nike had already started imagining a planet without waste and developing a roundabout procedure that would transform old shoes into recovered materials. Reuse-a-Shoe helps the consumers to recycle their used shoes. Nike has imposed more than 32 million pairs of shoes, and around £120 million of manufacturing and assembling waste have been converted into wearable and usable materials. In Europe and North America, you can drop your old sneakers at any of the 292 Nike showrooms situated around the country and give your sneakers a second life. If you have used a pair of sneakers, then Nike's Reuse-a-Shoe centre takes any brand of sneakers. The shoe businesses are additionally taking a shot at better approaches to manage the control of this pattern of waste generation and make it less complex to part it with shoes without any culpability. Various start-up associations like Everlane and Veja to industry stalwarts like Nike, Adidas, New Balance, Saucony and Converse and many more footwear companies are moving towards non-harmful, biodegradable, recycled and eco-friendly materials. It's better to give the used shoes a subsequent new life by posting them on resale platforms like The Real, Depop and Thred Up or on shoe trading platforms like Fight Club, Sole Supremacy and Stock X. The most profitable method to generate economy can be obtained by the resale of used footwear to the developing countries. If this is the most economical source of used footwear available in the market, then the profit will be generated through the sales of used shoes. The distribution of used shoes into the market of developing countries usually leads to the damage the economy of their country, and also the local footwear manufacturer's economy is impacted. In Uganda, the import of an enormous volume of utilized shoes has fundamentally influenced the market of the residential footwear industry as of late. Increasingly, 7–8 million pairs of used shoes are brought into Uganda every year. While just 240,000 pairs are made up by the local footwear industry. There's no magic bullet solution for disposing of old shoes by reusing them and reduce the footwear waste generated [7, 17, 57].

Similarly, waste reuse techniques are applied in the leather industries, where larger pieces of leather are stored from the end collection productions. The reuse of the footwear is planned from the time of production itself. The footwear designers are informed by the materials planning department regarding quantity, colour and the type of leather pieces that are in the warehouse. From this data, a specialized footwear model is generated from the left-over leather in which all cuts and colours are applied, and the maximum available leather waste is utilized, resulting in the minimum amount of waste generation. Later finishing of the footwears is done by the finishing department where varnish, opaque and texture is applied. Therefore, the footwear pairs that are produced may not have unity among them as every pair is made up of different types of leather. Hence, the scrap leather is used for the production of new footwear models and is reintroduced in the market as a new product [7–10].

6.3 Recycle

Recycling of leather waste means that the waste material obtained from the footwear and leather factories is reprocessed before it is used to make a brand-new product. There is minimal scope to recycle waste footwear/leather into new products. When the products arrive at the end of their valuable life, manufacturers do not play any role in the assortment, recycling or removal of those end-of-life products. A sustainable portable footwear recycling application relies vigorously upon building up a fruitful value recuperation chain. A pair of shoes is made up of different materials and composition, and it may contain various types of recyclable materials like leather which may be manufactured by chromium-free method or chromium tanned method, polymers like PU, PVC, etc. It may also contain natural or synthetic textiles. Therefore, it is essential to recycle the used materials after the functional life of footwear or shoe is finished. Recycling gives us the specialized achievability to reuse most of these used materials as a raw material as a chemical feedstock or can be used in energy generation in an ecological friendly manner which may otherwise have been disposed of as waste in landfills or incinerated. The first step applied in building up a footwear recycling process is to isolate post-consumer shoe waste into mono-fraction material streams, which can be usually done by using mechanical or chemical method. Some companies in the UK and Germany have started to recycle footwear, but none of them has revealed the details of what they are manufacturing from footwear. Separation of the leather from the upper sole is subjected to chrome recovery from the waste leather, though it is not an economical process.

Currently, research is undergoing to develop a straightforward technique for the recycling of shoes into usable products. Many leading shoe companies have come forward and have started supporting the recycling method. Nike began to the shredding of old Nike shoes and turning them into reusable materials from shoes collected at the finish of life footwear from customers and converts them into the Nike Grind, a collection of elite materials produced by using recycled shoes and assembling scrap materials [7, 57]. Nike presented sport surfaces all through the world and evaluated new recycling procedures which will prompt the organization's first customary Nike Grind association with Field Turf. Nike Grind materials are used to change spaces from basketball courts and play top surfaces to running tracks and numerous others. Footwear recycling cannot be regular as garments recycling on the grounds that shoes are progressively more complex material. Adidas, moreover, utilizes bits of rubber from sneakers into floor coverings.

Another association named Terracycle, a New Jersey-based recycler, discovers purchasers for leather shoe bits which can be converted into ground surface and furniture. Simultaneously, plastic is utilized in building compartments and sound insulating materials, among various different employments. At present, there has been significant research carried out in this area of recycling, and various strategies have additionally been developed [7]. These are isolated into two parts: those who pulverize the leather skin structure entirely and those who use leather fibres. The first group incorporates strategies, for example, chrome recuperation and gelatin extraction, while the latter includes the utilization of the leather fibres as fillers in plastics and rubber materials. Furthermore, legislative implementation can play a significant job in developing financially attainable value recuperation chains for recycling of post-consumed shoes and footwear. Various techniques of recycling are described below [50, 56].

6.3.1 Composites

Leather is a natural polymer which is made up of collagen fibres and is cross-linked together in a three-dimensional structure [58]. The essential structure of collagen is comprised of twined triple units of peptide chains, and the triple helices are bonded together by hydrogen holding. Around 50% of the unprocessed hides or skins are discarded as waste after going through various operations like trimming, shaving and cutting procedures. This leather waste is considered to be hazardous because of the presence of chromium in them [59]. Chroming of leather is done to improve the appearance, feel lustre, physical properties and chemical properties and improving natural resistance. The major concern is over the conversion of trivalent chromium into hexavalent chromium on oxidation. This leather waste particulates can be therefore used as a filler or reinforcement in the preparation of composites [60–65]. The most common resins used are epoxy resin, unsaturated polyester resin and vinyl ester resins. This resin acts as the matrix. Epoxy resin is easy to process and work with. The main advantages of epoxy resin are they have a low coefficient of thermal expansivity, high rigidity, elasticity and have good resistance towards moisture and chemicals. Epoxies usually outperform different resins which are available in the market in terms of mechanical properties and have high resistance towards natural degradation, which prompts their selective use in making aeroplane parts. Apart from the high price of epoxy resin, they also possess some disadvantages like it is brittle and has poor fire resistance which hinders its application in industries. Epoxies have their applications in marine, automotive and electrical industries. Epoxy-based leather composite is shown in Fig. 7. Similarly, unsaturated polyester resins are also used as a matrix in the preparation of leather waste composites. They are thermoset resins cured from the liquid state to a solid state when subjected to proper conditions. Unsaturated polyester resins are also known as "polyester resins" or basically as "polyesters". There are two kinds of polyester resins which are utilized in standard laminating frameworks in the composites sector. First is orthophthalic polyester resin, and second is the isophthalic polyester resin. The former is the standard economical resin utilized by numerous individuals, and the latter is currently turning into the

Fig. 7 Epoxy resin-based leather composite [60]

most favoured resin in the marine businesses where superior waterproofing property is required. UP matrix has the following advantages: cheaper cost, extreme processing versatility and long history of performance. It is majorly used in construction, marine industries transportation, etc. Also, vinyl ester resins have almost similar molecular structure as compared to polyester resin but have different reactive sites in the primary location; they are only positioned at the end of the molecular chains. This long molecular chain makes the vinyl ester resins tougher and more resilient than polyesters. This long chain of the resin is capable of absorbing heavy shock loadings. The vinyl ester molecule has fewer ester groups. They are susceptible to water degradation due to hydrolysis, which means they are resistant to water and many other chemicals. Because of this property, vinyl esters are frequently used in the applications such as pipelines and chemical storage tanks. The cured molecular structure of the vinyl ester shows that it is tougher than polyester resin. These properties can be achieved at an elevated temperature and proper post-curing [66]. Leather waste is also used in the preparation of biodegradable composites. Polylactic acid (PLA) is a very well-known biodegradable plastic material obtained from the fermentation process. Biomass is extracted from the corn, potato and sugar which are used for the fermentation. It can be used to replace plastics obtained from petroleum-based materials. There is a worldwide demand for agriculturally based and environmentally sustainable biomaterials. Polylactic acid is mostly used in tissue engineering and medical surgery, biodegradable plastic bags, food packaging, automobile parts, etc. There is an increasing demand for the environmentally friendly biocomposites which are made from the agriculturally based biodegradable material which will replace the usage of conventional composites. All biopolymers are sustainable. Biocomposites offer various advantages such as ease of renewability, biodegradability, cheap cost and density and good physical and chemical properties. Leather biocomposites are prepared by using the PLA matrix and reinforced with waste leather fibres. Leather is made up of collagen [67, 68]. Similarly, natural rubber/leather composites and polyvinyl alcohol/leather composites are also made to use them in antistatic flooring and coatings [69]. Leather wastes can also be used with bioresins obtained through epoxidation and acrylation of rice bran oil, soya bean oil, palm oil, etc. Natural oils consist of long-chain unsaturated fatty acids. These fatty acids have 4–22 carbons, and the double bonds of the allylic carbons (ester groups), as well as the alpha carbons which are adjacent to the ester groups, are highly amenable for carrying out chemical reactions and considered as the active sites of the triglycerides. These active sites are used to introduce different polymerizable structures within the chain. Different procedures can be applied to introduce different polymeric groups into natural triglycerides. One can synthesize different types of bioresins by utilizing or simply functionalizing the various fatty acids. These resins are called as bioresins or green resins, which can be used to make leather waste reinforced green composites [60, 63]. This is how leather waste is used as a filler or as a reinforcement with different types of resins, rubber, polymers, etc. to make composites [70, 71].

6.3.2 Biodiesel

Biodiesel has gained significance popularity from the year 1970. Energy supply extracted from fossils has become an unattractive method and is also not feasible economically and environmentally. Therefore, due to these concerns, alternative and renewable energy resources have gained a lot of attention and momentum. Earlier vegetable oils were utilized as liquid fuels due to the invention of diesel engines. Biodiesel is an alternative fuel obtained from renewable agricultural sources [72, 73] like vegetables and animal sources like fats and proteins. Studies have shown that biodiesel is non-toxic and biodegradable. It is termed as an environmentally friendly energy source. This fuel has low emission values. Biodiesel persists various significant advantages like it is free of sulphur or aromatics and it lowers the carbon dioxide emission in the atmosphere and also supports the greenhouse effect [1, 74–77]. It was observed that there was a reduction in carbon monoxide emission, unburned hydrocarbons and particulate matter when biodiesel was used in conventional diesel-fired engines. Looking at these advantages of biodiesel, scientists started looking for new sources for the production of biodiesel. This led to the utilization of prefleshing leather wastes obtained from leather industries in the production of biodiesel. Most of this waste generated was from the fleshing procedure, which is called as prefleshing. The fats which were extracted were transformed into biodiesel. The wastes obtained during prefleshing consist of reasonable amounts of fat and protein. Due to this considerable amount of fat content, these leather wastes fulfil the criteria for biodiesel production. Therefore, the usage of biodiesel in the diesel-fired engine vehicles can act as an alternative fuel which may also provide satisfactory economic benefits towards meeting the essential energy needs of the industry. It also gives slightly better fuel economy as compared to diesel fuel [78–84].

6.3.3 Protein Extraction

Leather processing generates a large amount of solid and liquid wastes. More than 50% of leather waste is made from the hide/skin processing. The leather waste comprises the high amount of proteins and fats in them. Therefore, leather waste obtained from the tanneries is subjected to thermal and enzymatic treatment to isolate valuable protein products. Thermally treated fats' yielded protein is not as economical as it consumes high amounts of energy. Enzymatic processing of leather waste produces a protein, which is more economical [34]. The enzymatic treatment is usually optimized at a very low temperature, over a short period, ultimately yielding in a reasonable amount of protein. Therefore, thermal and enzymatic treatments are implemented on the untanned leather waste to recover a maximum amount of hydrolysed protein with a minimum amount of residue. Enzymatic extraction is the most preferred procedure used as it is effortless as this method does not emit any harmful gases or odours, and the wastewater discharged does not contain any toxic chemicals. Therefore, due to this method, not only the ecological problem is solved, but also the valuable and useful products are produced from leather waste [48, 56, 64].

6.3.4 Chrome Recovery

Raw leather is made up of collagen fibres. These putrescible collagen fibres are converted into the non-putrescible leather matrix during the tanning process. Tanning is done by using basic chromium sulphate. During the tanning process of hiding/skin, only 60% of chromium salts get utilized, while rest of the chromium salts gets disposed of along with solid waste material some through wastewater. This chrome in leather waste is carcinogenic and can cause adverse effects on human health like ulcers, kidney malfunction, lung diseases, perforated nasal septum, etc. [85, 86]. Therefore, there is a need to dispose of these harmful waste materials through special techniques which are very expensive. Currently practised solid leather waste disposal methods are land disposal, thermal incineration and anaerobic digestion which have their respective disadvantages.

Usually, leather waste containing chromium is subjected to pyrolysis method at 450 and 600 °C using a fixed bed reactor loaded with 50–60 g. This yielded charred carbonaceous residue of ammonium carbonate along with oil and gas products [87]. The charred residue yield was in the range of 38–49% with a calorific value of 4300–6000 kcal/kg, which was suitable for the use of solid fuel. Further studies were carried on the management of chromium-containing leather solid wastes through double pyrolysis. In a single pyrolysis system, the percentage of charred carbonaceous residual material was 40% which is a major drawback. Therefore, a second pyrolysis method was used to reduce the mass residual carbonaceous material, gases and value-added products. The advantage of pyrolysis over combustion is a reduction in CO_2, which plays a vital role in the greenhouse effect. This method yielded a high energy content combustible renewable gas with a minimum residual mass of 24%. Double pyrolysis of chromium-containing leather solid waste disposal method is safe in recovering useful products without converting the oxidation state of trivalent chromium (Cr^{3+}) into hexavalent chromium (Cr^{6+}) form in the residual ash. The generated gases can replace the thermal energy requirement. The liquid by-products find numerous applications due to the presence of thiol groups and nitrogen content. The residual carbon ash, which contains trivalent chromium (Cr^{3+}), can be used in steel manufacturing industries [40, 88].

6.3.5 Papermaking

The raw leather consists of fibrous material which is therefore used as a co-material in the paper production. Around 10% of leather fibres are used in paper manufacturing. It was found to be that leather fibres enhanced the overall properties of paper, particularly due to the inter-fibre cohesion. The disadvantage is that a higher amount of leather is tough to process on simple equipment. The main difficulty is to remove water from the leather because of the hygroscopic nature of the leather. Card production requires higher levels of leather, which may not be possible on simple equipment. Some advantages are that the cellulose fibres gain in cohesion due to leather fibres, and the compound paper that is produced has a very attractive appearance and feel with relatively high absorbency. The paper produced from leather fibres has a very appealing appearance and could be sold in a very niche market. The utilization of these waste leather fibres as a source of raw material for the production of paper is only economically feasible if the price of the wood pulp is high [89, 90].

6.3.6   Absorbing Material

Leather industries and footwear industries produce a large amount of scrap leather and ground leather. Leather is hygroscopic in nature. Therefore, ground leather can be used as an absorbing material to clean up oil spills from the vessels on the beaches and oil industries (crude oil, diesel oil, etc.). The reason behind oil leakage sometimes consists of natural seeps, offshore exploration and production, the outflow from well, prohibited discharge from the container, tanker accident, terrestrial and atmospheric input, tanker ship, barge accident, prohibited discharge of water from the ballast, oil loading to the oil discharging terminal and oil loading/discharging put off. As per the analysis disbursed by the U.S. National Science, solely 2% of hydrocarbon pollution finding its approach into the ocean annually comes from tanker accidents, 11% comes from natural sources—tar sands and oilseeds, 13% comes from the atmosphere, 24% from all sources of transport, an astounding 50% comes from down drains and rivers to the ocean from cities and industries. Multiple causes each of natural (e.g. environmental condition factors, natural calamities) and somewhat by human action result in oil spill pollution. Some applications may require specialized equipment and sophisticated techniques. This may require a huge quantity of grounded leather. So, leather as oil absorbent is applied to a minimal oil spill area, as it may require up to 2000 tons/year of waste grounded leather [89–92].

7   Conclusion

Leather industry and footwear industries are one of the major solid waste-generating sectors. The quantification of the wastes generated is a tough task. Most of the waste generated is from the post-consumer shoe waste or end of life of footwear. There is a lot of pressure from the legislative council to stop the waste disposal in landfills as there is a lack of empty land left for dumping and also due to environmental concerns. The main objective is to reduce, reuse and recycle the waste that is disposed of from the leather and footwear industries. It was observed that Asia is first in the production of leather and footwear, which ultimately leads to the mass production of solid waste in the leather and footwear sectors. Almost 60% of the solid leather wastes in the world is generated by Asia. This has raised us to develop many technical, economic and environmental issues to recycle waste and highlighted the urgency to overcome the barriers which already exists in the currently established methods which are used to recover end-of-life products from the leather and footwear industry. Recycling and product recovery activities should be identified and studied to reduce the landfilling of waste footwear products and hazardous substances which could enter the environment and impact human health. This will help in recovering the economic value of the end-of-life waste materials, components and products. The 3R of the waste management hierarchy works on the research and development, the designing of the product, raw material requirements and the order of the operation carried out. These factors will help us in reducing, reusing and recycling the waste generated from footwear industries and lead us to a waste-free environment. Reduction and reusing techniques do not have much impact on waste deduction; therefore, recycling of waste is the most successful operation used by the industries. The recycling technologies need large quantities of wastes; therefore, leather sectors must organize the waste collection, transportation and also the recycling operations. Regarding the recycling of leather waste, there are only two options that are implemented on a large scale, i.e. preparation of fertilizers and next is chromium and energy recovery after the incineration process. These factors are very crucial in establishing a sustainable end-of-life product recovery from the waste in the footwear industry. Hence, to improvise the recycling method towards the generation leather waste and footwear wastes, leather industries and footwear sector will have to develop a new sustainable internal organization and develop various new responsibilities and parameters which will control the waste generation in the leather industries and footwear manufacturing factories. This will help in creating a well-structured organization which could deal on these environmental issues.

References

1. Colak SELIME, Zengin GÖKHAN, Ozgunay H, Sarikahya H, Sari O, Yuceer L (2005) Utilization of leather industry prefleshings in biodiesel production. J Am Leather Chem Assoc 100(4):137–141
2. Aten A, Innis KF, Knew E (1955) Flaying and curing of hides and skins as a rural industry. Food and Agricultural Organization of the United Nations, Rome
3. Taylor MM, Cabez LF, Marmer WN, Brown EM, Kolomaznik K (1998) Functional properties of hydrolysis products from collagen. J Am Leather Chem Assoc (USA)
4. Gaidau C, Ghiga M, Filipescu L, Stepan E, Lacatus V, Popescu M (2007) Advanced materials obtained from leather–by products. In: CEEX conference proceedings Romania, pp 252–252
5. Ferreira MJ, Almeida MF (2011) Recycling of leather waste containing chromium-a review. Mater Sci Res J 5(4):327
6. Kolomaznik K, Mladek M, Langmaier F, Janacova D, Taylor MM (2000) Experience in industrial practice of enzymatic dechromation of chrome shavings. J Am Leather Chem Assoc 95(2):55–63
7. Staikos T, Rahimifard S (2007) Post-consumer waste management issues in the footwear industry. Proc Instit Mech Eng Part B: J Eng Manuf 221(2):363–368
8. Soling for low cost footwear. World Footwear, July/August 2006. Shoe Trades, Cambridge, pp 18–20
9. The future of polyurethane soling. World Footwear, January/February 2005. Shoe Trades, Cambridge, pp 18–20
10. European Footwear Statistics (2005) http://europa.eu.int/comm/enterprise/footwear/statistics.html

11. EU market survey 2004: footwear, Centre for the Pro-motion of Imports from Developing Countries (CBI) (2004)
12. US Shoe Stats (2005) http://www.apparelandfootwear.org/data/shoestats
13. Footwear market predictions: forecasts for global foot-wear trading to 2009, SATRA Technology Centre (2003)
14. Council Directive 99/31/EC on the landfill of waste, OJ L 182, 26 April 1999
15. Landfill Allowances and Trading Scheme (England) Regulations, Statutory Instrument 2004 Number 3212, Department of Environment, Food, and Rural Affairs (2004)
16. Wicks R, Bigsten A (1996) Used clothes as development aid: the political economy of rags, http://ideas.repec.org/p/hhs/gunwpe/0017.html
17. Temsch R, Marchich M (2002) UNIDO programs funded by Austria to strengthen the leather sector in Uganda, Evaluation report UNIDO projects US/UGA/ 92/200, US/UGA/96/300, United Nations Industrial Development Organisation, http://www.unido.org
18. Council of Leather Exports, CLE (2008) Export of leather & leather products—facts & figures 2007-8'
19. Ganguly S (2008) Agra—Centre of the Indian footwear industry. World Footwear, March April
20. Kumar SC (1997) Indian leather industry: growth, productivity and export performance. APH Publication, New Delhi
21. Khan KAS, Special feature—28th international footwear conference, Guangzhou, China, 12 June
22. Kulkarni P (2005) Use of eco labels in promoting exports for developing countries to developed countries: lessons from Indian leather footwear industry. Centre for International Trade, Economics and Environment
23. Lynch OM (1969) The politics of untouchability. Columbia University Press, New York
24. Tesfor G, Lutz C, Ghauri P (2003) Comparing export marketing channels: developed versus developing countries. Int Mark Rev 21(4/5):409–422
25. El Boushy AK, Dieleman SH, Koene JIA, van der Poe AFB (1991) Tannery waste by-product from cattle hides, its suitability as a feedstuff. Bioresour Technol 35:321–323
26. Association of Official Agricultural Chemists (1975) In: Horwitz E (ed) Official methods of analysis, 12th edn. Washington, DC
27. Zuriaga-Agusti E, Galiana-Aleixandre MV, Bes-Pia A, Mendoza-Roca JA, Risueno-Puchades V, Segarra V (2015) Pollution reduction in eco-friendly chrome free tanning and evaluation of the biodegradation by compositing of the tanned leather wastes. J Clean Prod 87(15):874–881
28. Highberger JH (1956) The chemical structure and macromolecular organization of the skin protein. In: O'Flaherty F, Roddy WT, Lollar KM (eds) The chemistry and technology of leather, vol I. Rheinhold Publishing Corporation, New York, pp 65–193
29. Sivaprakash K, Maharaja P, Pavithra S, Boopathy R, Sekaran G (2017) Preparation of light weight constructional materials from chrome containing buffing dust solid waste generated in leather industry. J Mater Cycles Waste Manage 19(2):928–938
30. Bidermann K, Neck H, Neher MB, Wilhelmy V Jr (1962) A technical economic evaluation of four hide curing methods. Agricultural economic report, no. 16, marketing economics division, economic research service, USDA, Washington
31. FAO (1990) Production yearbook, vol 42. Food and Agricultural Organization of the United Nations, Rome
32. Gawecki K, Lipinska H, Rulkowski A (1981) Meal from residues of the tanning and meat industries in feeds for fattening chickens. Rocz Naukowi Zootechniki 8(1):185–192
33. Gustavson KH (1956) The chemistry of the tanning process. Academic Press, New York
34. Henrickson RL, Turgut H, Rao BK (1984) Hide protein as a food additive. J Am Leather Chem Ass 79:132–145
35. Sundar VJ, Gnanamani A, Muralidharan C, Chandrababu NK, Mandal AB (2011) Recovery and utilization of proteinous wastes of leather making: a review. Rev Environ Sci Bio/Technol 10(2):151–163
36. Fela K, Wieczorek-Ciurowa K, Konopka M, Woźny Z (2011) Present and prospective leather industry waste disposal. Pol J Chem Technol 13(3):53–55

37. Pawowa M (1995) Ecological aspects of hide processing and leather waste. Radom: Wyd. ITeE
38. Taborski W, Kowalski Z, Wzorek Z, Konopka M, Chojnacka K, Chojnacki A (2005) Thermal utilization of leather scrap after chrome tanning. J Am Leather Chem Assoc
39. Dalev PG, Simeonova LS (1992) An enzyme biotechnology for the total utilization of leather wastes. Biotech Lett 14(6):531–534
40. Kirk DW, Chan CCY, Marsh H (2002) Chromium behaviour during thermal treatment of MSW fly ash. J Hazard Mater B90:39–49
41. Nunes RR, Bontempi RM, Mendonça G, Galetti G, Rezende MOO (2016) Vermicomposting as an advanced biological treatment for industrial waste from the leather industry. J Environ Sci Health Part B 51(5):271–277
42. Choudhary RK, Swati A, Hait S (2019) Vermicomposting of primary clarified tannery sludge employing *Eisenia fetida*. In: Water resources and environmental engineering II. Springer, Singapore, pp 125–135
43. Allison L (1965) Organic carbon. In: Methods of soil analysis: part 2 chemical and microbiological properties, vol 9, pp 1367–1378
44. Dhayalan K, Fathima NN, Gnanamani A, Rao JR, Nair BU, Ramasami T (2007) Biodegradability of leathers through anaerobic pathway. Waste Manag 27(6):760–767
45. Beccari M, Carucci G, Lanz AM, Majone M, Papini MP (2002) Removal of molecular weight fractions of COD and phenolic compounds in an integrated treatment of olive oil mill effluents. Biodegradation 13(6):401–410
46. Borja R, Alba JSEG, Martinez LMPG, Monteoliva M, Ramos-ormenzana A (1995) Effect of aerobic pretreatment with *Aspergillus terreus* on the anaerobic digestion of olive-mill wastewater. Biotechnol Appl Biochem 22(2):233–246
47. Borja R, Banks CJ, Wang Z, Mancha A (1998) Anaerobic digestion of slaughterhouse wastewater using a combination sludge blanket and filter arrangement in a single reactor. Biores Technol 65(1–2):125–133
48. Sekaran G, Swarnalatha S, Srinivasulu T (2007) Solid waste management in leather sector. J Des Manuf Technol 1(1):47–52
49. Swarnalatha S, Ramani K, Karthi AG, Sekaran G (2006) Starved air combustion–solidification/stabilization of primary chemical sludge from a tannery. J Hazard Mater 137(1):304–313
50. Sekaran G, Shanmugasundaram KA, Mariappan M (1998) Characterization and utilisation of buffing dust generated by the leather industry. J Hazard Mater 63(1):53–68
51. Lima DQ, Oliveira LCA, Bastos ARR, Carvalho GS, Marques JGSM, Carvalho JG, De Souza GA (2010) Leather industry solid waste as nitrogen source for growth of common bean plants. Appl Environ Soil Sci
52. Love AHG (1983) Chromium-biological and analytical considerations. In: Chromium: metabolism and toxicity. CRC Press Boca Raton, pp 1–12
53. Leonard A, Lauwerys RR (1980) Carcinogenicity and mutagenicity of chromium. Mutat Res/Rev Genet Toxicol 76(3):227–239
54. Aravindhan R, Madhan B, Rao JR, Nair BU, Ramasami T (2004) Bioaccumulation of chromium from tannery wastewater: an approach for chrome recovery and reuse. Environ Sci Technol 38(1):300–306
55. Kirk DW, Chan CC, Marsh H (2002) Chromium behavior during thermal treatment of MSW fly ash. J Hazard Mater 90(1):39–49
56. Bhoyar RV, Titus SK, Bhide AD, Khanna P (1996) Municipal and industrial solid waste management in India. J IAEM 23:53–64
57. NIKE Reuse-A-Shoe, http://www.nike.com
58. Oliveira LCA, Van Zanten Coura C, Guimaraes IR, Gonclaves M (2011) Removal of organic dyes using Cr containing activated carbon prepared from leather waste. J Hazard Mater 192(3):1094–1099
59. Malek A, Hachemi M, Didier V (2009) New approach of depollution of solid chromium leather waste by the use of chelates. Econ Environ Impacts 170(1):156–162
60. Kale RD, Jadhav NC (2019) Utilization of waste leather for the fabrication of composites and to study its mechanical and thermal properties. SN Appl Sci 1(10):1231

61. Nair BU, Raghava Rao J, Sreeram KJ, Thanikaivelan P (2002) Green route for the utilization of chrome shavings. Environ Sci Technol 36(6):1372–1376
62. Henrickson RL, Ranganayagi MD, Ali A (1984) Age, species, breed, sex and nutrition effect of hide collagen. CRC Crit Rev Food Sci Nutritio 3:159
63. Kale RD, Jadhav NC, Pal S (2019) Fabrication of green composites based on rice bran oil and anhydride cross-linkers. Iran Polym J 28(6):471–482
64. SAstry TP, Sehal PK, Ramasami T, Value added eco-friendly products from tannery solid wastes. J Environ Sci Eng
65. Srinivasan TS, Nendrakumal MA, Krishnan TS, Scaria KJ (1985) Chrome shavings-a tannery waste, current practice and future trends for its utilization. In: Presented in the 3rd AAP animal science congress, Seoul, Korea
66. Jadhav AC, Pandit P, Gayatri TN, Chavan PP, Jadhav NC (2019) Production of green composites from various sustainable raw materials. In: Green composites. Springer, Singapore, pp 1–24
67. Lim LT, Auras R, Rubino M (2008) Processing technologies for poly (lactic acid). Prog Polym Sci 33(8):820–852
68. Ponsubbiah S, Suryanarayana S, Gupta S, Composite from leather waste
69. Al-Joulani NM, Effect of rubber and leather wastes on concrete properties
70. Dong Y, Ghataura A, Takagi H, Haroosh HJ, Nakagaito AN, Lau KT (2014) Polylactic acid (PLA) biocomposites reinforced with coir fibres: evaluation of mechanical performance and multifunctional properties. Compos A Appl Sci Manuf 63:76–84
71. Ambone T, Joseph S, Deenadayalan E, Mishra S, Jaisankar S, Saravanan P (2017) Polylactic acid (PLA) biocomposites filled with waste leather buff (WLB). J Polym Environ 25(4):1099–1109
72. Teli MD, Jadhav AC (2016) Extraction and characterization of novel lignocellulosic fibre. J Bionanosci 10(5):418–423
73. Pandit P, Singha K, Jadhav A, Gayatri TN, Dhara U (2019) Applications of composites materials for environmental aspects. Compos Environ Eng, 33–55
74. Alcantara R, Amores J, Canoira LT, Fidalgo E, Franco MJ, Navarro A (2000) Catalytic production of biodiesel from soy-bean oil, used frying oil and tallow. Biomass Bioenerg 18(6):515–527
75. Briggs MS, Pearson J (2005) Biodiesel processing handout for new hampshire science teachers association workshop. UNH Biodiesel Group http://www.unh.edu/p2/biodiesel. Last accessed 13(02), p 2012
76. Canakci M (2005) Performance and emissions characteristics of biodiesel from soybean oil. Proc Instit Mech Eng Part D: J Autom Eng 219(7):915–922
77. Graboski MS, McCormick RL (1998) Combustion of fat and vegetable oil derived fuels in diesel engines. Prog Energy Combust Sci 24(2):125–164
78. Guo Y, Leung YC, Koo CP (2002) A clean biodiesel fuel produced from recycled oils and grease trap oils (BAQ 2002). Better Air Qual Asian Pac Rim Cities 16:55–61
79. Kinast JA (2003) Production of biodiesels from multiple feedstocks and properties of biodiesels and biodiesel/diesel blends. Final report; Report 1 in a series of 6 (No. NREL/SR-510-31460). National Renewable Energy Lab., Golden, CO. (US)
80. Knothe G (2005) Dependence of biodiesel fuel properties on the structure of fatty acid alkyl esters. Fuel Process Technol 86(10):1059–1070
81. Ma F, Hanna MA (1999) Biodiesel production: a review. Biores Technol 70(1):1–15
82. McCormick RL, Graboski MS, Alleman TL, Herring AM, Tyson KS (2001) Impact of biodiesel source material and chemical structure on emissions of criteria pollutants from a heavy-duty engine. Environ Sci Technol 35(9):1742–1747
83. McGill R, Storey J, Wagner R, Irick D, Aakko P, Westerholm M, Nylund NO, Lappi M (2003) Emission performance of selected biodiesel fuels. SAE Trans, 1456–1473
84. Ramadhas AS, Muraleedharan C, Jayaraj S (2005) Performance and emission evaluation of a diesel engine fueled with methyl esters of rubber seed oil. Renew Energy 30(12):1789–1800
85. Chen JC, Wey MY, Chiang BC, Hsieh SM (1998) The simulation of hexavalent chromium formation under various incineration conditions. Chemosphere 36(7):1553–1564

86. Yılmaz O, Kantarli IC, Yuksel M, Saglam M, Yanik J (2007) Conversion of leather wastes to useful products. Resour Conserv Recycl 49(4):436–448

87. Zuriaga-Agustí E, Galiana-Aleixandre MV, Bes-Piá A, Mendoza-Roca JA, Risueño-Puchades V, Segarra V (2015) Pollution reduction in an eco-friendly chrome-free tanning and evaluation of the biodegradation by composting of the tanned leather wastes. J Clean Prod 87:874–881

88. Oliveira LC, Coura CVZ, Guimarães IR, Gonçalves M (2011) Removal of organic dyes using Cr-containing activated carbon prepared from leather waste. J Hazard Mater 192(3):1094–1099

89. Temsch R, Marchich M (2002) UNIDO programs funded by Austria to strengthen the leather sector in Uganda, evaluation report UNIDO Projects US/UGA/92/200, US/UGA/96/300. United Nations Industrial Development Organisation (UNIDO)

90. Wilford A (1997) Environmental aspects of footwear and leather products manufacture. UNIDO document for the Thirteenth Session of the Leather and leather products Industry Panel

91. Margesin R, Schinner F (1999) Biological decontamination of oil spills in cold environments. J Chem Technol Biotechnol 74(5):381–389

92. Wei QF, Mather RR, Fotheringham AF, Yang RD (2003) Evaluation of nonwoven polypropylene oil sorbents in marine oil-spill recovery. Mar Pollut Bull 46(6):780–783

93. National Hide Association (1979) Hides and skins. National Hide Association, Sioux City

Printed in the United States
by Baker & Taylor Publisher Services